Cellular Automata and Complex Systems: Methods for Modeling Biological Phenomena

Eleonora Bilotta, University of Calabria, Italy

Pietro Pantano, University of Calabria, Italy

Medical Information Science Reference

MEDICAL INFORMATION SCIENCE REFERENCE

Hershey · New York

Director of Editorial Content:	Kristin Klinger
Director of Book Publications:	Julia Mosemann
Acquisitions Editor:	Lindsay Johnston
Development Editor:	Julia Mosemann
Publishing Assistant:	Casey Conapitski, Travis Gundrum, Natalie Pronio
Typesetter:	Deanna Jo Zombro
Production Editor:	Jamie Snavely
Cover Design:	Lisa Tosheff
Printed at:	Yurchak Printing Inc.

Published in the United States of America by
Medical Information Science Reference (an imprint of IGI Global)
701 E. Chocolate Avenue
Hershey PA 17033
Tel: 717-533-8845
Fax: 717-533-8661
E-mail: cust@igi-global.com
Web site: http://www.igi-global.com

Library of Congress Cataloging-in-Publication Data

Bilotta, Eleonora, 1958-
Cellular automata and complex systems : methods for modeling biological phenomena / by Eleonora Bilotta and Pietro Pantano.
p. cm.
Summary: "The theme of this book is the use of Cellular Automatas (CAs) to model biological systems, describing 2-D CAs to create populations of "life-like agents" with their own genomes"--Provided by publisher. Includes bibliographical references and index.
ISBN 978-1-61520-787-9 (hardcover) -- ISBN 978-1-61520-788-6 (ebook)
1. Biocomplexity--Mathematical models. 2. Biology--Mathematical models. 3. Cellular automata. I. Pantano, Pietro, 1954- II. Title.
QH323.5.B4535 2010
570.1'51135--dc22

2009042357

British Cataloguing in Publication Data
A Cataloguing in Publication record for this book is available from the British Library.

Table of Contents

Foreword

"*Cellular Automata and Complex Systems. Method for Modeling Life-like Phenomena*" presents ten years of research by Eleonora Bilotta and Pietro Pantano. The book offers an innovative approach to the study of discrete dynamical systems, such as one-dimensional and two-dimensional Cellular Automata (CA). The first section introduces recent work by the authors on the use of Genetic Algorithms to discover complex rules for one-dimensional CA. It goes on to show how these CA can generate novel structures with language-like properties. The second section provides a detailed discussion of two-dimensional Cellular Automata, beginning with von Neumann's initial proposal of an artificial self-replicating system. The authors show that self-replication is a common feature not only of biological but also of computational systems. They go on to show how it is possible to analyze computational self-replicators as if each system were a biological agent with its own genetic code. On this basis, they develop a taxonomy of self-replicators. The third part of the book shows how cellular automata can be used to generate sound and music, rendering some of the beauty of complexity.
The authors present new techniques relevant to many important areas of CA research. More specifically:

1. they show how Genetic Algorithms can be used to detect complex rules both in one-dimensional and two-dimensional systems;

2. they create a formal language providing new insights into emerging CA structures (gliders and regular domains), and use sets of production rules to generate complex behavior;
3. they describe the current state of the art in Boolean CA;
4. they provide a taxonomy of self-reproducing systems accessible to a wide-audience;
5. they demonstrate how cellular automaton can be seen as life-like agents, each with its own genetic code;
6. they evolve populations of complex agents;
7. they translate the output of cellular automata into sounds and music, using media to gain a better understanding of CA behavior;
8. they present a gallery of patterns generated by one and two-dimensional CA.

The end product is a virtual laboratory which the authors use to explore the life-like processes set in motion by specific CA. Their simulations show how it is possible to change CA's form and function by varying their "genome". Their study of the relationship between the genome, structure and function of CA provides new ways of interpreting evolutionary trajectories in CA parameter space and demonstrate CA power as a model of life-like phenomena.

All told, I found this book to be most timely and challenging. Its high level of exposition should appeal to a broad readership. It is a gem to behold and contemplate.

Leon Chua
University of Berkeley, USA

***Leon Chua**, Ph.D., is a full professor at the Electrical Engineering and Computer Sciences Department of the University of Berkeley. Prof. Chua is the inventor of the Chua circuit (1983) and of Cellular Neural Networks (1988). He has made major contributions to several different fields of research including Cellular Neural/Nonlinear Networks, Nonlinear Circuits and Systems, Nonlinear Dynamics, Bifurcation theory and Chaos theory. His name is included in the ISI list of the most frequently cited authors in engineering. He is the founder and editor of the International Journal of Bifurcation and Chaos and has received numerous awards including several honorary degrees. Recently, a nano-technology research group in HP discovered a memristor (memory resistor), that Prof. Chua predicted in 1973. The discovery has opened interesting prospects for research. Prof. Chua is an expert in Cellular Automata, where he has worked to place the discipline rests on sound mathematical foundations. With World Scientific, he has recently published several books in this area.*

Acknowledgment

The research presented in this book is the result of more than ten years of work with several generations of students, who have always expressed strong interest in our work and enthusiasm for our discoveries. We would particularly like to thank Valerio Talarico, Antonio Lafusa and Stefano Vena who helped us to develop our simulation tools, together with Lorella Gabriele and Adriano Talarico, who assisted us in the preparation of diagrams, tables and graphs. Special thanks are due to Richard Walker who helped us to translate the book into English and who provided much precious advice. We should also mention Julia Mosemann, Development Editor of IGI Global, who provided us with valuable editorial suggestions. Last but very much not least, we would like to thank our colleague Pier Augusto Bertacchini who was crucial to creating the rich and stimulating academic environment in which we achieved our results.

Eleonora Bilotta
University of Calabria, Italy

Pietro Pantano
University of Calabria, Italy

Chapter 1
Outline

INTRODUCTION

Ever more frequently, contemporary science finds itself in situations in which the only way it can address the complexity of nature is to develop new methods. One of the most common models is the Cellular Automaton - a system in which large numbers of particles, distributed on a lattice, developing according to deterministic local rules, generate unpredictable large scale behavior. Cellular Automata (from now on CAs) provide important insights into a vast range of physical, biological, social, economic and psychological phenomena.

The theme of this book is the use of CAs to model biological systems. More specifically, we will describe how we have used 2-D CAs to create populations of "life-like agents" with their own genomes, whose development and morphology (the pattern generated by the CAs) are determined by the collective behavior of smaller units, comparable to the molecules and cells of biological organisms. Often, when we seek to explain complex processes in life forms, our tools are too weak or too difficult to use. Mathematical modeling with CAs allows us to go beyond traditional approaches. The ultimate goal is to eliminate some of the intrinsic complexity of biological systems, and find what this complexity hides: general laws of self-organization common to a wide range of natural phenomena. So how can we

DOI: 10.4018/978-1-61520-787-9.ch001

analyze the behaviors of these systems that resemble biological organisms? And what sort of biological problems can we address in this way? In this chapter, we describe the relationship between CAs and biological systems and identify the fundamental questions we will address in the rest of the book. To help readers in finding their way, we conclude with a short outline of what is to follow.

BACKGROUND

This book brings together two programs of research which, though distinct, have many tools and methods in common. The first is one more attempt by humans to compete with the Great Architect and build life-like properties into human artifacts. This is the program initiated by von Neumann and carried forward by Langton and followers. The second program, originating in the early 1980s, is an attempt to create "new kind of science", (Wolfram, 2002) rooted in concepts such as complexity, emergence and the collective behavior of automata. As we will see, this second program of research brings together empirical data (Wolfram's computer experiments), with solid mathematics: the description of the discrete behaviors of 1-D CAs, using systems of non-linear equations (Chua, 2006; 2007).

The Problem of Self-Replication

As early as the 1940s, researchers saw the study of artificial self-replicating systems as a way of understanding the properties of biological organisms. Today, we think of self-replication in far broader terms, as a fundamental property of many natural and artificial systems (Maynard Smith and Szathmàry, 1995; Langton, 1995).

To self-replicate a system needs to generate a structure with functionality similar to its own. In many systems, self-replication involves a cycle of growth followed by separation (the separation of the membrane during the replication of biological cells, a baby's separation from its mother after it is born, social separation leading to the creation of new social units, etc.). Other systems spread by producing systems similar to themselves, at distant points in time and space. This too is self-replication. In both cases, we observe complex, self-organizing development processes, allowing systems to grow and spread in time and space. If we view life as a machine for processing information, it follows that these processes will require the transport of information. This vision is shared by neurobiology, logic, automata theory, cognitive psychology and even molecular biology. It is closely connected to McCulloch and Pitt's use of mathematical logic to define the first neural networks (in the early 1940s) (McCulloch & Pitt, 1943; 1947) and to von Neumann's concept of a self-replicating machine (proposed in the early 1950s) (von Neumann, 1966). It was von Neumann, who, when thinking about the origins of life, first conceived the theoretical possibility of a machine containing a complete description of itself, and with the ability to construct a copy of itself, using the description (von Neumann 1966; Langton 1984; Derrida & Weisbuch, 1986). Since then, scientists have built many mathematical models of biological organisms and their functions, including the cognitive processes studied by Artificial Intelligence. Implemented in computers, these models have made it possible to simulate complex behaviors. This approach is rooted in the traditions of mechanistic biology. In the XVII and XVIII century, biology began to use mechanical metaphors to explain the functioning of the organs of the human body (Canguilhem, 1966). Thus the idea of organisms as information processing machines brings together two metaphors: life as a machine and life as language. This latter line of thought led to Weismann's idea that organisms' genetic material is organized as a code or the script of a code, carrying biological information about their hereditary characteristics (Maynard Smith & Szathmàry, 1995). Once the structure of DNA had been discovered and scientists had begun to study the genetic code, the biochemical facts appeared to justify the view that organisms were determined by a genetic program. What remained to be studied was the epigenetic relationship between genotype and phenotype about which very little was known.

In the meanwhile, Shannon and Weaver's concept of information content (1948) had a powerful influence on biology, leading biologists to see the genetic program as equivalent to instructions and algorithms, written in an artificial programming language. Many scientists criticized this way of applying information theory to biology (Apter and Wolpert, 1965; Waddington, 1968). But regardless of this criticism, the influence of cybernetics and computer science was too strong to resist. Very soon the powerful metaphor of a genetic program was to dominate the whole of molecular biology.

It was against this background that von Neumann sought to identify the mathematical foundations underlying the self-replication of living cells, giving rise to the new field of Artificial Life. Within the field, the study of self-replicating systems was of great importance. We can distinguish four main lines of research:

a. Attempts to build universal constructors based on von Neumann's self-replicating automaton; this work mainly dates to the 1950s and 1960s;
b. Attempts to create a minimal system capable of non-trivial self replication; this line of study began with Langton (1984) with later contributions in the 1980s and 1990s;
c. The enhancement of Langton's self-replicators with additional computational capabilities;

d. The evolution of self- replicators using Genetic Algorithms. This work began in the 1990s and is still in progress.

This research will be further discussed in Chapters 7, 8 and 10.

Complexity as a Science

Among researchers trying to create links between different scientific disciplines, there is growing awareness that biological, physical, psychological, social and economic systems are all extremely complex. New methodological and epistemological approaches are needed to:

a. Define what we mean by a complex system;
b. Identify the essential characteristics and components of a complex system;
c. Define complex systems' basic behaviors and dynamics;
d. Identify the structures and novel forms of organization (patterns) that emerge from these dynamics.

1-D CAs contain many interacting elements and give rise to unpredictable emergent behavior and can thus play a key role in this endeavor. The operation of a CA can be simulated on a computer. This means it is no longer necessary to reduce complex phenomena to simpler processes. On the contrary, we can work in the opposite direction, reconstructing the phenomenon from the underlying processes. This possibility allows researchers to conduct completely new forms of experimental manipulation and testing. In a way, it represents a completely new epistemological approach to science.

One of the key actors in the development of CAs has been Stephen Wolfram (1994). Wolfram's work, discussed in detail later in this book, provides a first set of principles for the classification of 1-D CAs. Following in his tracks, Langton and colleagues (Langton, 1986; 1990; Li, Packard & Langton, 1990; Li & Packard, 1990) have defined additional metrics for the analysis of complexity. This work has become the basis, first for studies of CA's computational mechanics (Crutchfield & Young, 1989; Crutchfield, 1994; Hanson & Crutchfield, 1997) and later for the development of evolutionary techniques (Packard, 1988; Mitchell et al., 1993; 1994).

Recently Chua (Chua 2006; 2007; 2009) has demonstrated that it is possible to formalize and prove Wolfram's empirical observations through mathematical analysis. To this end, he maps CA rule tables to the attractors of specially adapted versions of Cellular Neural Networks (CNN), which are non-linear dynamic systems (Chua, 1998). Chua projects the local rules for a CA onto a Boolean cube in 3-D Euclidean space. He then introduces the fundamental concept of *linear separabil-*

ity and defines a complexity index *k* for each rule. He goes on to identify a special class of local rules with k=2, which he identifies with what he calls the Wolfram complexity threshold. Applying these concepts to the 256 local rules in an ECA (Elementary CAs, which will be explained in next chapters), he identifies 16 distinct genetic families, each composed of 16 parent genes. This classification allows him to explain, but not predict, the essential features in the patterns generated by each local rule. He encodes these families of genes for CA rules using a representation that he calls a "Double helix Torus".

Chua goes on to prove not only that ECAs local rules 110, 124, 137 and 193 share identical behaviors, but they all have universal computational capabilities. Using Klein's method (Chua et al., 2004), he partitions the 256 local rules that Wolfram had studied in his experiments into 89 global equivalence classes, from which he derives 30 local equivalence classes. In these classes, there is always a mapping between the parameters of the non-linear difference equations for members of the same class.

In subsequent work, Chua characterized the long term behavior of 206 rules for 1-D CAs. Each of these rules can have a number of different attractors and invariant orbits. He uses a single, randomly generated, initial state to uniquely determine the precise characteristics of the attractor whose basin of attraction contains the random configuration used for the test. On the basis of these experiments, Chua makes a distinction between CAs that are *bilateral* or *non-bilateral* and between CAs that are *invertible* or *non-invertible.* This is a completely new approach to the study of CA behavior and to the prediction of their asymptotic long-term behavior. It is an approach that does not depend on knowing the initial state of the CA, but on a gallery of graphs showing the way the CA develops in time (*forward graphs)* and tracing back the behaviors leading to a certain attractor (*backward graphs).* He uses this technique to map all 256 rules, fifty of which are complex. In substance, Chua's idea is to develop a "rigorous *analytical* theory of 1-D CAs via a nonlinear dynamics perspective". In Chapter 4, we will extend Chua's work to the multidimensional case.

MAIN FOCUS

In the various sections of this book, we will discuss the importance of CAs for Complexity Science, the search for complex rules and self-replicating systems, the use of CAs to model biological systems, and the discovery of basic rhythms we can translate into sounds and music. In our presentation, we will be guided by our own research in which we have used evolutionary techniques to discover complex rules and self-replicators. Our findings show that reverse engineering is not the only way to find complex rules and that Genetic Algorithms are a valid alternative. We demonstrate that the rules space for multistate CAs is practically unlimited, accommodating a wide range of simple, complex and chaotic systems. We outline a first attempt to treat the general rules of CA behavior as a language system and show how we can identify the fundamental elements of the language, using production rules to compose them into more complex structures. Finally, we describe how we have used this approach as the basis for a rudimentary form of genetic engineering.

In our discussion of 2-D self-replicators, we begin by explaining our basic methodology based on Genetic Algorithms. We go on to demonstrate that self-replication is a fundamental characteristic of the artificial as well as the natural world. We then describe the huge range of behaviors produced by k-totalistic systems, and present an initial taxonomy of these behaviors, which we will discuss further in later chapters. As in the case of 1-D CAs, we analyze the organization of rules space. To conclude we identify several general behavioral models for 2-D CAs and show how these models generated fractal patterns, derived from the Sierpinski Triangle and the Sierpinski Arrowhead.

Modeling

Models are shorthand representations of aspects of reality that we believe to be useful or which are particularly salient from a cognitive viewpoint. In this sense, models are tools to gain a better understand of the systems they represent (Gurney & Nisbet, 1998). Every model consists of a set of components that interact with a well-defined environment (Bossel, 1994). For cognitive science, models are "structural analogues of the world" with "a structure that corresponds directly to the structure of the state of affairs that the discourse describes" (Johnson-Laird, 1983). Recent philosophy of science takes a very similar approach. (Falkenhainer, 1990; Nowak & Thagard, 1992; Thagard, 1992; Brewer, 2001Palmieri, 2003). When scientists think about a process, or use their imagination to produce scientific theory, they are "thinking in models". In fact, they use models not only to represent the final output of their work but to describe the processes underlying the analysis and consequent organization of theory. The way these models are used is based on intellectual procedures that date back to Galileo and the origins of the scientific method. In brief, the concept of a model has become a fundamental element in contemporary epistemological discourse.

There are, of course, many different kinds of model. The first distinction we have to make is between *quantitative* and *qualitative* models. Some of the latter are designed, not to obtain a quantitative response, but to gain insights into the mechanisms underlying a specific situation or phenomenon. These are what we call *strategic models* (Nisbet et al., 1983). Formal models of dynamic systems, expressed in mathematical language, are called *mathematical models*. When formal models of dynamic systems are simulated on computers, they are called *simulation models* (Humphreys, 2004). Over the last 25 years, this last kind of model has taken on a role of great importance. One of the reasons is that we can also use computers to simulate traditional mathematical models. In this setting, the computer makes it possible to hide some of the complexity of the calculation, which is handled at the computational level, without appearing in the output of the simulations.

Most scientific modeling begins by defining the system to be modeled. In many cases, this requires a long phase of observation and a vast body of prior knowledge to identify the basic principles underlying the working of the system. These are then translated into mathematical equations. Finally the equations are encoded in artificial languages. Many of the key issues in this process concern scientific visualization (Bilotta et al.,2006; 2007a; 2007b; 2007c; 2007d; 2007e; 2007f), the design of graphical output and the conventions used to make the output understandable to the scientific community. Simulations make it possible to visualize a system and its dynamics, and to manipulate variables; they allow us to make observations that would not be possible with the real system. Finally when they are over, the data they produce can be used to validate and, if necessary, reformulate the model. This way of proceeding represents an enormous contribution to the scientific method, opening new trails for scientists to follow.

Outline of the Book

The book is divided into 4 main sections. The first is dedicated to the role of CAs in complexity science. The section is composed of 4 chapters.

After this chapter, in which we describe the main themes of the book and outline its structure, Chapter 2 presents a formal treatment of CAs, organized in the same order in which Wolfram developed his original ideas. It goes on to introduce the use of Genetic Algorithms as a way of evolving complex rules. Chapter 3 introduces the basic assumption that artificial systems have life-like properties and that we can analyze them in the same way we would analyze biological organisms. In other words, we can look at artificial organisms as agents inhabiting an environment. We go on to describe the modeling procedures we use in the rest of the book. We outline our modeling paradigm, based on the identification of fundamental structures (gliders, regular domains, and particles in 1-D systems, self-replicators in k-totalistic CAs), systematic observation of their behavior and the extraction of general behavioral laws. We show how we can consider the rules for 1-D and 2-D CAs as their genetic code and how, by "mutating" the code, we can change the shape, structure and intrinsic nature of our systems. The computational tools we have developed allow us to precisely identify the genetic networks underlying the behavior of self-replicators as they develop in time. Chapter 4 reviews the methods we have used to analyze CAs behavior and to identify emergent structures in the patterns they produce. This quantitative and qualitative analysis allows us to detect the key elements determining the behavior of the system and makes it easier for us to model biological behavior. Chapter 4 also includes the first ever attempt to analyze patterns in terms of human perception. The method we have developed allows us to identify the rules underlying the visual organization of emergent structures and show that these rules depend in turn on general laws governing the organization of perceptual stimuli and human perception. Another approach, we discuss in this chapter, is the use of automatic procedures to detect emergent structures such as regular domains and gliders. We describe how we can use these procedures to introduce language constructs and to automatically generate new structures from those that already exist.

The theme of the second section of the book is the search for complex rules and the discovery of self-replicating systems. The section contains 4 chapters.

Chapter 5 opens with a brief review of Wolfram's pioneering work. It goes on to present the metrics we use in our own analysis, and in particular the Input Entropy function we use for 1-D and 2-D CAs. Chapter 6 demonstrates that when we use Input Entropy in a fitness function for a Genetic Algorithm, we can successfully find complex rules for 1-D multistate CAs. To make sure that other researchers can repeat our experiments, we provide a detailed description of our methods. In Chapter 7, we describe how we can apply a Genetic Algorithm to a special formalization of 2-D CAs and produce systems similar to the Game of Life (Gardner, 1970). We show that these simulations can give rise to an infinite range of self-replicators. In a recent paper (Bilotta & Pantano, 2006), we presented an initial taxonomy of these systems. Additional examples are described in (Bilotta e Pantano 2008). Chapter 8 presents a gallery of self-replicating systems, generated with our evolutionary techniques. We observe that they are highly diverse in terms of behavior, complexity and the number of rules they use during development. As in nature, we do not find just one replication mechanism but many. This suggests that self-replication is extremely common and that the ability to self-replicate is intrinsic to the very nature of a CA.

The theme of the third part of the book is biological modeling using emergent language structures in 1-D CAs. The basic assumption is that 1-D CAs are universal computational machines, with the ability to process, store and transport information. We see these machines as agents with an artificial genetic code that evolves when environmental conditions change. These mutations produce lineages of self-replicators, each of which follows a specific trajectory in parameter space. We observe the geometrical structure of these systems. Our observations allow us to identify a special class of self-replicator, with the ability to return to the initial conditions of the simulation after a certain number of steps. We call these systems *universal constructors.*

As self-reproducers have different forms and behavior, we have classified all these systems in classes. The last part of the section provides a taxonomy of the various classes of self-replicators we have found in our work.

Chapter 9 demonstrates that it is possible to compose individual elements in 1-D CAs into larger structures, using mechanisms similar to those we find in natural language. When we analyze the number of regular domains in a system and the period of these domains, we discover that some domains are the result of mergers between two pre-existing domains. Just as in natural language, this way of proceeding allows us to generate an infinite number of words (neologisms) and, as a consequence, an infinite number of sentences. The system (and its elementary particles) can be seen as a structured set of elements (comparable to the elements of a natural language), including:

a. Phonemes (individual cells, with their own specific state and characteristics);
b. Composition rules allowing phonemes to combine in structured particles similar to morphemes (structures that interact to produce emergent complex behavior);
c. A lexicon (domains, gliders, emergent particles);
d. Syntactical composition rules allowing the creation of higher level structures from lower level elements such as words (emergent structures) in the system.

The main goals of Chapter 10 are to:

a. demonstrate that self-replicators produce structures with fractal symmetries in time and space;
b. demonstrate the presence of precise mathematical patterns (the Sierpinski Triangle and other models) both in 1-D and 2-D CAs.
c. show that the artificial systems display the logic discussed by von Neumann and that mathematical modeling of life-like artificial organisms such as 2-D CAs can provide insights into the properties of biological organisms.
d. demonstrate the presence of developmental synchronies;
e. show that synchronization between basic models like the Sierpinski Triangle and other models of development can produce chaotic and complex patterns unlike any model previously seen.

Chapter 11 discusses the association between the development of emergent artificial organisms in 2-D CAs and their underlying genetic code, and shows how changes in the genome affect the organism's developmental dynamics. One consequence of our approach is that self-replicators, in any given generation, are always more complex than those in the previous generation.

Chapter 12 proves that it is possible to create structures with the ability to replicate any kind of pattern. These are what we call universal constructors. In this setting, any macro-structure created though arbitrary merging of simpler systems is itself a self-replicating structure.

The same phenomenon repeats on higher scales. In this way, even higher level structures (compositions of second or higher level self-replicators) become self-replicators.

Our experiments show that there exists not just one universal constructor, but a whole class of self-replicators with this property. The members of this class display a diverse range of development and development based on the Sierpinski model.

Chapter 13 presents visual representations of self-replication models previously described in Bilotta and Pantano (2006). The chapter has two basic goals. On the one hand, it illustrates the diversity of our self-replication models: many of which may have equivalents in nature. On the other, it demonstrates the productive, compositional nature of these systems. The chapter shows clearly that the development of these systems is not limited to symmetrical models, such as the Sierpinski Triangle. If we change the initial data, we find tightly coupled, synchronous combinations of structures that enter into collision with other elementary structures. We observe that, in some cases, the symmetries are perfectly conserved through this process: probably because the two structures are well synchronized in time and space or because they are in phase. In other cases, replication proceeds perfectly for a few steps but then degenerates into chaos. In yet other cases, symmetries are partially preserved–particularly on the outer surface of the organism–but the internal components collapse into chaotic behavior with no visible symmetries. Here too, we hypothesize the presence of quasi-linguistic structures, this time expressed not only in space but also in time.

In the fourth and final section of the book, we identify the rhythms displayed by 1-D and 2-D CAs and compare them with the rhythms we find in biology. The comparison allows us to describe CA behavior in time in terms of a small set of categories: periodic rhythms, rhythms that synchronize specific components of the structure, rhythms that synchronize the behavior of groups of organisms and their population dynamics, chaotic rhythm, and rhythms that resemble noise. We see how genetic changes can lead to changes in these rhythms and the way they are synchronized. Finally we show how we can transform these rhythmic structures into sounds and music.

CONCLUSION

In this chapter, we have presented the main topics we are going to address in the rest of the book and outlined the way in which we will organize our presentation.

Today's scientists have access to an enormous pool of knowledge, accumulated over centuries of scientific progress and new technologies that provide them with almost unlimited computational and information management capabilities. It is with these tools that they are addressing the challenges raised by complexity science. The most important problems they still have to face concern the epistemological foundations underlying the modeling process. In recent decades, modeling has changed the course of modern scientific thought. At the beginning of the 1980s, a group of physicists–Stephen Wolfram among them–argued that when computers became sufficiently powerful it would be possible to explain complex behaviors without resort to numbers or equations. This was the rationale behind the development of CAs, tiny computational systems that operate on a simple playing field, similar to a chess board. Unlike differential equations–that are hard to use–the rules governing the behavior of CAs are extremely simple. A typical rule might specify that a cellular automaton continues to move in a certain direction on the board until it collides with another automaton. Other rules describe the way the automaton interacts with other automata, what happens when they collide, the way they start from given initial conditions and organize themselves into new structures.

Scientific laws give algorithms, or procedures, for determining how systems behave. The computer program is a medium in which the algorithms can be expressed and applied. Physical objects and mathematical structures can be represented as numbers and symbols in a computer, and a program can be written to manipulate them according to the algorithms. When the computer program is executed, it causes the numbers and symbols to be modified in the way specified by the scientific laws. It thereby allows the consequences of the laws to be deduced.

Executing a computer program is much like performing an experiment. Unlike the physical objects in a conventional experiment, however, the objects in a computer experiment are not bound by the laws of nature. Instead they follow the laws embodied in the computer program, which can be of any consistent form. Computation thus extends the realm of experimental science: it allows experiments to be performed in a hypothetical universe. Computation also extends theoretical science. Scientific laws have conventionally been constructed in terms of a particular set of mathematical functions and constructs, and they have often been developed as much for their mathematical simplicity as for their capacity to model the salient features of a phenomenon. A scientific law specified by an algorithm, however, can have any consistent form. The study of many complex systems, which have resisted analysis by traditional mathematical methods, is consequently being made possible through computer experiments and computer models. Computation is emerging as a major new approach to science, supplementing the long-standing methodologies of theory and experiment (Wolfram, 1984).

This long quotation from Wolfram points to some of the key themes of this book.

New technologies, and especially computers, have changed the way we do science. Today we do not just have the traditional scientific method; we also have simulation. And simulation does not just provide new capabilities for analysis and data collection; it gives us new ways of visualizing scientific data, multiplying the power of our intellect and imagination and setting in motion virtuous circles of scientific creativity.

In this chapter, we have described the processes we use to model biological behaviors with discrete dynamics systems such as 1-D and 2-D CAs. Together with this description, we have also introduced the main intellectual issues we are going to address. Each is rooted either in von Neumann's original model, which is still a milestone in modern science, or in those of his successors. We hope this outline will guide readers from different backgrounds as they navigate the remaining chapters of the book.

REFERENCES

Apter, M. J., & Wolpert, L. (1965). Cybernetics and development 1: Information theory. *Journal of Theoretical Biology*, *8*, 244–257. doi:10.1016/0022-5193(65)90075-5

Bilotta, E., Di Blasi, G., Pantano, P., & Stranges, F. (2007d). A Gallery of Chua Attractors, Part IV. *International Journal of Bifurcation and Chaos in Applied Sciences and Engineering*, *17*(4), 1017–1078. doi:10.1142/S0218127407017665

Bilotta, E., Di Blasi, G., Pantano, P., & Stranges, F. (2007e). A Gallery of Chua Attractors, Part V. *International Journal of Bifurcation and Chaos in Applied Sciences and Engineering*, *17*(5), 1383–1511. doi:10.1142/S0218127407018099

Bilotta, E., Di Blasi, G., Pantano, P., & Stranges, F. (2007f). A Gallery of Chua Attractors, Part VI. *International Journal of Bifurcation and Chaos in Applied Sciences and Engineering*, *17*(6), 1801–1910. doi:10.1142/S0218127407018105

Bilotta, E., & Pantano, P. (2006). Structural and Functional Growth in Self-Reproducers. *Complexity*, *11*(6), 12–29. doi:10.1002/cplx.20131

Bilotta, E., & Pantano, P. (2008). Biological traits in artificial self-reproducing systems . In Orsucci, F., & Sala, N. (Eds.), *Complexity Science in Information Communication Technologies: Interaction Between Living Systems and Information Technologies* (pp. 109–123). Hershey, PA: IGI Global.

Bilotta, E., Pantano, P., & Stranges, F. (2006a). Computer Graphics meets Chaos and Hyperchaos. Some key problems. *Computer Graphics, 30*(3), 359–367. doi:10.1016/j.cag.2006.02.003

Bilotta, E., Pantano, P., & Stranges, F. (2007a). A Gallery of Chua Attractors, Part I. *International Journal of Bifurcation and Chaos in Applied Sciences and Engineering, 17*(1), 1–60. doi:10.1142/S0218127407017161

Bilotta, E., Pantano, P., & Stranges, F. (2007b). A Gallery of Chua Attractors, Part II. *International Journal of Bifurcation and Chaos in Applied Sciences and Engineering, 17*(2), 293–380. doi:10.1142/S0218127407017343

Bilotta, E., Pantano, P., & Stranges, F. (2007c). A Gallery of Chua Attractors, Part III. *International Journal of Bifurcation and Chaos in Applied Sciences and Engineering, 17*(3), 657–734. doi:10.1142/S0218127407017537

Bossel, H. (1994). *Modelling and simulation.* Wellesley, MA: A.K. Peters Ltd.

Brewer, W. F. (2001). Models in science and mental models in scientists and non-scientists. *Mind & Society, 2*(2), 33–48. doi:10.1007/BF02512358

Canguilhem, G. (1991). *The Normal and the Pathological.* New York: Zone Books.

Chua, L. O. (1998). *CNN: a Paradigm of Complexity.* Singapore: World Scientific Publishing Company.

Chua, L. O. (Ed.). (2006). *A Nonlinear Dynamics Perspective of Wolfram's New Kind of Science* (*Vol. 1*). Hackensack, NJ: World Scientific Publishing Company.

Chua, L. O. (Ed.). (2007). *A Nonlinear Dynamics Perspective of Wolfram's New Kind of Science* (*Vol. 2*). Hackensack, NJ: World Scientific Publishing Company.

Chua, L. O. (Ed.). (2009). *A Nonlinear Dynamics Perspective of Wolfram's New Kind of Science* (*Vol. 3*). Hackensack, NJ: World Scientific Publishing Company.

Chua, L. O., Sbitnev, V. I., & Yoon, S. (2004). A nonlinear dynamics perspective of Wolfram's new kind of science, Part III: Predicting the unpredictable. *International Journal of Bifurcation and Chaos in Applied Sciences and Engineering, 14*, 3689–3820. doi:10.1142/S0218127404011764

Crutchfield, J. P. (1994). The Calculi of Emergence: Computation, Dynamics, and Induction. *Physica D. Nonlinear Phenomena*, *75*, 11–54. doi:10.1016/0167-2789(94)90273-9

Crutchfield, J. P., & Young, K. (1989). Inferring Statistical Complexity. *Physical Review Letters*, *63*, 105–108. doi:10.1103/PhysRevLett.63.105

Derrida, B., & Weisbuch, G. (1986). Evolution of overlaps between configurations in random Boolean networks. *Journal of Physics*, *47*(8), 1297–1303. doi:10.1051/jphys:019860047080129700

Falkenhainer, B. (1990). A Unified Approach to Explanation and Theory Formation . In Shrager, J., & Langley, P. (Eds.), *Computational Models of Discovery and Theory Formation* (pp. 157–196). San Mateo, CA: Morgan Kaufman.

Gardner, M. (1970). Mathematical games: the fantastic combinations of John Conway's new solitaire game "Life." . *Scientific American*, *223*, 120–123. doi:10.1038/scientificamerican1070-120

Gurney, W. S. C., & Nisbet, R. M. (1998). *Ecological Dynamics*. New York: Oxford University Press.

Hanson, J. E., & Crutchfield, J. P. (1997). Computational Mechanics of Cellular Automata: An Example. *Physica D. Nonlinear Phenomena*, *103*(1-4), 169–189. doi:10.1016/S0167-2789(96)00259-X

Humphreys, P. (2004). *Extending Ourselves: Computational Science, Empiricism, and Scientific Method*. New York: Oxford University Press.

Johnson-Laird, P. N. (1983). *Mental models: Towards a cognitive science of language, inference, and consciousness*. Cambridge, MA: Harvard University Press.

Kelso, J. A. S., Mandell, A. J., & Shlesinger, M. F. (Eds.). (1988). *Dynamic Patterns in Complex Systems*. Singapore: World Scientific Publishing Company.

Langton, C. G. (1984). Self-Reproduction in Cellular Automata. *Physica D. Nonlinear Phenomena*, *10*, 135–144. doi:10.1016/0167-2789(84)90256-2

Langton, C. G. (1986). Studying artificial life with cellular automata. *Physica*, *22D*, 20–149.

Langton, C. G. (1990). Computation at the edge of chaos. *Physica*, *42D*, 2–37.

Langton, C. G. (1995). *Artificial life: an overview*. Cambridge, MA: The MIT Press.

Li, W., & Packard, N. H. (1990). The structure of the elementary cellular automata rule space. *Complex Systems*, *4*(3), 281–297.

Li, W., Packard, N. H., & Langton, C. G. (1990). Transition phenomena incellular automata rule space. *Physica D. Nonlinear Phenomena*, *45*, 77–94. doi:10.1016/0167-2789(90)90175-O

Maynard Smith, J., & Szathmàry, E. (1995). *The Major Transitions in Evolution.* New York: W. H. Freeman Press.

McCulloch, W., & Pitts, W. (1943). A Logical Calculus of Ideas Immanent in Nervous Activity. *The Bulletin of Mathematical Biophysics*, *5*, 115–133. doi:10.1007/BF02478259

McCulloch, W., & Pitts, W. (1947). On how we know universals: The perception of auditory and visual forms. *The Bulletin of Mathematical Biophysics*, *9*, 127–147. doi:10.1007/BF02478291

Mitchell, M., Crutchfield, J. P., & Hraber, P. T. (1994). Dynamics, Computation, and the "Edge of Chaos": A Re-examination. In G. Cowan, D. Pines, & D. Melzner (Eds.), Complexity: Metaphors, Models, and Reality. (Vol. 19, pp. 497-513). SFI Series in the Sciences of Complexity. Reading, MA: Addison-Wesley.

Mitchell, M., Hraber, P., & Crutchfield, J. P. (1993). Revisiting the Edge of Chaos: Evolving Cellular Automata to Perform Computational Tasks. *Complex Systems*, *7*, 89–130.

Nisbet, R. M., McKinstry, J., & Gurney, W. S. C. (1983). A strategic model of material cycling in a closed ecosystem. *Mathematical Biosciences*, *64*, 99–113. doi:10.1016/0025-5564(83)90030-5

Nowak, G., & Thagard, P. (1992). Copernicus, Ptolemy, and explanatory coherence. In R. N. Giere (Ed.), Minnesota Studies in the Philosophy of Science, Volume XV: Cognitive Models of Science (pp.274-309). Minneapolis, MN: University of Minnesota Press.

Packard, N. H. (1988). Adaptation toward the edge of chaos . In Kelso, J. A. S., Mandell, A. J., & Shlesinger, M. F. (Eds.), *Dynamic Patterns in Complex Systems* (pp. 293–301). Singapore: World Scientific Publishing Company.

Palmieri, P. (2003). Mental models in Galileo's early mathematization of nature . *Studies in History and Philosophy of Science*, *34*, 229–264. doi:10.1016/S0039-3681(03)00025-6

Shannon, C., & Weaver, W. (1948). *A Mathematical theory of communication.* Urbana, IL: University of Illinois Press.

Thagard, P. (1992). *Conceptual Revolutions.* Princeton, NJ: Princeton University Press.

von Neumann, J. (1966). *Theory of Self-Reproducing Automata (edited and completed by AW Burks).* Urbana, IL: University of Illinois Press.

Waddington, C. H. (1968). The Basic Ideas of Biology. In C.H. Waddington (Ed.), *Towards a Theoretical Biology: Prolegomena* (Vol. 1, pp. 1-32). International Union of Biological Sciences Symposium.

Wolfram, S. (1984). Computer software in science and mathematics. *Scientific American, 251*, 188–203. doi:10.1038/scientificamerican0984-188

Wolfram, S. (Ed.). (1994). *Cellular Automata and Complexity: collected papers.* Reading, MA: Addison-Wesley.

Wolfram, S. (2002). *A New Kind of Science.* Champaign, IL: Wolfram Media.

Chapter 2
Basic Definitions

INTRODUCTION

Cellular Automata (CAs) are discrete dynamic systems that exhibit chaotic behavior and self-organization and lend themselves to description in rigorous mathematical terms. The main aim of this chapter is to introduce CAs from a formal perspective. Ever since the work of von Neumann (1966), von Neumann & Burks (1970), Toffoli & Norman (1987), Wolfram (1983; 1984), and Langton (1984; 1986; 1990), specialists have recognized CAs as a model of crucial importance for complexity studies. Like other models used in the investigation of complex phenomena (Chua & Yang, 1988; Chua, 1998), a CA consists of a number of elementary components, whose interactions determine its dynamics. In this and the following chapters, we will sometimes refer to these elementary components as *cells,* sometimes as *sites.* The cells of a CA can be positioned along a straight line or on a 2 or 3-dimensional grid, creating 1-D, 2-D and 3-D CAs. Automata consisting of cells whose only possible states are 0 or 1, are Boolean Automata; automata whose cells can assume more than 2 states are multi-state CAs. In both cases, the CA contains "elementary particles" whose dynamics are governed by simple rules. These rules determine sometimes unpredictable, emergent behaviors, ranging from the simple, through the complex to the chaotic. The basic parameters determining these behaviors are

DOI: 10.4018/978-1-61520-787-9.ch002

the number of "neighbours" affecting the behavior of a cell, the size and geometry of the area surrounding the cell (its "neighborhood"), the states cells are allowed to assume and the local rules determining their dynamics. All the cells in a CA are governed by the same rules which they use to interpret their own state and that of their neighbours and to decide the next step in their development. In the models considered here, each cell reads the state of its neighbours and determines it own state, simultaneously with the other cells. This kind of CA is thus a kind of parallel computer. Boolean 1-D CAs, in which each cell has a neighbourhood of three cells, are known as Elementary Cellular Automata (ECAs). Recent fine work from Chua (2006; 2007) and his co-workers (Chua et al., 2002; Chua et al. 2004) shows that although ECAs have been studied for a long time (Wolfram, 1984; 2002; Wuensche & Lesser 1992), they can still offer surprises. Wolfram (2002), has argued that ECAs already show the most important features of the complex behavior of higher dimensional and multi-state CAs. We believe, however, that this is not entirely true. In particular, ECAs do not show the same kinds of complex behavior seen in multi-state automata. Even more importantly, they do not display self-replication except in a rudimentary form. As we will see in later chapters, ECAs cannot generate the complex structures and geometries generated by self-reproducing CAs operating in higher dimensions. In the final part of this chapter, we describe the way these patterns are organized in time and space, and give examples of the extreme complexity and beauty to which they often give rise (Bilotta & Pantano, 2006).

A number of authors have analyzed CAs with continuous valued states, CAs that evolve asynchronously, non-homogeneous CAs, CAs with probabilistic rules, etc. (Sipper et al., 1997; Nehaniv, 2002; Fatès & Morvan, 2005; Fatès et al., 2006). For a review of the literature see Wolfram (2002). A number of researchers have used generalizations of CAs to model physical phenomena on different scales (Hoekstra et al., 2008). In this book, we will describe models of biological phenomena based on CAs with discrete states that evolve synchronously. By contrast, asynchronous are highly complex and difficult to control. We will touch on these processes only in Chapter 13, when we compare replication in a synchronous CA with emergent replication in an asynchronous process.

Another critical issue in the study of CAs is the *rule space*, a concept first introduced by Langton (1986; 1990). Unlike Wolfram (1984), Li & Packard (1990) and other authors (Oliveira et al., 2001), who focused on the patterns they produced, Langton classified CAs on the basis of the rules governing their dynamics. His goal was to produce an exhaustive study of high-dimensional and non-Boolean CAs, similar to earlier studies of ECAs. Unfortunately, however, the high dimensionality of the rule space made this impossible. Thus, to contain the size of the problem, Langton restricted his study to *totalistic* and *outer-totalistic* rules. Our own approach is completely different. Rather than seeking a complete classification of CA behavior, like Wolfram, or a parameter-based categorization of rule space, we have tried to find (usually complex) rules that produce "interesting" behavior. We have restricted our search to a subspace of rules we have already considered in previous work (Bilotta et al., 2002; 2003b). It is these *k-totalistic* rules, we introduce in this chapter, together with examples of the way they operate.

We begin with a formal treatment of CAs and of the 1- 2- and 3-D patterns they produce. We then proceed to examine the subspace of k-totalistic rules and describe how we can use genetic algorithms (Holland, 1994; Mitchell, 1996) to discover complex rules. We also see how k-totalistic CAs can be used in discussions of self-replication, an issue we examine in later chapters. Finally we discuss CAs' global dynamics and the related concepts of "attractors" and "basins of attraction". To this end, we study the development of 2- and 3-D CAs in time and space and show how these can be represented graphically, using novel forms of tessellation and 3-D patterns to complement standard representations.

BACKGROUND

Formal Definition of a Cellular Automaton

A CA is a *discrete dynamic system* (Hilborn, 2000; Strogatz, 2000) in the sense that cells can only assume discrete values, time proceeds in discrete units and the state of the system (that is the state of its cells) varies with time. Formally, any CA can be represented as a 4-tuple (Wolfram 1983; 2002):

$$A = (D, S, N, f)\ 1) \qquad (2.1)$$

where D is a positive integer that indicates the CA *dimension* (1, 2, 3 or more), S is a set of *finite states*, $N = (x_1, \ldots, x_n)$ is a *neighbourhood vector* consisting of n elements in Z^D (where Z indicates the set of relative numbers), and f is a *local rule* defined by:

$$f : S^n \rightarrow S \qquad (2.2)$$

Each cell in the CA has a neighbourhood N, with a local interaction radius r.

Figure 1a shows the case for $D = 1$, $r = 1$, and $n = 3$; Figure 1b the case for $D = 1$, $r = 2$, and $n = 5$ and Figure 1c, the case for $D = 1$, $r = 3$, and $n = 7$. In general, for $D = 1$,

$$n = 2r + 1 \quad (2.3)$$

Furthermore, for $d = 1$, (2.2) associates n values of S, with a new value for S. Thus:

$$(s_{i-r}(t-1),\ldots,s_i(t-1),\ldots s_{i+r}(t-1)) \rightarrow s_i(t) \quad (2.4)$$

with $s_i \in S$, $S = \{0,1,\ldots,k-1\}$, $\forall i \in M$, where M is the number of cells in the CA.

Wolfram (1984) provides an alternative formulation for (2.4) namely:

$$s_i(t) = F(s_{i-r}(t-1),\ldots,s_i(t-1),s_{i+r}(t-1))$$

The expression indicates that state s_i of the i-th cell at time t depends on the state of the cell and on that of its neighbours at time $t - 1$. Applying (2.4), we can compute the state of the cell as a function of the state of its neighbours:

$$s_i(t) = f\left[\sum_{j-r}^{j=r} \alpha_j s_{i+j}(t-1)\right] \quad (2.5)$$

where f is an integer function and the a_j are integer constants.

Figure 1d shows the case in which $D = 2$, $r = 1$, $n = 5$. This is called the *von Neumann neighbourhood*. Figure 1e illustrates the case with $D = 2$, $r = 1$, and $n = 9$. This is the *Moore neighbourhood*. Figure 1f depicts another example of a *Moore neighbourhood* in which $D = 2$, $r = 2$, and $n = 27$. In general, for $D = 2$: $n = (2r + 1)^2$.

(2.4) thus becomes:

$$\begin{pmatrix} s_{i-r,j-r}(t-1),\ldots,s_{i,j-r}(t-1)\ldots,s_{i+r,j-r}(t-1) \\ s_{i-r,j}(t-1),\ldots,s_{i,j}(t-1)\ldots,s_{i+r,j}(t-1) \\ s_{i-r,j+r}(t-1),\ldots,s_{i,j+r}(t-1)\ldots,s_{i+r,j+r}(t-1) \end{pmatrix} \rightarrow s_{ij}(t) \quad (2.6)$$

with $s_{ij} \in S$ for all cells in the CA.

The concepts just described can be generalized to three dimensions (see Figure 1g) and for all *Moore neighbourhoods*. We can thus write:

Figure 1. The first row represents a 1-D CA with (from left to right, a, b, c) $r = 1, r = 2, r = 3$ *and* $n = 3, n = 5$ *and* $n = 7$ *. The second row represents 2-D CAs with (from left to right, d, e, f) a von Neumann neighbourhood with* $n = 5$*, a Moore neighbourhood with* $n = 9$ *and* $r = 1$ *and a Moore neighbourhood with* $n = 25$ *and* $r = 2$*. The first image on the third row (g) shows a 3-D CA with* $n = 27$ *and* $r = 1$*. The center image (h) shows a 2-D CA with hexagonal cells. In this case,* $n = 7$ *and* $r = 1$*. This kind of image can be drawn for any figure that covers the plane and for any 3-D space-filling shape, for example the dodecahedron depicted in the third image (i).*

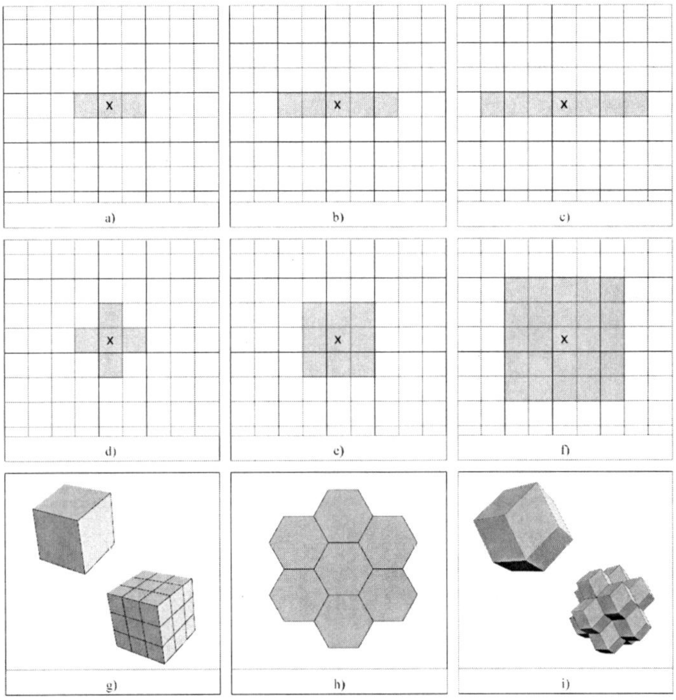

$$n = (2r + 1)^d. \tag{2.7}$$

We can extend this treatment to other ways of covering the plane, for example with hexagons (see Figure 1h) and of filling space, for instance with dodecahedrons (see Figure 1i). In these cases, cells evolve as a function of their own state and that of their neighbours.

In Boolean CAs, $S = \{0,1\}$; in multi-state CAs, $S = \{0,1,..,k-1\}$, with k>**2**. In elementary CAs, $S = \{0,1\}$, D = 1 and r=**1**. If M is the total number of cells,C, the set of all possible configurations of the CA has a cardinality of M^k, where each $p \in P$ defines the state of all cells belonging to the CA. Thus, in the 1-D case (D = 1), p can be represented by a string c', giving the state of each individual cell in the CA, as follows:

$$c' = s_1 s_2 \ldots s_i \ldots s_{M-1} s_M \tag{2.8}$$

where $s_i \in S, \forall i = 1..M$.

The neighbourhood of an ECA has $2^3 = 8$ possible configurations. These configurations can be enumerated using a single number n_l in the range 0 to 7. If we define $\alpha_{-1} = 2^0 = 1, \alpha_0 = 2^1 = 2, \alpha_1 = 2^2 = 4$, we obtain $n_l = \sum_{j=-r}^{j=r} \alpha_j s_{i+j}(t-1)$ the argument of function f in (2.5).

The local rule (2.2), associated with a neighbourhood, can be represented explicitly in a table, known as the *rules table*. Each table consists of eight expressions (2.4) known as *elementary rules* (see Table 1).

Table 1. Rules table for an ECA. $a_i \in S, \forall \mathrm{i} = 1 \ldots 8$

$111 \rightarrow s_1$
$110 \rightarrow s_2$
$101 \rightarrow s_3$
$100 \rightarrow s_4$
$011 \rightarrow s_5$
$010 \rightarrow s_6$
$001 \rightarrow s_7$
$000 \rightarrow s_8$

The results can be represented in a string:

$$l = s_1 s_2 s_3 s_4 s_5 s_6 s_7 s_8 \tag{2.9}$$

which provides a concise representation of the rules table in Table 1 (Figure 2).

In the case of an ECA, the rules table can be represented concisely as a sequence of 8 bits - a byte. The length of string (2.9) is thus 2^3 and the number of possible strings is $2^{2^3} = 256$. Any ECA can thus be identified by a decimal number in the range 0 to 255. The string for a given rules table will be the same number in binary format. To convert the number to decimals we apply the formula:

$$n_d = 2^7 * s_1 + 2^6 * s_2 + 2^5 * s_3 + 2^4 * s_4 + 2^3 * s_5 + 2^2 * s_6 + 2^1 s_7 + s_8 .$$

If we consider a 1-D multi-state CA, with a neighbourhood of n elements $N = (x_1, \ldots\ldots, x_n)$, the number of possible configurations is k^n. In this case, we can write a rules table similar to Table 1, this time with k^n rows, each corresponding to an elementary rule (2.4). We can thus construct a string similar to (2.9):

$$l = s_1 s_2 \ldots \ldots \ldots \ldots s_{k^n} . \tag{2.10}$$

The number of possible local rules (2.2) is equal to the number of possible strings (2.10), that is k^{k^n}.

Table 2 shows the number of possible rules for specific cases.

A rule that discriminates between every possible configuration of the neighbourhood is said to be *exhaustive.* When the last character in the string is 0, the rule is *regular*. In the general case, the number of α_j in (2.5) is given by $\alpha_j = k^{r-j}$.

Figure 2. The top row shows every possible configuration of an ECA. The bottom row shows the development of the ECA as determined by a local rule (2.2), or rules table. The rule used here can be represented by the string l = 10000001

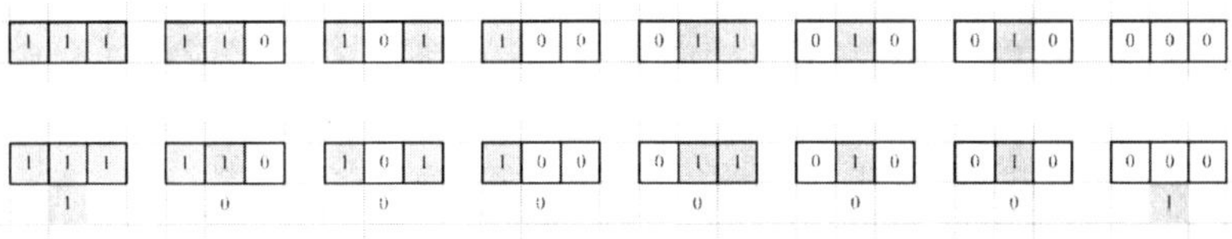

Table 2. Number of characters in the rules string and number of rules for different dimensions, different numbers of states and different radii of interaction. As we can see, the number of rules rapidly becomes extremely large. We consider Moore neighbourhood for D = 2.

Dimension (d)	States (k)	Radius (r)	Length of string (k^n)	Number of rules (k^{k^n})
1	2	1	8	256
1	2	2	32	4294967296
1	3	1	27	7625597484987
1	4	1	64	$3.403*e^{38}$
2	2	1	512	$1.3407*e^{154}$
2	3	1	19683	3^{19683}

Totalistic Rules

In an ECA, the rule $l = s_1s_2s_3s_4s_5s_6s_7s_8$ can be represented by a string of 8 bits. Consider the case in which $s_2 = s_5$ and $s_4 = s_7$. In this instance, configurations 110, 011, 100 and 001 give the same result, creating a left-right symmetry. The string for these rules can thus be rewritten:

$$l = s_1s_2s_3s_4s_2s_6s_4s_8 \tag{2.11}$$

For regular ECAs, we can rewrite (2.11) in the form $l = s_1s_2s_3s_4s_2s_6s_4 0$. In this case, the number of rules N_R will depend on just 5 parameters. Thus $N_R = 2^5 = 32$ and the rules for the system are 254, 250, 236, 232, 222, 218, 204, 200, 182, 178, 164, 160, 150, 146, 132, 128, 126, 122, 108, 104, 94, 90, 76, 72, 54, 50, 36, 32, 22, 18, 4, and 0. With this kind of symmetric rules table, an initially symmetric pattern will maintain its symmetry for the whole life of the CA.

In 1983, Wolfram introduced a new class of what he called "totalistic rules". With totalistic rules, the state of an individual cell depends on the *number* of neighbouring cells in a given state, regardless of their position in the neighbourhood. In the case of ECAs, totalistic rules are symmetrical, even when (110) gives the same results as (101) and (100) the same result as (010). In these cases, known as 2-totalistic rules, $s_2 = s_3$ and $s_4 = s_6$. With these rules, we can rewrite (2.11) as follows:

$$l = s_1 s_2 s_2 s_4 s_2 s_4 s_4 s_8 . \tag{2.12}$$

(2.12) depends on just 4 parameters: $s_1 s_2 s_4 s_8$. The number of totalistic rules N_T is given by $N_T = 2^4 = 16$. In the case of regular rules, $s_8 = 0$. In this case, there are just 8 rules (rules 254, 232, 150, 128, 126, 104, 22, 0). If we set $a_j = 1 \forall j$, we can rewrite (2.5) in the form:

$$s_i(t) = \mathrm{f}\left[\sum_{j=-r}^{j=r} a_j s_{i+j}(t-1)\right] = \mathrm{f}(s_{i-1}(t) + s_i(t) + s_{i+1}(t))$$

This formulation shows that the evolution of the system depends exclusively on the sum of the values of the states of neighbouring cells. If we want to be explicit, we can write the rules as in Table 3. The Table shows us that the local rule depends on just 4 parameters. Thus, any totalistic rule can be uniquely represented by a single decimal number in the range 0 to 15.

The total number of rules is given by s $2^4 = 16$.

In the case $k = 2$, $r = 2$, there are $2^5 = 32$ possible configurations of neighbouring cells. However, with totalistic rules, different configurations with the same number of active cells give the same result. So we can rewrite (2.10) in the form:

$$l = s_1 s_2 s_2 s_4 s_2 s_4 s_4 s_2 s_4 s_4 s_8 s_4 s_{16} s_2 s_4 s_4 s_8 s_4 s_8 s_8 s_{16} s_4 s_8 s_8 s_{16} s_8 s_{16} s_{16} s_{32} .$$

As can be seen from this expression, the number of totalistic rules, N_T, depends on 6 parameters $s_1 s_2 s_4 s_8 s_{16} s_{32}$. As a result, $N_T = 2^6 = 64$. Half of these rules have

*Table 3. Rules table for totalistic CAs with r=**1**,k=**2**. In this case, there are 4 elementary rules.*

$1+1+1=3 \rightarrow s_1$
$1+1+0=2 \rightarrow s_2$
$1+0+0=1 \rightarrow s_4$
$0+0+0=0 \rightarrow s_8$

$S_{32} = 0$ and are thus regular. These rules, listed exhaustively in the Appendix , can be enumerated using a single decimal number in the range 0 to 63.

For CAs with $D = 1$, $k = 2$, $r = 2$, 2.5 becomes

$$s_i(t) = \mathrm{f}(s_{i-2}(t) + s_{i-1}(t) + s_i(t) + s_{i+1}(t) + s_{i+2}(t))$$

Table 4 shows the rules table, now composed of 6 elementary rules.

These results can be extended to 1-D CAs with $k > 2$. The number of independent parameters is given by:

$$N' = \left[1 + (k-1)(2r+1)\right] \qquad (2.13)$$

For example, for $k = 3$ and $r = 1$, $N' = \left[1 + 2 * 3\right] = 7$. The total number of rules is given by $N_R = 3^7 = 2187$. For $D = 1$, $k = 2$, *and* $r = 1$ (2.13) can be rewritten in the form: $N' = 3k - 2$. These results can be easily extended to the two-dimensional case. In this case, (2.13) becomes:

$$N' = \left[1 + (k-1)n\right] \qquad (2.14)$$

Table 4. Rules table for CA with D = 1, k = 2, and r = 2. Each rule depends on 6 parameters. Individual rules can be enumerated using a decimal number in the range 0 to 63.

$1+1+1+1+1=5 \rightarrow s_1$
$1+1+1+1+0=4 \rightarrow s_2$
$1+1+1+0+0=3 \rightarrow s_4$
$1+1+0+0+0=2 \rightarrow s_8$
$1+0+0+0+0=1 \rightarrow s_{16}$
$1+0+0+0+0=0 \rightarrow s_{32}$

where n is the number of neighbouring cells. With a *von Neumann neighbourhood* with D=**2**,k=**2**, n=**5** and $N' = 6$; with a Moore neighbourhood, n=**9** and $N' = 10$

Totalistic rules have the advantage that they drastically reduce the size of the rule space. The disadvantage is that they make no distinction between the cell whose state is being calculated and the other cells in its neighbourhood. Introducing this distinction, we obtain so-called outer totalistic rules. As an example, consider a Boolean CA with D=2,r=2 and a Moore neighbourhood. According to the state of the cell, we have 2 neighbourhoods abcd0efgh and abcd1efgh, with the constraint that a+b+c+d+e+f+g+h=s<9.Each neighbourhood can be specified with 9 parameters. To construct the rules table, we thus need 18 parameters. The most famous example of this kind of rule is Wolfram's rule 224, (Wolfram 1985), used in the Conway's game of Life (Gardner, 1970). The rules table for the game is described in Table 5.

Development of 1-D CAs in Space and Time

The rules table, governing the development of a CA, is composed of a certain number of elementary rules. Each cell in the CA reads the state of the cells in its neighbourhood and determines its own state using the corresponding elementary rule in the rules table. In what follows, we limit our discussion to 1-D CAs.

In general, one-dimensional CAs, indicated by $D = 1$, are represented as a line of cells along an axis. Each cell contains a letter, showing the state of the cell. The subscript indicates the position of the cell on the line. Thus a_9 indicates the state of cell 9. Alternative ways of representing the CA are to show the state of each cell as a number (Figure 3a) or to use a colour code. This latter technique provides a compact representation for CAs with large numbers of cells.

Table 5. The rules table for "Life". The table can be represented by the binary string 000000000011100000 or the decimal code 224.

State=1	State=0
$1+1+1+1+1+1+1+1=8 \to 0$	$1+1+1+1+1+1+1+1=8 \to 0$
$1+1+1+1+1+1+1+0=7 \to 0$	$1+1+1+1+1+1+1+0=7 \to 0$
$1+1+1+1+1+1+0+0=6 \to 0$	$1+1+1+1+1+1+0+0=6 \to 0$
$1+1+1+1+1+0+0+0=5 \to 0$	$1+1+1+1+1+0+0+0=5 \to 0$
$1+1+1+1+0+0+0+0=4 \to 0$	$1+1+1+1+0+0+0+0=4 \to 0$
$1+1+1+0+0+0+0+0=3 \to 1$	$1+1+1+0+0+0+0+0=3 \to 1$
$1+1+0+0+0+0+0+0=2 \to 1$	$1+1+0+0+0+0+0+0=2 \to 0$
$1+0+0+0+0+0+0+0=1 \to 0$	$1+0+0+0+0+0+0+0=1 \to 0$
$0+0+0+0+0+0+0+0=0 \to 0$	$0+0+0+0+0+0+0+0=0 \to 0$

Figure 3. Shows the spatial organization of a 1-D CA. The number in the subscript indicates the position of the cell; the number in the next row shows its state (s={0,1}). The following row provides a visual representation of the state of the CA, in which white indicates a state of 0 and black a state of 1. Figures 3b and 3c represent the development of the CA in time and space. The rule for the CA is listed in the Appendix of the Chapter, under code C2.1.

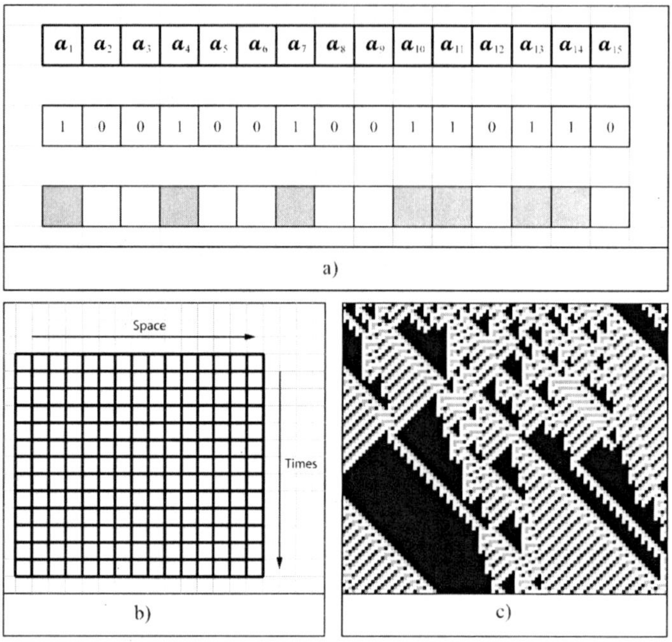

Each row of cells represents the state of the CA at a given moment in time. Thus, the x-axis shows the spatial organization of the cells while the y-axis shows their development over time, beginning at the top and finishing at the bottom of the diagram.

Development of 2-D CAs in Space and Time

To represent 2-D CAs, we can replace the linear sequence of cells we used in Figure 3, with a sequence of grid diagrams. As with 1-D CAs, the state of individual cells is represented with color codes. In Figure 4, cells with a state of 0 are white and those with a state of 1 are black. In the first diagram, all except 1 cell has a state of 0. The following diagrams show the development of the cell at successive time steps. The final diagram depicts the cell after 15 steps. The cell's development is governed by a k-totalistic rule with $k = 4$ and $r = 1$, and a *Moore neighbourhood.* As in Figure 4, the initial seed expands in a fairly regular fashion until it occupies the entire space. The rule for the CA is defined in the Collection of Rules, with code C2.2.

If we use the z-axis to represent the CA's development over time, we can construct 3-D spatial-temporal diagrams for 2-D CAs, similar to those used in the 1-D case. Figure 5 shows a 3-D representation of the CA. In this case, the picture shows 20 steps in the development process. The regularities in the patterns are even more evident than in the previous diagram. Figure 5b shows the spatial-temporal diagram for a CA with two initial seeds, evolving over 80 time steps and rapidly becoming extremely complex.

It is still possible to make out structures, but the overall picture is extremely complex. The CA depicted here consists of 80*80 cells. As in the one-dimensional case, the cell is organized as a torus ensuring that the neighbourhood of cells on the boundaries of the space is identical to that of the other cells. The results become evident, when the CA fills all the available space, as in the last stages of the developmental process, depicted in Figure 5b.

As can be seen in Figure 6, different CAs have very different developmental dynamics. This diagram depicts the growth of a seed using totalistic Boolean rules with *Moore neighbourhood* 169, 994, 674. As previously shown, the total number of rules for this CA is $2^{10} = 1024$.

Figure.4. Development of a CA with $k = 2, k = 4$ and $r = 1$. *The k-totalistic rule for the CA is given in the Appendix, with code C2.2. Initially only one cell is active (state=1). Figure 4 shows the CA's development in successive time steps. The final picture shows the CA after 15 time steps. By this stage, the CA has occupied all the space available.*

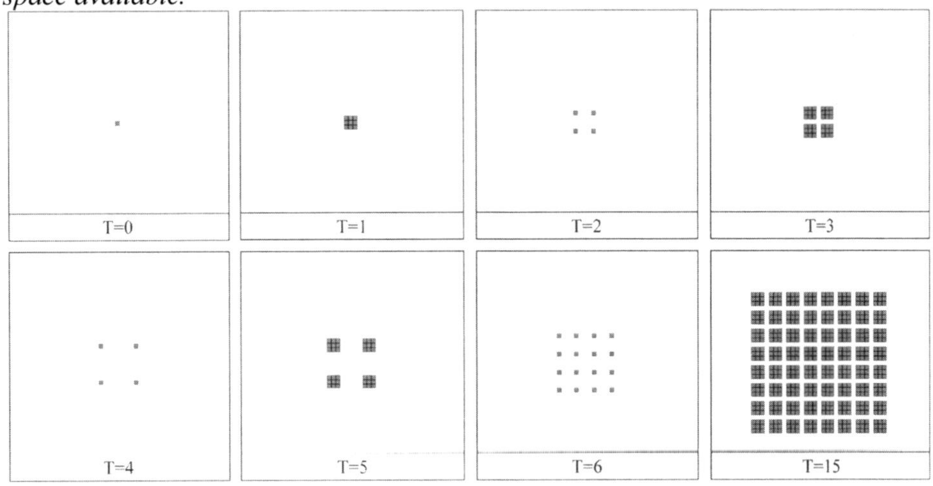

Figure 5. 3D diagrams showing 20 (a) and 80 (b) time steps in the development of the CA with rule C2.2, as described in the Collection of rules. The developmental process begins at the top of the diagram and proceeds towards the bottom. The result is a structure similar to the Sierpinski Arrowhead . The structure fills all available space.

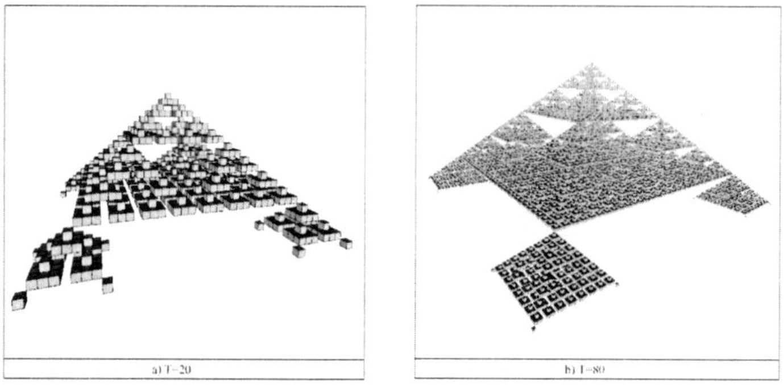

Development of 3-D CAs in Space and Time

The situation does not change when we consider 3-D CAs. These CAs can use several alternative geometries. To illustrate the possibilities, the first two rows of Figure 7 shows 8 steps in the development of a CA with cubic cells, applying totalistic rule 198. The third row shows 4 steps in the development of a second CA with cubic cells, this time applying outer-totalistic rule 23406. The next row shows the application of the same rule to cells that are truncated octahedrons. The fourth row depicts 4 steps using totalistic rule 12434 and cells that are rhombic dodecahedrons.

In the CAs with cubic cells, each cell's neighbourhood comprises 27 cells. There are thus $2^{28} = 268435456$ totalistic rules for the CA. In the CA with rhombic dodecahedrons, the neighbourhood includes 13 elements and there are $2^{14} = 16384$ rules. The cells that are truncated octahedrons have a neighbourhood of 9 elements. Given that this CA uses outer-totalistic rules, the number of rules is $2^{16} = 65536$. We chose the symmetries shown here because they are examples of space-filling CAs. However, so long as we define neighbourhoods in a suitable way, we can easily construct alternative symmetries. The underlying CA formalism is unaffected by the choice of cell shape and neighbourhood.

Figure 6. The first three rows (a, b, c) illustrate the first 7 and the 10th time step in the development of totalistic CAs (the CAs with decimal codes 169, 994, 674) for $D = 2, k = 4$ *and* $r = 1$. *The fourth row (d, e, f) provides a 3-D diagram showing ten steps of development with the same rules.*

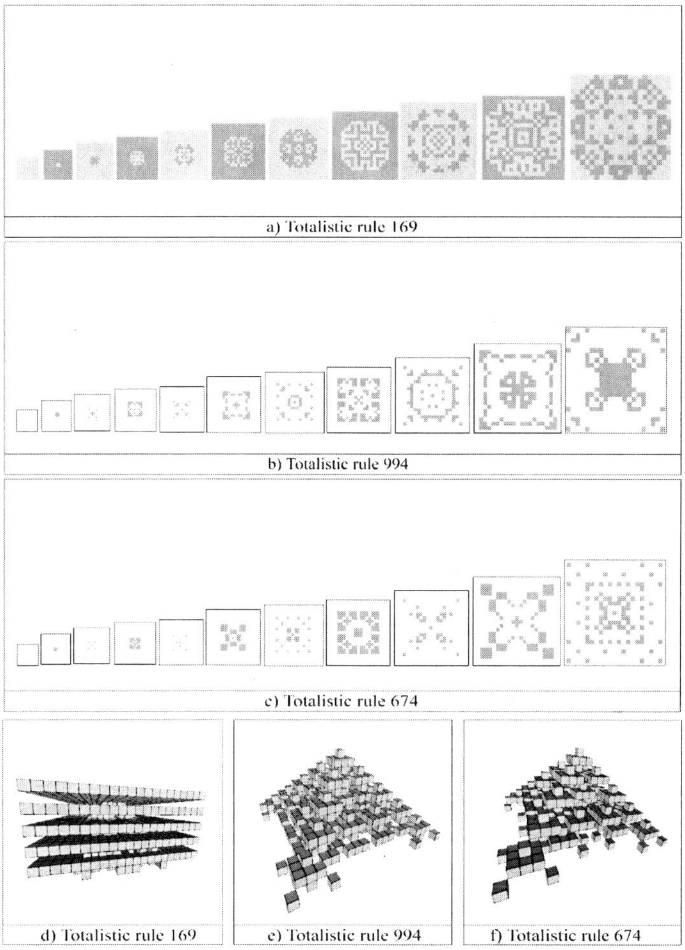

Emergent Structures

The CAs we have considered so far all consist of identical cells with non-linear local interactions. The most feature of their behavior is the emergence of structures that stand out clearly against the background. Although interactions among individual cells are purely local, the structures they produce can be extremely large, often displaying large scale symmetries in time and space. Later, we will examine

Figure 7. The first two rows show the development over time of a CA with cubic cells whose development is governed by totalistic rule 198. The third row shows the temporal development of a CA whose cells are rhombic dodecahedrons and whose development is governed by totalistic rule 12434; the fourth row shows the development of a CA based on outer-totalistic rule 23406. The cells of this CA have the form of a truncated octahedron.

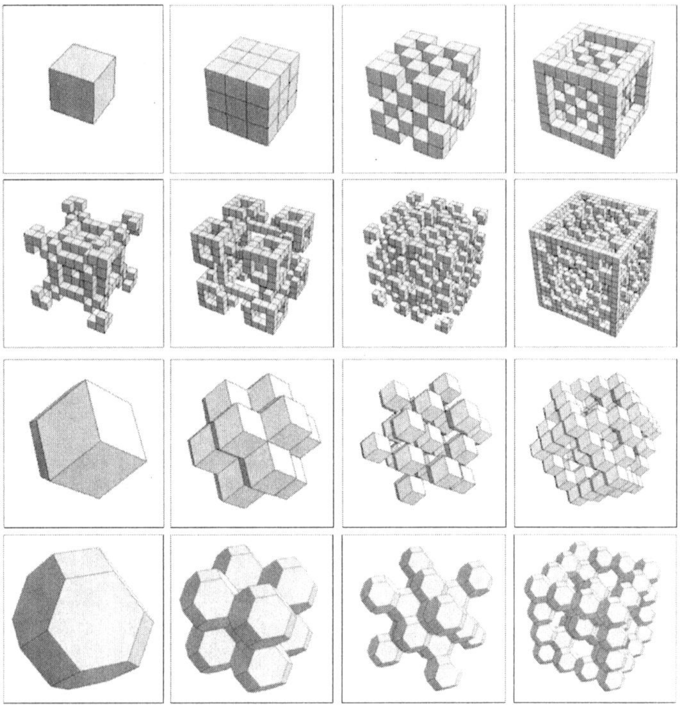

many examples of such structures, both from 1-D and 2-D CAs. Together display a wide variety of shapes and behaviors. Some of the most interesting are gliders, glider-guns and self-replicators.

Gliders are structures that remain unaltered as they move through space. When they collide with other gliders they transform, sometimes disappearing and sometimes generating novel structures. Glider-guns are highly complex structures that over time generate whole swarms of glider-guns. By contrast, self-replicators are structures that replicate over time, in well-defined self-replication processes. Different self-replicators can cohabit within the same system. Their Complex interactions sometimes lead to the transformation of one kind of self-replicator into another. In the rest of this book, we will see the key features of these emergent structures and study how they behave.

MAIN FOCUS OF THE CHAPTER

One of the fundamental problems in research on CAs is the search for rules of self-organization relevant to the study of physical and biological systems. Such rules lead to complex behaviors with long-distance correlations that are not directly linked to local interactions among neighbours. These are what Wolfram (1984) have called "complex rules". Wolfram looked for complex rules by systematically exploring specific rule spaces, for example ECAs, or the space of totalistic rules with $D = 1$, $r = 2$ and $r = 2$. However, the very high cardinality of rule spaces with $D > 1$ or $k > 2$, means that a systematic approach is no longer feasible for these CAs. An effective alternative, which we explore in the coming chapters, is to use Genetic Algorithms.

Genetic Algorithms (GAs) were first proposed by Holland (1994). The idea is to reproduce the evolution of a population subject to natural selection. In a GA, an initial population of randomly generated "artificial organisms" improves its ability to survive in its ecosystem (its "fitness") via a process of mutation, cross over, and selection. Given a population of "rules", we can use GAs to evolve rules of increasing complexity. As discussed in Chapter 5, the formal definition of complexity is an open issue much debated among experts. In the work presented here, we base our fitness function on Wuensche's (1999) concept of Input-Entropy. Using this concept, we have conducted experiments, demonstrating the effectiveness of GAs as a tool for exploring rule space and identifying complex rules (Bilotta et al. 2002; 2003a; 2003b). In what follows, we present results for $D = 1$ and $D = 2$. To handle this last case, we introduce a new category of what we call "k-totalistic rules".

Looking for CAs with Interesting Behaviors

If we apply Genetic Algorithms to 1-D CAs with $k>2$, we generate a large number of rules. Figure 8 illustrates a few examples. The CAs shown in the figure are based on the rule strings (2.10)

Let us now analyze the patterns in Figure 8, and the kinds of self-organization seen in the diagrams. The main structures we observe are uniform domains, regular domains, gliders, and other "glider-gun" structures. Uniform domains are regions in a pattern in which all cells present in the region have the same state; regular domains are regions in which cells are not necessarily in the state but nonetheless exhibit well-defined and easily visualized periodic behavior. Gliders are mobile structures that exhibit well defined periodic behavior as they move (Hanson & Crutchfield, 1997). More rigorous definitions of these structures will be introduced in Chapter 4.

Providing a rigorous analysis of the patterns shown in Figure 8 would be a difficult and lengthy task, outside the scope of this book. Here we will limit ourselves to the analysis of just some of the rules, showing how we can identify emergent

Figure 8. Examples of complex rules. The first 5 are 1-D CAs with $k = 4$ and $r = 1$ (a, b, c, d, e). In the sixth example $k = 4$ and $r = 2$, (f), in the seventh and eighth $k = 5$ and $r = 1$ (g, h), in the ninth, tenth and eleventh $k = 6$ and $r = 1$ (i, j, k) In the last example (l), $k = 8$ and $r = 1$. In the Collection of rules, these systems are assigned codes C2.3-C2.14.

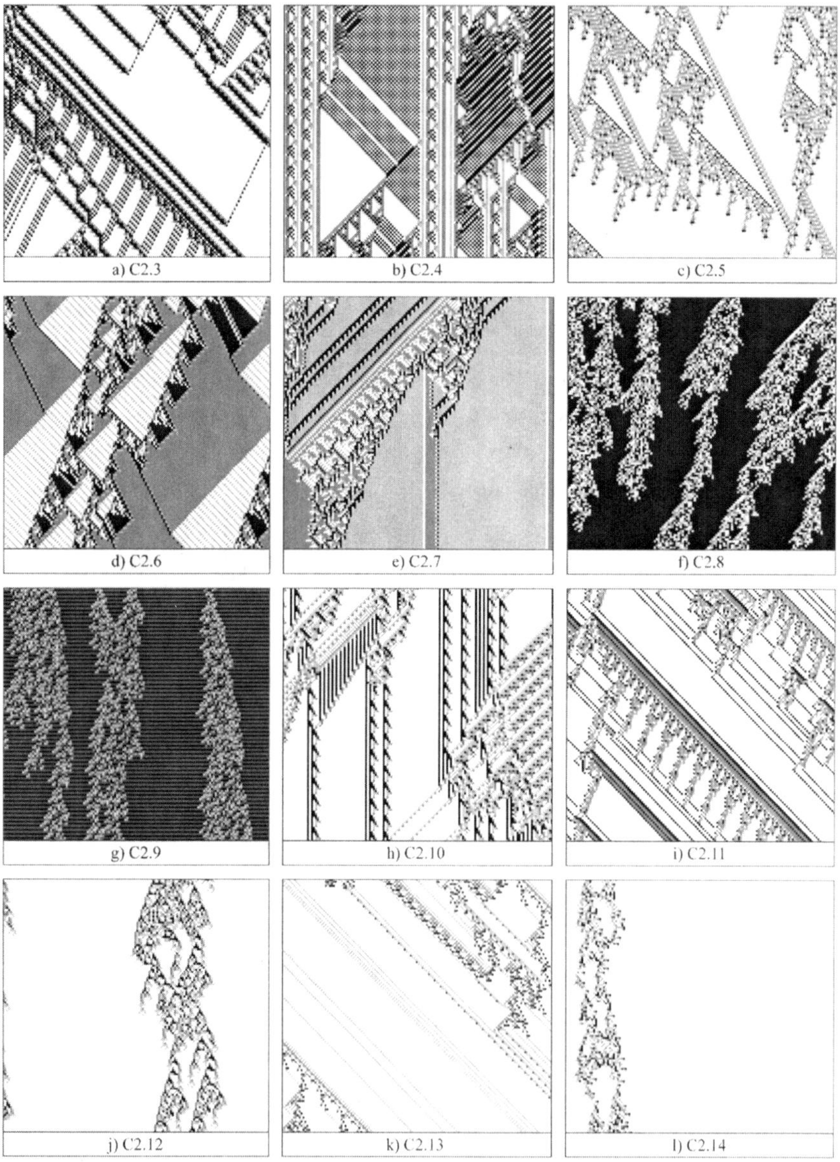

structure. Consider, for example, rule C2.3. A preliminary analysis of the pattern produced by this rule shows a uniform domain of cells with state 1, and a series of gliders moving within the uniform domain (1).

a) The first glider, denoted here as α (code 0002) travels towards the right.
b) The second,β (code3103), travels towards the left.
c) The third, γ (code 310310) travels towards the left at the same speed as β . γ is constructed by adding an additional component (with state 01) to β In other words, γ=β01. Chapter 9 describes how we can use this composition as the basis for a construction language.
d) The fourth glider, δ, consists of a single periodic cell which oscillates between state 3 and state 0 and which moves leftwards at the same speed as the previous two gliders.

Careful examination shows other interesting phenomena, represented in Figure 9 (Bilotta & Pantano, 2001). For example, there are two gliders α **and** β. β disappears and α shifts towards the right (formally $\alpha+\beta=\alpha)_{s+}$). This behavior is not uniform. In reality, different phases of the glider produce different behaviors. Another composition is $\alpha+\beta=\beta)_{s-}$. In this case, the surviving glider, β shifts towards the left. In yet another class of interaction, we observe α+β=δ.

Without attempting a detailed discussion of individual rules, a number of observations are in order.

a) Rules presented in Figure 8 may produce both uniform and regular domains. In many cases, they generate elementary gliders that cross the uniform domains leaving a dense path behind them.
b) Rule C2.4 produces a glider-gun. Interaction among gliders (300131003022) produces a regular domain (3001310030220) (Figure 9b). If we insert a (0) between two gliders we effectively couple them.
c) Rules C2.11 e C2.13 also allow coupling between gliders. These interactions lead to the formation of regular domains or new gliders of gradually increasing complexity. For example the very large glider (10000000000000424200000000001241002000010015101241100110020100000013340001320003401 3231) has the ability to couple with itself (Figure 9c). These rules can also produce couplings between different gliders as when glider (14000010001451500000 0014145500000225001313224124110001) couples with glider (255100100 040000101552151520134110521002551). We will see how this mechanism can be encoded to produce linguistic structures.
d) The very large gliders described in point (c) are not the only ones. We observe the same phenomenon for rules C2.8 and C2.9. For example, glider (21202)

Figure 9. The first image (a) illustrates the interaction between two gliders governed by rule C2.3. The second (b, rule code C2.4) shows a regular domain created by composing gliders of the same type. The third (c, rule code C2.11) shows the composition of two gliders of the same type and the simultaneous appearance of other gliders. The fourth (d, rule code C2.9) shows a glider with a very long period that divides the diagram into two homogeneous regions.

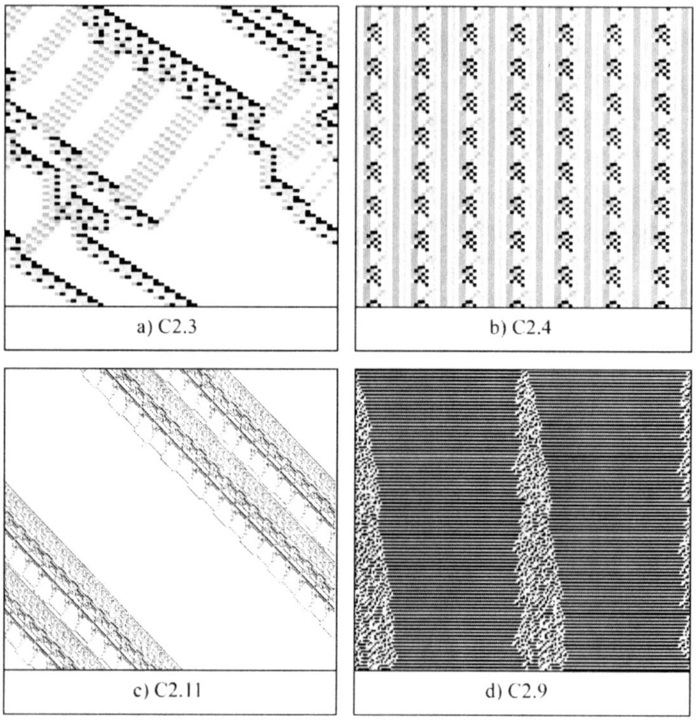

produced by rule C2.8, has a period of 82, a maximum size of 21 characters (133000130300012120211) and a minimum size of a single character (3). C2.9 produces a glider at the point of contact between two domains (0) and (2). The new structure then gives rise to a highly complex glider with a period of more than 200 steps (see Figure 9d) that continues to act as a separator and link between two uniform domains.

From what we have seen, it is clear that analyzing CAs at the level of individual rules is a very time-consuming task. Complex rules of interaction and composition, and the presence of very large structures, covering large areas of space and time, are symptoms of highly variable behaviour that does not lend itself to easy classification. In later chapters, we adopt a "genetic" approach to the description of these characteristics, identifying features of CAs behavior that are in some way similar to those of biological systems.

K-Totalistic Rules

Exhaustive approaches to the exploration of rule space are unfeasible for CAs with multiple states and large neighbourhoods. The k-totalistic rules, we introduce below, avoid this problem by reducing the size of the rule space. The totalistic and outer-totalistic rules introduced by other authors avoid this problem by considering only the sum of state values, independently of the state of individual cells, thereby reducing the size of the rule space. This means, however, that two distinct neighbourhoods of type (111) are both associated with the same rule (102). This leads to an enormous reduction in the number and variety of emergent structures, such as self-replicators [Bilotta et al., 2003b]. k-totalistic rules avoid an explosion in the size of the rules space, while still taking account of the number of neighbouring cells in specific states.

Consider a new class of CAs, whose development is a function of the total number of neighbouring cells, in a given state. We will refer to these CAs as k-totalistic CAs. From a formal point of view, we can describe a k-totalistic CA as follows.

A neighbourhood $N = (x_1, ..., x_n)$ is a set of *n* different elements $x_i \in Z^D$. Let the number of elements in *N*, with state 0 be h_0, the number of elements with state 1, h_1, and so on up to h_{k-1}. In this way, we can associate any $N = (x_1, ..., x_n)$ with a string $(h_0 h_1 ... h_{k-1})$, where the values of h_i are not arbitrary, but satisfy the constraints:

$$h_0 + h_1 + ... + h_{k-1} = n \text{ and } h_i \geq 0 \text{, for } i = 0, 1, ..., k-1. \tag{2.15}$$

Let *V* be the number of possible configuration of *N*. Totalistic rules associate each of these configurations with a numeric string $(h_0 h_1 .... h_{k-1})$. Let *G* be the set of all these configurations. We define a k-totalistic rule *T* as an application that associates any configuration $v \in V$ with an element in *G* that satisfies constraints (2.15):

$$T : V \rightarrow G \tag{2.16}$$

Application (2.16) can be represented explicitly as a table that associates each of the first k-1 totals with a number between 0 and k-1. Given constraints (2.15), the last total depends on the others. For example, consider string (240) for a CA with $k = 4$, $D = 2$ and $r = 1$, and with a Moore neighbourhood, composed of 9 cells, in which 2 are in state 0, 4 are in state 1, 0 in state 2 and 3 in state 3. This latter value is obtained by applying (2.15). In other words, $h_4 = 9 - 2 - 0 - 4 = 3$.

String (240) corresponds to one element $q \in G$. To construct the rules table, we need to precisely count the number of elements in *G*. The presence of constraints (2.15) substantially lowers the number of possible configurations. Computing this value is the same as computing the number of ways in which we can place *n* non-distinguishable objects in *k* containers.

From combinatorial analysis we obtain:

$$N_T = \binom{k+n-1}{n} = \frac{(k+n-1)!}{(n)!(k-1)!} \tag{2.17}$$

Where N_T is the number of elementary rules.

In the case considered earlier, k=4 and n=9. Thus $N_T = \frac{12!}{9!3!} = \frac{12x11x10}{6} = 220$

Appendix 1 provides an example of a rules table associating each element of *G* with a particular configuration $v \in V$ of neighbourhood *N*. We can represent this kind of table as a string of numbers of length N_T.

$$l = \left(s_1, s_2, \ldots\ldots\ldots, s_{N_T}\right) \tag{2.18}$$

with $s_i \in S$.

To match a configuration $(h_0 h_1 \ldots . h_{k-1}) \in V$ with a specific value s_i in the string (2.18) and vice versa, we fix the convention that s_i corresponds to the ith configuration in lexicographical order.

Table 6 compares values of N_T for k-totalistic and exhaustive rules. All the examples refer to a 2-dimensional CA, with a Moore neighbourhood.

With $k = 2$, the k-totalistic rules are identical to traditional totalistic rules for binary CAs.

Table 6. Numbers of k-totalistic and traditional exhaustive rules for different values of .k and r

k	r	L (k-totalistic rules)	L (exhaustive rules)
2	1	10	512
	2	26	33554432
	3	50	$\sim 1.1 * 10^{15}$
3	1	55	19683
	2	351	$\sim 8.5 * 10^{11}$
	3	1275	$\sim 2.4 * 10^{23}$
4	1	220	262144
	2	3266	$\sim 1.1 * 10^{18}$
6	1	2002	10077696
8	1	11440	134217728

Complex K-Totalistic Rules for 2-Dimensional CAs

Complex rules for 2-D CAs with $k > 2$, produce a vast number of highly differentiated emergent structures, making analysis a challenge. By using k-totalistic rules, we can significantly reduce the cardinality of the rule space and use genetic algorithms to search for complex rules. As in the 1-D case, we use fitness functions based on Input Entropy. This method is highly effective in discovering rules that produce complex emergent structures including self-replicators. Figure 10, for example, shows 40 steps in the development of a complex k-totalistic CA, with random initial conditions. The rule for the CA (C2.15) has $d = 2$, $k = 4$, and $r = 1$. Using (2.18), we can encode the string as follows:

200033300001310300020200200000032000320000000213002200013030003
0000330302003000010020000000000023021033130000000021000003000213122
22030000000000000000100300030300000000020033000003021022022130030120
0000000000000003102000000.

In the pattern produced by this rule, we observe a large number of complex structures, including gliders, glider-guns, self-replicators, etc.

Table 7 summarizes the features of some of these structures and their respective codes.

In the following chapters, we provide an exhaustive treatment of these emergent structures including a careful analysis of their behavior and self-replication capabilities.

Figure 10. Pattern generated by k-totalistic rule C2.15 with $D = 2, k = 4$ and $r = 1$ *listed in the Appendix. We observe a number of emergent structures including gliders, glider-guns, self-replicators, etc.*

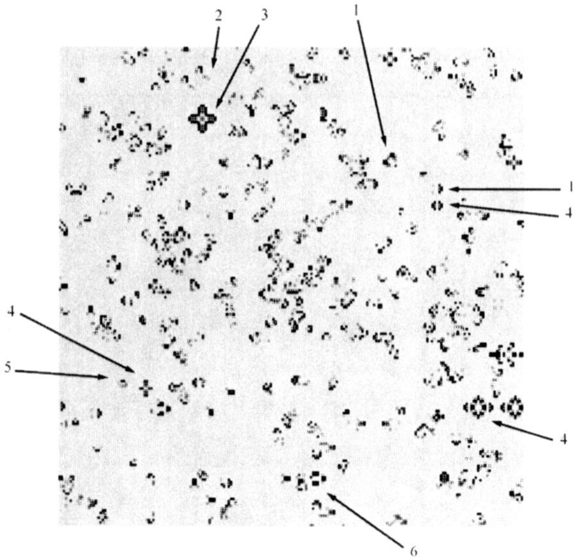

FUTURE TRENDS

The search for complex rules with the ability to generate self-organizing structures is a fundamental challenge for CA research. In this chapter, we have seen how Genetic Algorithms can facilitate the search for such rules and demonstrated GAs ability to find complex rules. We have shown, furthermore, that although the high cardinality of rule space would normally prevent us from using GAs for multi-state CAs, we can avoid this difficulty by using k-totalistic rules.

Despite these achievements, many problems remain to be resolved. In particular:

a) How should we measure the complexity of a rule?
b) What are the best fitness functions when we apply genetic algorithms?
c) How should we choose the parameters for the Genetic Algorithm?

When we want to analyze individual rules additional questions arise:

a) How should we analyze the patterns we observe?
b) How should we classify these patterns?
c) What kinds of emergent structure do we observe?

Table 7. Emergent structures produced by k-totalistic rule C2.15 with $D = 2, k = 4$ and $r = 1$.

Emergent Structures	Features	Emergent structures' codes
1	Structure 1 is a transient form that degenerates into two gliders. The two gliders travel in opposite directions.	00000330000 00001331000 00012021000 00330001000 00332000000 00011103000
2	Structure 2 is highly complex and behaves like a glider gun. The structure is involved in a process of self-replication. The code for th process is shown bottom right.	0000330000000 0001030000000 0001030000000 0000000000000 0000001100000 0000000010000 Code for the self-replication process 00000001000000 00000003300000 00000003300000 00030301000000 00033300000000
3	Structure 3 is highly complex. The development of the structure includes the emergence of glider-guns and self-replication.	00000000330000000 00000003223000000 00000003223000000 00000003223000000 00003333003333000 00032220220222300 00032220220222300 00003333003333000 00000003223000000 00000003223000000 00000003223000000 00000000330000000
4	Structure 4 is a self-replicator with linear behavior. The picture provides a snapshot of one moment during the self-replication process.	00000330000 00000220000 00000220000 00033003300 00000220000 00000220000 00000330000
5	Glider	000003300000 000010010000 000001110000
6	This is a very complex structure that behaves as a glider-gun. The gun generates self-replicating gliders with a structure reassembling structure 4.	0033300000 0033301000 0300003300 0000003300 0033301000 0033300000

d) What are the fundamental features of these structures?
e) What kind of automated procedures can we use to find these structures?
f) When should we consider these structures as self-replicators?
g) What are the characteristics of these self-replicators?

CONCLUSION

These are the questions to which we try to respond in the rest of this book, fully aware that in some cases, the responses we offer are not complete. The main themes will be the search for complex rules and artificial self-replication, topics to which we will dedicate many chapters of the book. Our findings suggest that CAs can be a useful tool for modeling biological behavior and identifying new computing paradigms (Adamatzky et al., 2007; Paun et al., 2006; Roska et al., 2006; Roska & Chua, 1993).

In this chapter we have offered the reader a formal treatment of Cellular Automata, of the way they develop in time and space and of certain specific categories of systems based on totalistic and k-totalistic rules. The former, first introduced by Wolfram in 1984, are well known in the CA literature. The latter were first introduced by the authors of this book, in 2002. The motivation behind the introduction of k-totalistic rules is the need to reduce the size of the rules space, so as to allow the kind of exhaustive search used by Wolfram or at least a GA-based search for complex rules, as in our own work.

Complex CAs give rise to emergent self-organizing structures, including self-replicators. In later chapters we will discuss these patterns and behaviors in greater depth. In some cases we propose complete answers to the questions we have raised; in others, our answers are only partial.

REFERENCES

Adamatzky, A., Bull, L., De Lacy Costello, B., Stepney, S., & Teuscher, C. (Eds.). (2007). Unconventional Computing. Frome, UK: Luniver press.

Bilotta, E., Lafusa, A., & Pantano, P. (2002). Is self replication an embedded characteristic of Artificial/Living matter? In Standish, R. K., Bedau, M. A., & Abbass, H. A. (Eds.), *Artificial Life VIII* (pp. 38–48). Cambridge, MA: The MIT Press.

Bilotta, E., Lafusa, A., & Pantano, P. (2003a). Life-like self-reproducers. *Complexity*, *9*(1), 38–55. doi:10.1002/cplx.10118

Bilotta, E., Lafusa, A., & Pantano, P. (2003b). Searching for complex CA rules with Gas. *Complexity*, *8*(3), 56–67. doi:10.1002/cplx.10063

Bilotta, E., & Pantano, P. (2001). Observations on complex multi-state CAs. Paper presented at Advances in Artificial Life: 6th European Conference, ECAL 2001. In J. Kelemen, & P. Sosik (Eds.), *Lecture Notes in Computer Science* (pp. 226–235). Heidelberg: Springer-Verlag.

Bilotta, E., & Pantano, P. (2005). Emergent Patterning phenomena in 2D Cellular automata. *Artificial Life Journal*, *11*(3), 339–362. doi:10.1162/1064546054407167

Bilotta, E., & Pantano, P. (2006). Structural and Functional Growth in Self-Reproducers. *Complexity*, *11*(6), 12–29. doi:10.1002/cplx.20131

Chua, L. O. (1998). *CNN: A Paradigm for Complexity*. Singapore: World Scientific Publishing Company.

Chua, L. O. (2006). A Nonlinear Dynamics Perspective Wolfram's New Kind of Science: *Vol. 1. World Scientific Series on Nonlinear Science*. Singapore: World Scientific Publishing Company.

Chua, L. O. (2007). A Nonlinear Dynamics Perspective of Wolfram's New Kind of Science: *Vol. 2. World Scientific Series on Nonlinear Science*. Singapore: World Scientific Publishing Company.

Chua, L. O., Sbitnev, V. I., & Yoon, S. (2004). A nonlinear dynamics perspective of Wolfram's new kind of science. Part III: Predicting the unpredictable. *International Journal of Bifurcation and Chaos in Applied Sciences and Engineering*, *14*, 3689–3820. doi:10.1142/S0218127404011764

Chua, L. O., & Yang, L. (1988). Cellular neural networks: Theory and Applications. *IEEE Transactions on Circuits and Systems*, *35*, 1257–1290. doi:10.1109/31.7600

Chua, L. O., Yoon, S., & Dogaru, R. (2002). A nonlinear dynamics perspective of Wolfram's new kind of science. Part I: Threshold of complexity. *International Journal of Bifurcation and Chaos in Applied Sciences and Engineering*, *12*, 2655–2766. doi:10.1142/S0218127402006333

Fatès, N., & Morvan, M. (2005). An Experimental Study of Robustness to Asynchronism for Elementary Cellular Automata. *Complex Systems*, *16*(1), 1–27.

Fatès, N., Morvan, M., Schabanel, N., & Thierry, E. (2006). Fully asynchronous behavior of double-quiescent elementary cellular automata. *Theoretical Computer Science*, *362*, 1–16. doi:10.1016/j.tcs.2006.05.036

Gardner, M. (1970). Mathematical games: the fantastic combinations of John Conway's new solitaire game Life. *Scientific American, 223*, 120–123. doi:10.1038/scientificamerican1070-120

Hanson, J. E., & Crutchfield, J. P. (1997). Computational Mechanics of Cellular Automata: An Example. *Physica D. Nonlinear Phenomena, 103*, 169–189. doi:10.1016/S0167-2789(96)00259-X

Hilborn, R. C. (2000). *Chaos and Nonlinear Dynamics: An Introduction for Scientists and Engineers*. New York: Oxford University Press.

Hoekstra, A. G., Falcone, J., Caiazzo, A., & Chopard, B. (2008). Multi-scale Modeling with Cellular Automata: The Complex Automata Approach. In Cellular Automata (LNCS 5191, pp. 192-199).

Holland, J. (1994). *Adaptation in natural and artificial systems*. Cambridge, MA: The MIT Press.

Langton, C. G. (1984). Self-Reproduction in Cellular Automata. *Physica, 10D*, 135–144.

Langton, C. G. (1986). Studying artificial life with cellular automata. *Physica, 22D*, 20–149.

Langton, C. G. (1990). Computation at the edge of chaos. *Physica, 42D*, 2–37.

Li, W., & Packard, N. (1990). The structure of the elementary cellular automata rule space. *Complex Systems, 4*, 281–297.

Mitchell, M. (1996). *An Introduction to Genetic Algorithms*. Cambridge, MA: The MIT Press.

Nehaniv, C. L. (2002). Evolution in Asynchronous Cellular Automata . In Standish, R. K., Bedau, M. A., & Abbass, H. A. (Eds.), *Artificial Life VIII*. Cambridge, MA: The MIT Press.

Oliveira, G. M. B., De Oliveira, P. P. B., & Omar, N. (2001). Definition and Applications of a Five-Parameter Characterization of One-Dimensional Cellular Automata Rule Space. *Artificial Life Journal, 7*(3), 277–301. doi:10.1162/106454601753238645

Paun, G., Rozenberg, G., & Salomaa, A. (2006). *DNA Computing: New Computing Paradigms*. New York: Springer-Verlag.

Roska, B., Molnara, A., & Werblin, F. S. (2006). Parallel Processing in Retinal Ganglion Cells: How Integration of Space-Time Patterns of Excitation and Inhibition Form the Spiking Output. *Journal of Neurophysiology, 95*, 3810–3822. doi:10.1152/jn.00113.2006

Roska, T., & Chua, L. O. (1993). The CNN Universal Machine: an analogic array computer. *IEEE Transactions on Circuits and Systems-II*, *40*, 163–173. doi:10.1109/82.222815

Sipper, M., Tomassini, M., & Capcarrere, M. S. (1997). *Evolving Asynchronous and Scalable Non-Uniform Cellular Automata.* Paper presented at the International Conference on Artificial Neural Networks and Genetic Algorithms (ICANNGA97). Heidelberg: Springer-Verlag.

Strogatz, S. H. (2000). *Nonlinear Dynamics and Chaos: With Applications to Physics, Biology, Chemistry, and Engineering*. New York: Perseus Books.

Toffoli, T., & Norman, M. (1987). *Cellular Automata Machines: A New Environment for Modeling*. Cambridge, MA: MIT Press.

von Neumann, J. (1966). *Theory of Self-Reproducing Automata* (Burks, A. W., Ed.). Urbana, IL: University of Illinois Press.

von Neumann, J., & Burks, A. W. (1970). von Neumann's self-reproducing automata. In A. W. Burks (Ed.), Essays on Cellular Automata (pp. 3-64). Urbana, IL: University of Illinois Press.

Wolfram, S. (1983). Statistical Mechanics of Cellular Automata. *Reviews of Modern Physics*, *55*, 601–644. doi:10.1103/RevModPhys.55.601

Wolfram, S. (1984). Universality and Complexity in Cellular Automata. *Physica*, *10D*, 1–35.

Wolfram, S. (1985). Undecidability and intractability in theoretical physics. *Physical Review Letters*, *54*(8), 735–738. doi:10.1103/PhysRevLett.54.735

Wolfram, S. (2002). *A New Kind of Science*. Champaign, IL: Wolfram Media.

Wuensche, A. (1999). Classifying Cellular Automata Automatically: Finding gliders, filtering and relating space-time patterns, attractors basins and the Z parameter. *Complexity*, *4*(3), 47–66. doi:10.1002/(SICI)1099-0526(199901/02)4:3<47::AID-CPLX9>3.0.CO;2-V

Wuensche, A., & Lesser, M. (1992). *The Global Dynamics of Cellular Automata: An Atlas of Basin of Attraction Fields of One-Dimensional Cellular Automata.* Reading, MA: Addison-Wesley.

APPENDIX

Totalistic Rules:Table 8

Table 8. Case $D = 1, k = 2, r = 2$

0 . - 00000000000000000000000000000000	32. - 10000000000000000000000000000000
1. - 00000000000000000000000000000001	33. - 10000000000000000000000000000001
2. - 00000000000000010000000100010110	34. - 10000000000000010000000100010110
3. - 00000000000000010000000100010111	35. - 10000000000000010000000100010111
4. - 00000001000101100001011001101000	36. - 10000001000101100001011001101000
5. - 00000001000101100001011001101001	37. - 10000001000101100001011001101001
6. - 00000001000101110001011101111110	38. - 10000001000101110001011101111110
7. - 00000001000101110001011101111111	39. - 10000001000101110001011101111111
8. - 00010110011010000110100010000000	40. - 10010110011010000110100010000000
9. - 00010110011010000110100010000001	41. - 10010110011010000110100010000001
10. - 00010110011010010110100110010110	42. - 10010110011010010110100110010110
11. - 00010110011010010110100110010111	43. - 10010110011010010110100110010111
12. - 00010111011111100111111011101000	44. - 10010111011111100111111011101000
13. - 00010111011111100111111011101001	45. - 10010111011111100111111011101001
14. - 00010111011111110111111111111110	46. - 10010111011111110111111111111110
15. - 00010111011111110111111111111111	47. - 10010111011111110111111111111111
16. - 01101000100000001000000000000000	48. - 11101000100000001000000000000000
17. - 01101000100000001000000000000001	49. - 11101000100000001000000000000001
18. - 01101000100000011000000100010110	50. - 11101000100000011000000100010110
19. - 01101000100000011000000100010111	51. - 11101000100000011000000100010111
20. - 01101001100101101001011001101000	52. - 11101001100101101001011001101000
21. - 01101001100101101001011001101001	53. - 11101001100101101001011001101001
22. - 01101001100101111001011101111110	54. - 11101001100101111001011101111110
23. - 01101001100101111001011101111111	55. - 11101001100101111001011101111111
24. - 01111110111010001110100010000000	56. - 11111110111010001110100010000000
25. - 01111110111010001110100010000001	57. - 11111110111010001110100010000001
26. - 01111110111010011110100110010110	58. - 11111110111010011110100110010110
27. - 01111110111010011110100110010111	59. - 11111110111010011110100110010111
28. - 01111111111111101111111011101000	60. - 11111111111111101111111011101000
29. - 01111111111111101111111011101001	61. - 11111111111111101111111011101001
30. - 01111111111111111111111111111110	62. - 11111111111111111111111111111110
31. - 01111111111111111111111111111111	63. - 11111111111111111111111111111111

Example of a Rules Table: Table 9

Table 9.

n.1 0009>>2	n.38 0432>>2	n.75 1224>>2	n.112 2133>>1	n.149 3150>>2	n.186 5004>>1
n.2 0018>>3	n.39 0441>>2	n.76 1233>>2	n.113 2142>>2	n.150 3204>>1	n.187 5013>>1
n.3 0027>>0	n.40 0450>>2	n.77 1242>>1	n.114 2151>>0	n.151 3213>>2	n.188 5022>>2
n.4 0036>>1	n.41 0504>>2	n.78 1251>>2	n.115 2160>>2	n.152 3222>>2	n.189 5031>>2
n.5 0045>>2	n.42 0513>>2	n.79 1260>>3	n.116 2205>>2	n.153 3231>>3	n.190 5040>>2
n.6 0054>>3	n.43 0522>>2	n.80 1305>>2	n.117 2214>>2	n.154 3240>>2	n.191 5103>>1
n.7 0063>>2	n.44 0531>>2	n.81 1314>>2	n.118 2223>>3	n.155 3303>>2	n.192 5112>>2
n.8 0072>>2	n.45 0540>>2	n.82 1323>>2	n.119 2232>>0	n.156 3312>>2	n.193 5121>>2
n.9 0081>>2	n.46 0603>>2	n.83 1332>>2	n.120 2241>>2	n.157 3321>>2	n.194 5130>>0
n.10 0090>>2	n.47 0612>>2	n.84 1341>>3	n.121 2250>>2	n.158 3330>>1	n.195 5202>>2
n.11 0108>>0	n.48 0621>>2	n.85 1350>>3	n.122 2304>>3	n.159 3402>>0	n.196 5211>>2
n.12 0117>>2	n.49 0630>>2	n.86 1404>>3	n.123 2313>>0	n.160 3411>>2	n.197 5220>>2
n.13 0126>>0	n.50 0702>>2	n.87 1413>>1	n.124 2322>>1	n.161 3420>>0	n.198 5301>>2
n.14 0135>>2	n.51 0711>>2	n.88 1422>>2	n.125 2331>>2	n.162 3501>>2	n.199 5310>>1
n.15 0144>>2	n.52 0720>>2	n.89 1431>>2	n.126 2340>>1	n.163 3510>>2	n.200 5400>>2
n.16 0153>>2	n.53 0801>>2	n.90 1440>>0	n.127 2403>>3	n.164 3600>>2	n.201 6003>>1
n.17 0162>>2	n.54 0810>>2	n.91 1503>>2	n.128 2412>>2	n.165 4005>>2	n.202 6012>>2
n.18 0171>>2	n.55 0900>>2	n.92 1512>>2	n.129 2421>>2	n.166 4014>>3	n.203 6021>>2
n.19 0180>>3	n.56 1008>>2	n.93 1521>>0	n.130 2430>>2	n.167 4023>>2	n.204 6030>>2
n.20 0207>>2	n.57 1017>>2	n.94 1530>>2	n.131 2502>>2	n.168 4032>>3	n.205 6102>>1
n.21 0216>>2	n.58 1026>>2	n.95 1602>>2	n.132 2511>>2	n.169 4041>>2	n.206 6111>>2
n.22 0225>>2	n.59 1035>>2	n.96 1611>>3	n.133 2520>>3	n.170 4050>>1	n.207 6120>>3
n.23 0234>>3	n.60 1044>>2	n.97 1620>>2	n.134 2601>>1	n.171 4104>>2	n.208 6201>>2
n.24 0243>>0	n.61 1053>>3	n.98 1701>>2	n.135 2610>>2	n.172 4113>>2	n.209 6210>>2
n.25 0252>>2	n.62 1062>>2	n.99 1710>>3	n.136 2700>>3	n.173 4122>>2	n.210 6300>>0
n.26 0261>>2	n.63 1071>>3	n.100 1800>>3	n.137 3006>>1	n.174 4131>>2	n.211 7002>>2
n.27 0270>>0	n.64 1080>>2	n.101 2007>>2	n.138 3015>>1	n.175 4140>>2	n.212 7011>>3
n.28 0306>>2	n.65 1107>>2	n.102 2016>>1	n.139 3024>>0	n.176 4203>>2	n.213 7020>>0
n.29 0315>>0	n.66 1116>>0	n.103 2025>>2	n.140 3033>>2	n.177 4212>>0	n.214 7101>>1
n.30 0324>>2	n.67 1125>>2	n.104 2034>>1	n.141 3042>>2	n.178 4221>>2	n.215 7110>>2
n.31 0333>>2	n.68 1134>>2	n.105 2043>>3	n.142 3051>>2	n.179 4230>>1	n.216 7200>>2
n.32 0342>>2	n.69 1143>>2	n.106 2052>>2	n.143 3060>>1	n.180 4302>>2	n.217 8001>>2
n.33 0351>>3	n.70 1152>>2	n.107 2061>>0	n.144 3105>>0	n.181 4311>>2	n.218 8010>>2
n.34 0360>>2	n.71 1161>>1	n.108 2070>>1	n.145 3114>>2	n.182 4320>>2	n.219 8100>>3
n.35 0405>>2	n.72 1170>>2	n.109 2106>>2	n.146 3123>>2	n.183 4401>>2	n.220 9000>>2
n.36 0414>>2	n.73 1206>>2	n.110 2115>>1	n.147 3132>>3	n.184 4410>>2	
n.37 0423>>2	n.74 1215>>0	n.111 2124>>2	n.148 3141>>2	n.185 4500>>2	

$d = 2$, $k = 4$ *e* $r = 1$

In this case, the string associated with (2.18) is: 230123222202022222322230 22020222322222222222222222222222232322022221220221232222333122 02202232233212132012121202223022301213222222312311022210223221 2232 2221020222232321222222021222222112221220222212122212322023012 22232

Rule Codes

C2.1: $D = 1, k = 3, r = 1$

001220000100011010200211112

C2.2: $D = 2, k = 4, r = 1$

0001000000000000000000000000300100222000000300000100000332020000
000000120000300100000000030000000000001000030000020000000000001000000
0000000000000000000000001030000020200002020310202000110200001000100
0001020020020202000003010

C2.3: $D = 1, k = 4, r = 1$

1103310121130010311100032023131121012031121310120002002000103020

C2.4: $D = 1, k = 4, r = 1$

1003303223132010301303122220011022012112301000110203022121100220

C2.5: $D = 1, k = 4, r = 1$

2121233221123213331113012232011212113021131233232113332323011231

C2.6: $D = 1, k = 4, r = 1$

2212121112330000220222121131202323122020111013012022233101002300

C2.7: $D = 1, k = 4, r = 1$

3130031131300220011112232322020302113221122033023011330221230101

C2.8: $D = 1, k = 4, r = 2$

1000331130312222103330201000032032303000330302002020003000033002
0200100103313100321133123203200113122200000030223230000101330201 32
2111132300330311020022013001302000332003002010310220300230200001 20
1003001233200103000202100102120123301200331000000330210232310202 01
3120032103310303020210202302310120010110021322112033202211002101 01
2000112021110020321022100102323303002000212002302022003010331221 10
2301331320012321100323313101200002121210021220001212120223211120 02
0310103301210333000313301321101220001210100212002001012203103303 20
2310200302321001210303001103122231300012003331211110100023320001 02
0001213003202122033001102000120001320002210021030033212201130033 01

0303131231200302320110000213112000231330002201012003211210010000110
0300110010033302100031023220100000332233303021002100220322322210 1
01221301320001311000103120313030000110030123320101021333030321002 1
020013322203023000202220310130210321200300330222002310100333213202
123202220032203101332030200013313220030133131111003012200300001320
23020201022003101321013110220221000000

C2.9: $D = 1, k = 5, r = 1$

020033102003130030103034243000220040002000013210130032200000 0
0013021100000124123100000003000030401420214320201401313213321242

C2.10: $D = 1, k = 5, r = 1$

120234134144030423004220014043423014034214324400004004031003 4
001402331442003001012022231210010110102240403003234330113113440 0

C2.11: $D = 1, k = 6, r = 1$

320020250102000242105000405133504010313253220203451041235325 05
4445140100345020250535020214055053142425502050141503435523231550 1
2431234534134130052030520053130440001210350332114210145403500032 3
2120500143001040504241 00

C2.12: $D = 1, k = 6, r = 1$

320100003510005320043201224200341052313310001503500130022434 43
5000510210504200001204431404250155315104500200004404001503010122 5
5345150400010550421230200025121300151520222304144102200450455030 0
2003002230115100440040 20

C2.13: $D = 1, k = 6, r = 1$

034202050023204503052042032420040301003004014220320045010402 10
1030032111514005034010533024530000300013013303230302130005100220 0
5240450000042250521210505004541233040140002005015004002003004004 5
5505100100230343440002 30

C2.14: $D = 1, k = 8, r = 1$

020000654053604157051022270500521504103711000400250604713146063
4374006103100323021064011634160206204000536045040307402607005500 22
7440002500001156507600637014012574510531311310070106007316760133 46

03306620561707306607771302552300703677656003260005070350050714200100053074464106670612761210664266041134600470010304340200010370010200670420060527214750020122040007025172600227020033005103706740004732135566776006313014001452036616102300272170267106454703440055057757003320760635203160247005000620025007700710130300670

C2.15: $D = 2, k = 4, r = 1$

200033300001310300020200200000032000320000000021300220001303000300003303020030000100200000000000230210331300000000021000003000213122220300000000000000001003000303000000000200330000030210220221300301200000000000000003102000000.

Chapter 3
Modelling Biological Systems

INTRODUCTION

What are the mechanisms underlying biological systems' ability to transform themselves: the ability of structures to replicate for their own goals, or to meet the specific goals of the system or environment to which they belong? What kind of evolutionary process underlies the emergence of the simple structures which joined together and replicated to produce the life on earth? How can we reproduce these functions in digital machines? More generally, what are the common elements shared by complex physical, biological, social and artificial systems?

In this chapter, we will examine the fundamental problem of how we can use 1 and 2-D Cellular Automata to model biological phenomena. Modeling is today a fundamental method in many sciences, ranging from the neurosciences and nano-technology to genetic engineering (for a review see Freitas, 2004).

As we saw in Chapter 2, 1-D Cellular Automata consist of a large number of elementary units. Interactions among the units are determined by a single, deterministic, local rule that gives rise to forms of self-organization and a wide range of emergent shapes. Inside the automaton, we distinguish particles which self-organize and propagate through time and space. These are so-called gliders. The dynamics of these emergent structures can lead to the extinction of other particles, the generation

DOI: 10.4018/978-1-61520-787-9.ch003

of new chaotic or periodic structures (new gliders) and sudden transitions from an ordered to a chaotic state.

Whether we start from Wolfram's qualitative classifications (Wolfram, 1984), Crutchfield's analytical computational mechanics (Crutchfield & Young, 1989; Crutchfield, 1994; Hanson & Crutchfield, 1997) or Wuensche's systematic analysis of overall dynamics (Wuensche & Lesser, 1992), we are limited to CAs with a small number of cells. However, there exist alternative methods that allow us to consider systems with large numbers of cells and multistate CAs (Bilotta et al., 2003a). In our own work, we have used Genetic Algorithms to identify complex rules for 1-D multistate automata, applied new techniques to analyze the emergent, spatio-temporal patterns they produce, developed computational methods to identify these patterns automatically, and studied the way in which local or deterministic rules are translated into genetic structures. As reported in several chapters of this book, we have used quantitative and qualitative analysis to identify the organizational laws underlying emergent structures, focusing on a small set of production rules shared by all these systems (Bilotta & Pantano, 2001). We also adopted an alternative way of modeling 1-D CAs, using the Universal Neuron approach first proposed by Chua and co-workers (Chua 2006; 2007; 2009). This method allows us to identify similarities in the behavior of different systems and to identify associations between the characteristics of the automaton and the dynamic system to which it gives rise. The models we have created make it possible to generalize from elementary CAs (ECAs) to multistate systems. The results show how universal neurons can generate rules with emergent properties (Bilotta & Pantano, 2008).

As far as concerns self-replication, several areas of modern science have offered interpretations that go beyond traditional approaches. According to these interpretations, self-replication is a fundamental property of a wide range of natural and artificial systems. The first scientist to view self-replication in mathematical terms was von Neumann. Von Neumann's goal was to discover an underlying "logic of life" providing insights into biological phenomena and allowing their recreation in digital systems. In this program of research, Cellular Automata's universal computational capabilities make them a good tool for representing the internal mathematical essence of biological organisms (von Neumann, 1966; Wolfram, 1984; Langton, 1990). It is an approach that fits well with the idea that biological life is based on information processing.

Towards the end of the 1940s, on a suggestion from Stanley Ulam's, von Neumann invented the concept of a Cellular Automaton and produced a design for a mechanical, self-replicating system. He proposed an architecture based on three key components: a universal constructor (A), with the ability to build a copy of another machine (X), based on a description of the machine; a universal copier (B), with the ability to copy the instruction tape; and a controller (C) with the ability to activate first A then B and then to associate B with its description. A system comprising A, X and B and a tape of instructions has the ability to self replicate. So long as the functions and descriptions of the components remain unchanged, the result is always the same. The machine replicates itself including any "mutations" on the tape. Von Neumann pointed out that with the universal copier in place, the system can evolve machines of ever increasing complexity. The basic design of B ensures it can copy any machine, *however complicated* (von Neumann, 1966). One of Von Neumann's main achievements was to clarify the logical relationship between the description of the system (the instructions tape, or genotype) and its construction (the execution of the instructions needed to produce a new individual or phenotype). What he had in mind was a large class of self-replicating machines. These include kinematic models, cellular machines, machines similar to neurons, continuous machines, and probabilistic machines. As far as concerns the kinematic model, von Neumann limited himself to an informal description, writing much more about cellular machines (though only in note form). After his premature death in 1957, the work was completed by his student, Arthur Burks. In this way, von Neumann became the first scientist to propose CA models of biological phenomena - the theme of this book.

In this chapter, we begin by reviewing the history of CA-based biological modeling, focusing on the modeling of evolution and on the tools we use to represent and understand evolutionary processes. We then introduce our own models of 1-and 2-D CAs, preparing the ground for the more detailed analyses present in Chapter 6 (for 1-D CAs) and Chapter 7 (2-D CAs).

BACKGROUND

What are the main characteristics of evolution? What are its recurring features? Which features vary over time? Does evolution have a direction? Organic and inorganic chemical systems change continuously. Do these changes follow specific trajectories? These are unanswerable questions. Always, our observations are partial and incomplete; always, we are forced to examine evolutionary phenomena from just a few of the possible viewpoints. And so, our doubts on the complexity of evolutionary processes remain unresolved. The only apparent constant in evolution is the unending flow of matter, energy and information. It is a process that resists any attempt at analysis. Evolutionary features depend on information. Information refers to something that is important or significant in a specific context. An analysis of the literature (Kauffman, 1993; 1995; Kimura, 1983; Schuster & Swetina, 1988; Crutchfield & Schuster, 2003) suggests that there are three possible kinds of evolution.

a. Evolution without information or blind evolution: in which the evolutionary process proceeds randomly.
b. Natural selection–the process described by Darwin–in which random variations are selected by information present in the environment this is called.
c. Self-organized evolution in which some variants are more likely to exist than others and are thus more likely to be selected. These variants are dense with information.

In brief, evolution consists of a flow of matter and energy in which different configurations produce the huge number of living species we find on earth. When a system experiences a variation in this flow of matter and energy, it reacts. If it consists of a large number of units (as in the case of cellular automata) they react as if they were under the influence of an external force. As a result, the elements-which are intrinsically independent - move in a coordinated way until the variation ceases to exert its effect. The behaviors of the individual elements combine to create a collective movement within the group and a movement of the group itself in which all the particles move in unison. This requires a form of communication among the elements that distributes information through the environment. Similar situations determine similar patterns of communication, producing similar behaviors. The probability that this kind of behavior will emerge from the spontaneous dynamics of the system is high and we observe that it is very common in nature. Patterns of dynamic behavior with higher than average probability of occurrence are called "attractors". Many studies have simulated these behaviors and many have used CAs. Mathematical modeling of biological behavior is driven by the need to represent evolutionary processes, data and problems in a way that offers insights into the general laws of evolutionary behavior. The literature describes several different approaches.

a. Genetic-molecular models of the origin of life: Eigen with his quasi-species model (Eigen, 1971; Schuster & Swetina, 1988), Schuster with his hypercycles (Eigen & Schuster, 1979), and Ratner and Shamini (1980; 1983) with their "sysers" have all produced mathematical models, describing hypothetical stages in the emergence of pre-biotic self-replicating systems based on large molecules.
b. General models of evolution such as M. Kimura's theory of neutral evolution (Kimura, 1983) and S. Kauffman's random Boolean networks (Kauffman, 1993; 1995), that consider the role of information in evolutionary processes
c. Artificial models of the evolution of life that aim to understand the formal laws of life and evolution. These models–that lie at the heart of the research

effort we describe in this book–allow us to analyze the evolution of artificial organisms in an artificial biology, implemented *in silicio*.

d. Computational systems that use evolutionary optimization methods to resolve practical problems. Examples include J. H. Holland's Genetic Algorithms (Holland, 1975; 1994) as well as L. Fogel and collaborators' techniques of Genetic Programming (Fogel et al., 1966). Both are examples of systems that simulate evolutionary processes using fitness functions and natural selection. This is the approach we have used to look for 1-D multistate CAs and to evolve 2-D CAs with the ability to self-replicate.

It is important to realize, however, that there is always a huge gap between the complexity of the biological world and our mathematical models. To apply these models, and to run simulations, we have to reduce the complexity of the problem, considering just a few of the hundreds or maybe thousands of variables that influence evolution in the real world.

If we want to create models, we have to start with data from physical and biological experiments. Only later we can begin to generalize and represent our generalizations in mathematical models. This was the way R.A. Fisher, J.B.S. Haldane, and S. Wright proceeded when they created the mathematical theory of population genetics, providing a quantitative description of the dynamics underlying the distribution of genes in a population. Of necessity, they based their work on intensive experimental research in genetics in the first half of the 20th century. The theory, sometimes called the "synthetic theory of evolution", brings together Darwin's concept of natural selection and Mendel's genetics (Haldane, 1924). Experiments with fruit flies *(Drosofila)* attempted to reconcile Darwin's view of evolution as a continuous process with the discrete variations that characterized Mendelian genetics. According to population genetics, the main mechanism underlying gradual evolution is selection of organisms with beneficial mutations. Deterministic models consider populations of infinite size; stochastic models refer to finite populations. Subsequent work in molecular biology during the 1950s and 1960s gave rise to M. Eigen et al.'s models of the origin of life, and to S. A. Kauffman's models of genetic regulatory systems (de Jong, 2002). In particular, evolution began to be modeled as an intrinsically stochastic, non-linear dynamic system, in which a population of individuals, each with its own structure and genotype, follows a trajectory through the space of possible genotypes. The trajectory is anything but random. On the contrary, it is regulated by developmental constraints and natural selection of phenotypes.

In these models, genotypes are represented in a genotype space, consisting of sequences of genes, proteins or polynucleotides (Crutchfield e Schuster, 2003). However, in 1968, Kimura (Kimura, 1968) discovered that certain variations in DNA sequences had no apparent effect on fitness. This implied the possibility of neutral networks spanning genotype space. Such networks were obviously non-adaptive, at least in Darwin's sense. Neutrality modifies the shape of genotype spaces and the associated evolutionary dynamics. What we observe is that while phenotypes remain apparently constant over long periods of time, the corresponding genotypes vary continuously, as the population moves across large scale, selectively neutral networks in genotype space. In this view, the main role of selection is to remove deleterious mutations. As a result, the majority of mutations that reach fixation have neutral phenotypes. By contrast, advantageous mutations are rare and have little influence over the genetic sequences found in the population. The constraints created by the development (ontogenesis) of the individual organism are a second factor strongly influencing evolutionary dynamics. Haeckel attempted to make ontogenesis a component in a broader picture of evolution. In his view, phylogenesis (i.e. the birth of races - from the Greek "phylum" - meaning tribe or race and "genesis") depends neither on an intelligent creator nor on a magical life force. Rather it is a mechanically determined, physiological process, depending on a number of different processes. Ontogenesis recapitulates the long evolutionary history of the species to which the organism belongs. Haeckel argued that the development of an embryo is determined by general laws that force it to "recapitulate" the different stages in the evolution of its ancestors, with each stage completely determining the stage which follows. The result is a tree of life in which each branch shares a common ancestor.

Current theoretical studies of evolutionary dynamics look at evolution from two different perspectives. On the one hand, there is the long term view that explains the evolution of populations and their respective gene pools; on the other there are the shorter, faster processes that determine the development of individual organisms. If we are to understand the interaction between the two levels, it is essential we should first examine the non-linear patterns that emerge during development. In this way we can identify the emergent behaviors available to natural selection.

In the physical world, we have identified emergent phenomena and evolution in observations and models of phenomena such as spin, animal pattern generation (Murray, 1989) and Turing's reaction-diffusion model (Turing, 1952). Turing's model uses the theory of non-equilibrium thermodynamic systems. This includes the study of dynamic systems on the basis of their classification in phase space and bifurcation theory. According to the literature, "bifurcation theory is a branch of dynamic systems that analyses and classifies the structural changes that occur when there is a transition from one kind of behavior to another, due to a variation in a control parameter" (Crutchfield e Schuster, 2003). Work in this area has described evolution and emergence using Cellular Automata and Kauffmann's Random Boolean Networks (RBN). Other phenomena have been modeled using studies of systems displaying Self-Organized Criticality (SOC) (Bak, 1996). Simulations and related theoretical work have led to the idea that systems displaying self-organized criticality or behavior leading to structural complexity inhabit a border zone between ordered and chaotic regions of parameter space (Langton,1986; 1990). Evolutionary and developmental processes can generate systems with an internal structure or with complex, adaptive, behavior. Similar ideas have given new life to the analysis of structures and networks. Highly structured networks lie in an intermediate zone between networks with ordered and networks with random topologies. These observations suggest that complexity is an intermediate state between order and chaos. In other words, complexity is emergent self-organization. CAs ability to process information becomes apparent in their emergent structures and this is why they are called complex systems. Until a few years ago, we were aware of only a relatively small number of such systems. Today, however, researchers such as Langton, Mitchell, Wuensche and many others have developed highly effective methods for generating these systems. One of these methods is to use Genetic Algorithms (Mitchell et al., 1993; Mitchell, 1996). Particularly important is Wuensche's method (1999), which measures order and chaos in CA rules by measuring Shannon Entropy in the frequency of use of specific elementary rules.

On the basis of this work, we have expanded the scope of traditional mathematical modeling, introducing computational models and Genetic Algorithms that make it possible to find complex rules for 1 and 2-D CAs.

Currently, the main fields of application for evolutionary methods are in interdisciplinary areas such as Artificial Life. The goal, inspired by ideas from cybernetics, is to replicate typical features of intelligent biological behavior. Key concepts include learning, neural networks and adaptive behavior that changes as the system interacts with the external world. The outcome has been a better understanding of biological evolutionary processes. At the same time, models developed in this setting provide us with a new way of looking at these processes in abstract terms. They are no longer constrained by reality "as it is" but show us reality "as it could be" (Langton, 1995). In this sense, mathematical modeling is an eminently theoretical activity. Before we can create actual artificial organisms with traits similar to those of biological systems, much additional work will be needed.

MAIN FOCUS

This chapter examines developmental and evolutionary processes in 1-D and 2-D CAs. The basic underlying assumption is that these kinds of system are in some way life-like proto-organisms. We thus consider them in exactly the same way we would consider biological agents - as agents living in an environment, albeit an artificial one. This is the approach we will adopt not only in our analysis of interactions among agents, but also at a higher level when we consider phenomena such as agents dynamics and group formation. Our experiments have allowed us to find a huge set of complex rules as well as 2-D CA rules that generate self-replicators. To model these phenomena, we analyze the basic structures to which they give rise (in 1-D systems: gliders, regular domains, and particles; in k-totalistic CAs self-replicating structures), systematically observe their behavior and extract the general laws determining this behavior. Considering the rules for these CAs as their "genetic code", we observe how changes to the code give rise to a huge range of morphogenetic processes, changing the shape, structure and internal characteristics of the system.

We have discovered that self-replicators have a genetic structure. Our computational model allows us to precisely identify the genetic networks the self-replicators use during their development.

Phenotypes and Genotypes for 1-D Multistate CAs

Cellular automata are discrete dynamic systems, whose individuals components develop simultaneously under the control of a shared local rule. In addition to local interactions among their components, CAs display complex behavior that brings to mind some of the features of living systems. As already mentioned, one of their defining features is the ability of certain CAs to generate self-organizing structures such as gliders. These structures persist in time, carrying information from one side of the lattice to the other. Wolfram classified CAs with this ability as Class IV CAs and hypothesized that they had universal computational capabilities. This hypothesis–and the idea that life is essentially information processing–has inspired la large body of research into CAs and their ability to represent the "logic" of life.

In this chapter (and in the whole of this book), we consider CAs as a mathematical tool for the study of complex biological systems. In our methodological approach, we view CAs as systems with a genome that is transmitted from one generation to the next. To analyze these specific aspects of CAs, we have introduced a number of novel analogies. In the chapters which follow, they will become the basis for new computational models.

Phenotype: given a 1-D CA and a set of initial conditions, it is possible to generate a pattern in time and space that we consider as the CA's phenotype. Changes in the initial conditions will lead to substantial changes in this pattern. We thus observe a typical feature of chaos (sensitivity to initial conditions) combined with the presence of persistent self-organizing structures such as gliders and regular domains. In this setting, it is useful to analyze how the features of the CA are affected by changes to the rules (the control parameters for the dynamic system). We also observe a life cycle, during which the phenotype *grows* and *develops*.

Genotype: the rule governing the evolution of the CA can be represented by a string of characters. This is the CA's genotype. The same scheme can be applied to 1 and 2-D CAs. The rules table is represented as a string of characters in which each rule is identified by a prefix R (signifying rule) and a number. Take for example a 1-D CA with k=4 and r=1. This CA has 220 elementary rules. We encode the rules as R1, R2, R3, ...R220. In addition to this encoding scheme, we have built a computational tool that allows us to precisely identify the elementary rules used by the CAs as they construct patterns in time and space (1-D CAs) or 3-dimensional structures (2-D CAs).

If we generate a family of CAs, each with its own genotype, we can apply evolutionary operators such as mutation and cross-over. We can then compute fitness metrics for the phenotypes and use this information to guide an evolutionary process. These Genetic Algorithms allow us to evolve complex CAs.

Genes: in complex CAs (including CAs generated by evolutionary processes), specific self-replicating structures are associated with specific CAs. Some of these structures are associated with subsets of the genotype. These are the CA's *genes.* Some of the structures are *monofactorial*, in the sense they are clearly associated with a specific gene. Others are *multifactorial*, depending on the concurrent effects of multiple genes. Gliders seem to belong to this second category.

Genetic Transmission: When we analyze a CA's genes, we see how traits are transmitted from one generation to the next. This allows us to analyze increases in complexity and to see how certain genes are "dominant" with respect to others.

In brief, CAs provide us with a formal model we can use to examine some of the typical characteristics of life. It is a model in the spirit of von Neumann, simple enough to capture the essence of the phenomena we seek to understand yet sufficiently complex to conserve defining characteristics such as complexity, self-organization and emergence. In what follows, we introduce some of the formal concepts we use in the rest of this book.

Chapter 2 has already described the way in which we can represent rules tables for 1-D CAs. In our modeling approach, we consider the rules table for a CA as its genetic code.

Given a rule table and an initial configuration or initial state $c\left(0\right)$, it possible to obtain a list in the form $c\left(0\right), c\left(1\right), c\left(2\right), \ldots, c\left(t\right)$. We can represent this list as a matrix or a spatio-temporal diagram. This is the CA's phenotype (Figure 1).

If we change the initial conditions, we obtain completely different patterns. This is because these systems are extremely sensitive to initial conditions. Figure 2 shows some of the different phenotypes obtained by changing the initial conditions for a multistate 1-D CA. We observe not only the sensitivity to initial conditions, typical

Figure 1. Spatio-temporal diagrams (phenotypes). The rules are given in the Appendix of this Chapter, with codes C3.1-C3.6.

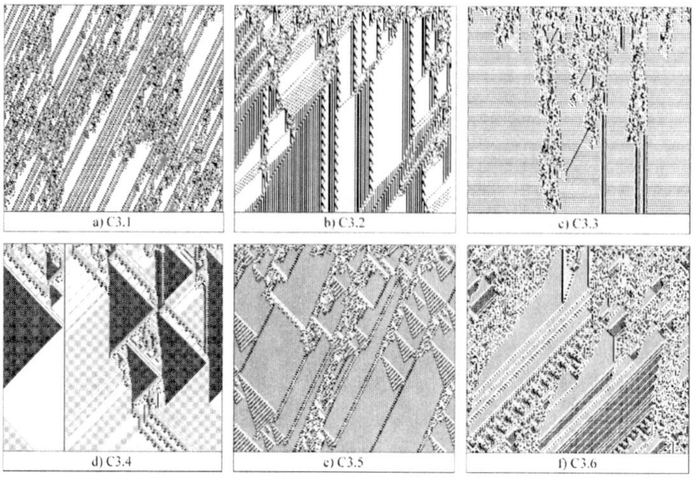

of chaotic systems, but the presence of persistent self-organizing phenomena such as gliders and regular domains.

Other more obvious changes can be obtained by changing the genetic code for the rules. This allows us to obtain completely different patterns (see Figure 3).

Thanks to the method we have created, we can produce a huge number of 1D CA systems. These systems are highly diverse in terms of behavior, complexity and the number of rules they use during development. As in nature, we do not find just one patterning mechanism but many.

Figure 2. Changes in the initial data of a 1-D, CA (C2.1) leading to minor changes in its spatio-temporal pattern (or phenotype)

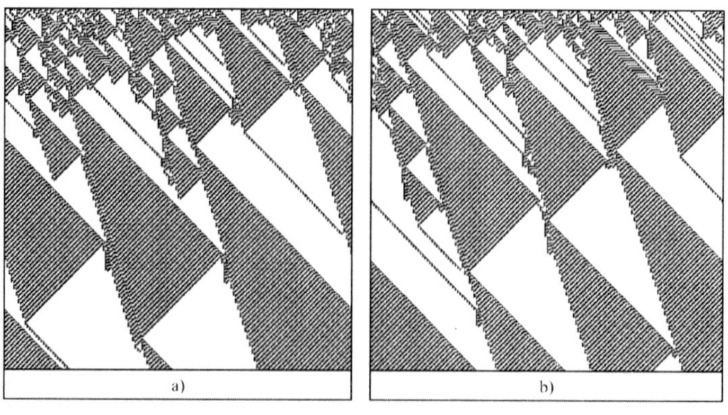

Figure 3. Changes in phenotype due to changes in the genetic code. Figure (a) shows the pattern generated by the rule, without changes. Figure (b) shows slight variations in non critical points. Figure (c) shows a larger variation in the pattern, due to a change in a critical point. The last Figure (d) shows a radical change due to the variations of the genetic code in two supercritical points, one leading to a chaotic pattern and the other to a periodic pattern.

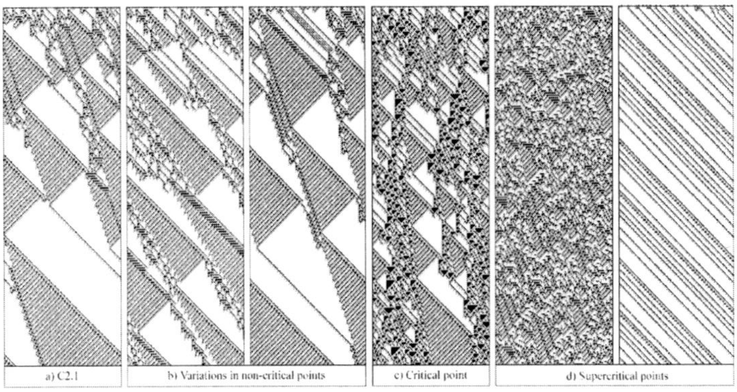

Dynamic Networks for Multistate 1-D CAs

To conclude, the spatio-temporal pattern is determined by the presence of a certain number ***s*** of elements such as gliders and regular domains. We can map these elements to *y genes* $g1, g2, \ldots. gy$. The set of these genes, $G = g1, g2, \ldots. gy$, represents the CA's genome. Ea*ch gene* corresponds to a single element in the spatio-temporal pattern and a single genetic network σg defined by the elementary rules used as the element develops in time and space. These are the n bases $b1_g, b2_g \ldots bn_g$ making up gene g. The order in which the rules are used defines a dynamic network $\sigma_g(t)$. The dynamic behavior of the *gene is determined by the fr*equency with which it uses specific rules. It is important to observe that different genes may use the same rules.

As an example, consider the glider (0102), produced by a k=3, r=1 system, with rule C2.1, and an initial state of 1. We will denote the gene for the glider by g. The corresponding genetic network σg is composed of the following elementary rules:

$$\sigma_g = \{R5, R14, R15, R16, R17, R18, R20, R24, R25\}$$

The corresponding dynamic networks are:

$$\sigma_g(1) = \{R5, R14, R15, R18, R20, R25\}$$

$$\sigma_g(2) = \{R5, R14, R15, R16, R17, R20, R24\}$$

Figure 4 shows the genetic network for the glider and the corresponding dynamic networks.

Figure 4. The genetic network corresponding to the gene for glider (0102) in the state 1, generated by rule C3.1, reported in the Collection of Rules. The first panel in the first column provides a graphical representation of the 9 rules used (a), compared to the complete set of rules (whose total number is 27) (d). The second column shows the times at which different rules are used (b, e). The central column shows the dynamic network for time 1 and for time 2 (respectively b, e). The third column provides an alternative representation of the dynamic networks showing only rules that are actually used (c, f).

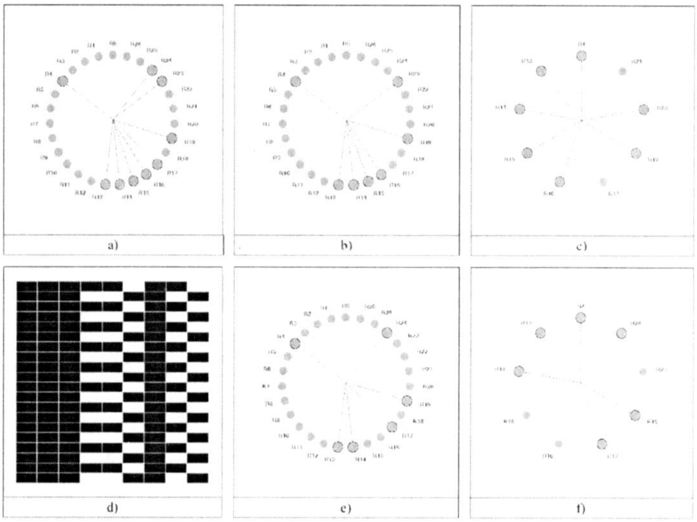

This approach allows us to analyze the detailed behavior of the genes and the way they are expressed. Figure 4, for instance, shows that R5, R14, R15 and R20 are always active. By contrast, the other bases switch on and off as the system evolves.

A careful analysis of the genetic networks and the corresponding bases (elementary rules) shows that certain local rules are re-used in different phenotypes. By identifying basic (monofactorial) traits we can identify the individual genes making up the genome and the associated structures in the CA. This work provides a starting point for the analysis of:

a. Multifactorial traits of the system;
b. Genetic networks;
c. The large scale phenotype, as represented by spatio-temporal patterns;
d. The genetic design intrinsic to any specific genome;
e. The language system created by the particles and structures (Bilotta & Pantano, 2006), with their multiple phonetic, morphological, lexical and syntactic levels of organization (Chomsky, 1965).

f. The production rules underlying the huge range of phenotypes to which the gene pool can give rise

Here there is a clear analogy with one the most important problems in biology. In its biological version, the problem is far too complex for direct analysis. In this setting, CA-based mathematical models can provide useful insights.

Once we have decoded a system we can extract pieces of the system and use them to create new systems, as a language. This approach, discussed in detail in Chapter 9, shows that it is possible to define procedures to produce combinations of gliders or combinations of gliders and regular domains.

Phenotypes and Genotypes for 2-D CAs

In previous work, we have used Genetic Algorithms to discover k-totalistic 2-D CAs with intrinsic self-replication capabilities and life-like properties (Bilotta et al., 2001; 2003b). We hypothesized that it might be possible to apply the same modeling approach we had used with 1-D multistate CAs. In this case too, we considered the rule for the CA as its genome. The figure 5 shows an example.

In the example, we show a rule for a 2-D CA (C3.7) with k=4 and r=1. As shown in the figure, the rules table is represented as a string of characters in which each rule is identified by a prefix R (signifying rule) and a number (in this case a number between 1 and 220). The gene sequence for the CA thus reads R1, R2, R3, ...R220. In addition to this encoding scheme, we have built a computational tool which allows us to precisely identify the rules used by the self-replicator (333) as it constructs its "body" (a 3-D spatio-temporal structure). In our analysis, we treat self-replicators as though they were living organisms with the ability to adapt to different environmental conditions. The genome for the self-replicator contains four

Figure 5. The genome for a 2-D CA (C3.8) with k=4, r=1, listed in the appendix of this chapter

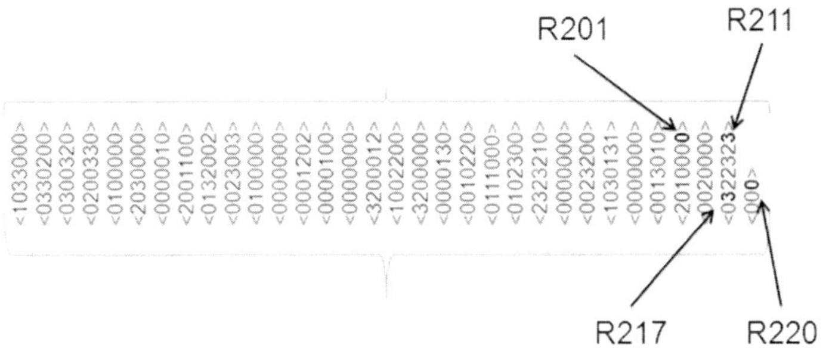

major classes of gene, each occupying a specific region of the genome and exerting a specific influence over the development process. Genes in region IFG (Indifferent Genes) have no effect on the overall behavior of the self-replicator; genes in region IG (Important Genes) have large effects; genes in region VIG (Very Important Genes) have very large effects; finally genes in region CG (Catastrophic Genes), of which there are several, produce catastrophic effects, completely changing the behavior of the self-replicator. (Bilotta et al., 2003b). In biological terms, some mutations are adaptive, some less so; some are lethal to the organism. What we observe is that each individual agent has its own characteristics, and adapts in its own specific way. It appears that there is no way of predicting these dynamics a p*riori.*

Dynamic Networks and the Development of Self-Replicators

CAs can be seen as implementing a RNA-like mechanism that translates the genotype into the phenotype. Given a rule for a 2-D CA and a set of initial conditions, the process operates as follows. On each step, the RNA consults the general rule, activates a gene sequence (a group of the set of elements making up the rules table) and translates the data into a specific spatial pattern, containing well-defined structures (equivalent to proteins). Step by step, the process produces substructures that contribute to the final configuration. In this way, the final pattern is determined by the interactions among three distinct elements.

a. **The genetic structure:** this contains the information necessary to describe the system;
b. **The gene expression structure:** this is monitored by a clock whose beats drive the simulation, determining which rules are consulted at each step;
c. **The phenotype structure:** the final pattern created by the procedure just described, and
d. **The protein structure:** the molecular components that determine the pattern that emerges on each step.

Each of these elements can be analyzed individually.

In general terms, we can distinguish an analytical level and an organizational level. When we observe self-replication, we distinguish between phenomena that concern the whole phenotype and those that involve the proteins responsible for individual steps in the production of a sub-structure. We can look at this problem at both levels. On the one hand, we can study the influence of individual groups of genes on the self-replicator; on the other, we can approach the problem in more general terms, looking at the complete set of substructures. In 2-D CAs, the two levels are linked. Both at the analytical and the organizational level, change produces a modification of the whole system. Like biological organisms, Cellular Automata display structures organized on different scales: proteins, phenotypes, dynamic gene networks, static gene networks, collections of rules and organized systems. This implies the existence of different levels of implementation - each with its own specific task. In other words, each level has its own algorithm describing how it should carry out its function. The global task - self-replication and the expression of self-replicating phenotypes - emerges from the implementation of the algorithms for the individual levels.

The processes that concern us may be cellular in nature (as in the case of local interactions among cells) or involve the whole system. What they show is that we can use Cellular Automata to simulate biological organisms and to understand evolutionary processes in artificial environments. The combination of global dynamics with cooperative dynamics during cell to cell interactions suggests general rules of pattern formation. These are problems we will discuss in greater detail in Chapter 11. However, we believe readers will benefit from seeing our genetic models straight way.

The first of our models (Figure 6) consists of a "glyph" showing the dynamic network created by a self-replicator during its development. Starting from the basic network for this CA (rule C2.2), the glyph has a configuration composed of 7 rules in the genetic code.

Figure 6. A glyph showing the dynamic network created by a self-replicator during its development

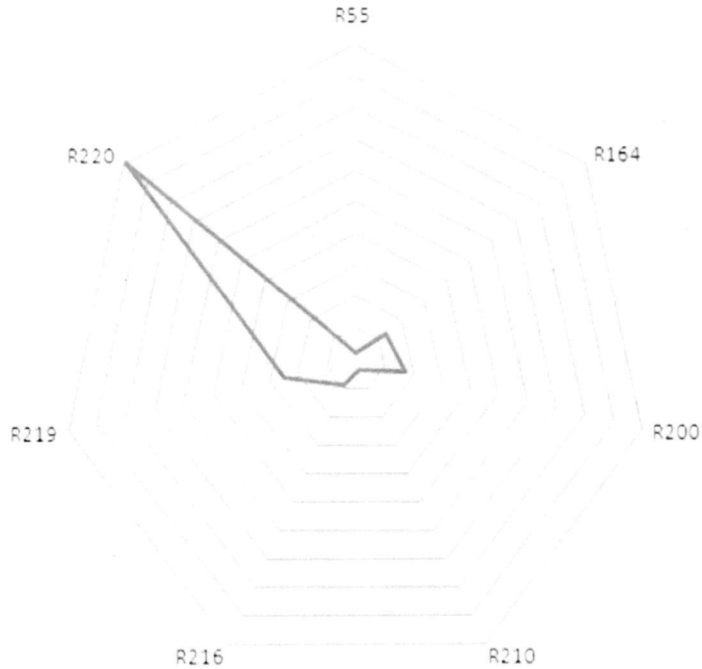

Figure 7 a) provides a schematic representation of structure formation in a self-replicator. In the left panel (image a), the top row shows the genetic code, the first column shows the simulation step and the rules used at each step. Figure 7 b) provides a virtual representation of the development process.

In Chapter 11, we will give a formal definition of the genetic networks underlying the self-replication process and show how these networks operate as fundamental structures for the emergence of patterns. We will then analyze various forms of mutation, illustrating sequences of changes in simple self-replicators and suggesting ways in which it is possible to represent these evolutionary processes. Finally, we will describe the changes in the shape of self-replicators that these processes determine.

Simulated Evolutionary Processes with 2-D CAs

In our work, we used Genetic Algorithms to simulate long term evolutionary processes and to analyze the relationship between organisms in the initial population and those evolved by the algorithm. To this end, we conducted a phylogenetic analysis. Our

Figure 7. Figure (a) shows the genetic code as it develops in time. In the illustration, the vertical column shows the simulation steps, while the horizontal line shows the rules used at each step. Figure (b) provides a virtual representation of the growth of the self-replicator, which graphically highlights the self-replication model this CA contains.

	R10	R512	R611	R612	R646	R647	R648	R650	R684	R701	R702	R703	R711	R712	R715
T1	0	0	0	0	0	0	0	0	0	0	0	0	0	9	2500
T2	1	4	0	0	0	0	0	4	4	0	0	8	0	4	2500
T3	0	0	5	4	4	12	0	0	0	4	16	0	0	4	2500
T4	0	0	0	0	0	0	0	0	0	0	0	0	0	36	2500
T5	4	16	0	0	0	0	0	16	16	0	0	32	0	16	2500
T6	0	0	20	16	20	48	16	1	0	8	32	4	0	4	2500
T7	0	0	0	0	0	0	0	0	0	0	0	0	45	36	2500
T8	4	16	0	0	0	0	0	16	16	0	0	32	0	16	2500
T9	0	0	20	16	16	48	0	0	0	16	64	0	0	16	2500
T10	0	0	0	0	0	0	0	0	0	0	0	0	0	144	2500
T11	16	64	0	0	0	0	0	64	64	0	0	128	0	64	2500
T12	0	0	80	64	88	192	96	9	0	16	64	12	0	4	2500
T13	0	0	0	0	0	0	0	0	0	0	0	0	189	36	2500
T14	4	16	0	0	0	0	0	16	16	0	0	32	0	16	2500
T15	0	0	20	16	16	48	0	0	0	16	64	0	0	16	2500
T16	0	0	0	0	0	0	0	0	0	0	0	0	0	144	2500
T17	16	64	0	0	0	0	0	64	64	0	0	128	0	64	2500
T18	0	0	80	64	80	192	64	4	0	32	128	16	0	16	2500
T19	0	0	0	0	0	0	0	0	0	0	0	0	180	144	2500
T20	16	64	0	0	0	0	0	64	64	0	0	128	0	64	2500

a)

	R10	R512	R611	R612	R646	R647	R648	R650	R684	R701	R702	R703	R711	R712	R715
T1	0	0	0	0	0	0	0	0	0	0	0	0	0	9	2500
T2	1	4	0	0	0	0	0	4	4	0	0	8	0	4	2500
T3	0	0	5	4	4	12	0	0	0	4	16	0	0	4	2500
T4	0	0	0	0	0	0	0	0	0	0	0	0	0	36	2500
T5	4	16	0	0	0	0	0	16	16	0	0	32	0	16	2500
T6	0	0	20	16	20	48	16	1	0	8	32	4	0	4	2500
T7	0	0	0	0	0	0	0	0	0	0	0	0	45	36	2500
T8	4	16	0	0	0	0	0	16	16	0	0	32	0	16	2500
T9	0	0	20	16	16	48	0	0	0	16	64	0	0	16	2500
T10	0	0	0	0	0	0	0	0	0	0	0	0	0	144	2500
T11	16	64	0	0	0	0	0	64	64	0	0	128	0	64	2500
T12	0	0	80	64	88	192	96	9	0	16	64	12	0	4	2500
T13	0	0	0	0	0	0	0	0	0	0	0	0	189	36	2500
T14	4	16	0	0	0	0	0	16	16	0	0	32	0	16	2500
T15	0	0	20	16	16	48	0	0	0	16	64	0	0	16	2500
T16	0	0	0	0	0	0	0	0	0	0	0	0	0	144	2500
T17	16	64	0	0	0	0	0	64	64	0	0	128	0	64	2500
T18	0	0	80	64	80	192	64	4	0	32	128	16	0	16	2500
T19	0	0	0	0	0	0	0	0	0	0	0	0	180	144	2500
T20	16	64	0	0	0	0	0	64	64	0	0	128	0	64	2500

b)

initial idea was that a group of agents with a common behavior is likely to have a common evolutionary history and that organisms belonging to the group will thus be more closely related to each other than to other organisms. In our work, we were guided by three basic assumptions. First, any group of artificial organisms must necessarily share a common ancestor. Second, evolution involves specific sequence of bifurcations. Third, changes in traits take place over time. Once we had found a self-replicator we began to modify its genetic networks, trying to identify universals and fundamental schemas and looking at possible mutations. Later we performed simulations in which we evolved self-replicators for many generations and analyzed their emergent evolutionary dynamics.

The evolution of 2D Cellular Automata can be represented as a dynamic model of change at different scales, or as a model comprising several interconnected layers. The starting point is always a self-replicating system. In our simulated model, this corresponds to a network (Figure 8). The evolutionary dynamics influence the instantiation and generation of species (Figure 9).

At the first level, evolution acts as a primitive force acting on the gene pool for a population of organisms. As the population evolves over the generations, natural selection of mutations creates new species with their own networks. Over time, these networks generate variants that are ever more distant from the networks observed at the beginning of the simulation: so much so that they seem to belong to different species. The presence of variants complicates the picture (Figure 10). Similar networks give rise to dissimilar self-replicators; different networks generate self-replicators that are very similar. At times, the same organism creates different competing or collaborating networks (Figure 11), setting in motion creative processes that generate new networks, more complex than their predecessors, in which the shapes, functionality and behaviors of their predecessors come together in new combinations.

Figure 8. The genetic network for a self-reproducer

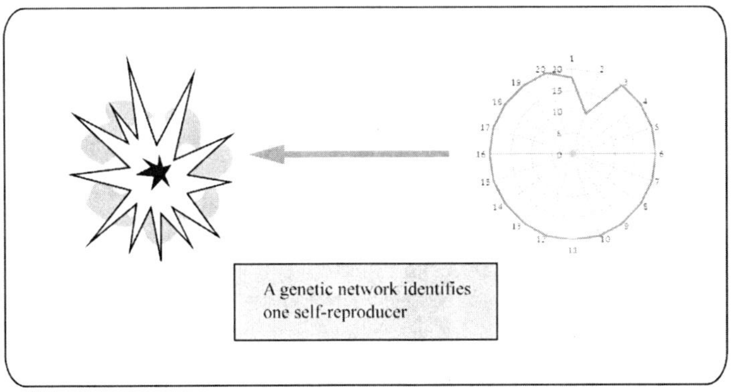

Figure 9. Simulated evolutionary model with self-replicators

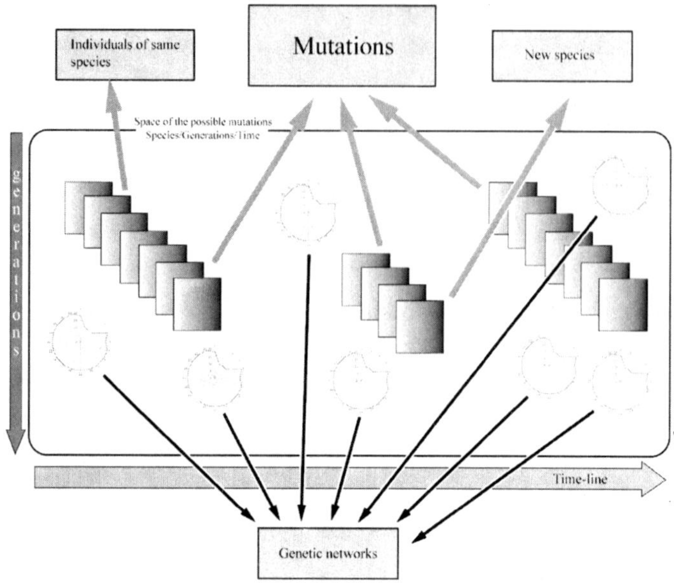

Figure 10. Variants complicate the picture for the first level

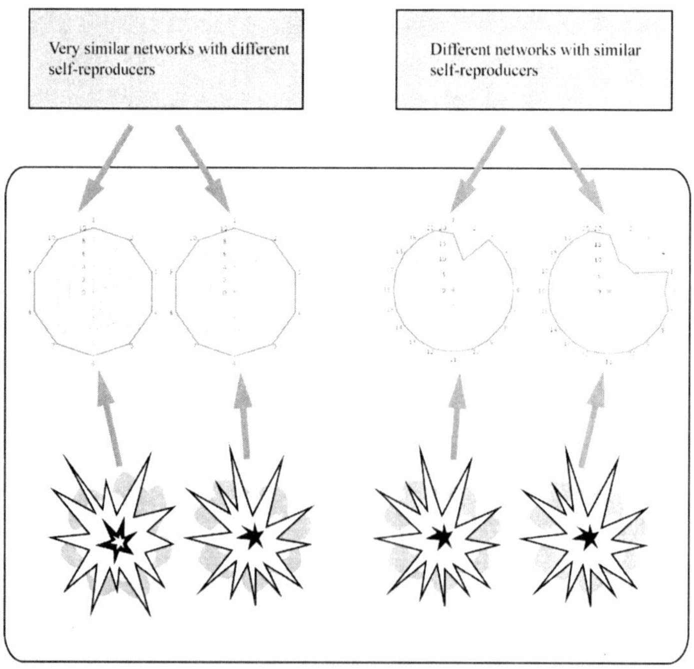

Figure 11. The same organism can give rise to different genetic networks producing new networks that are more complex than their predecessors, in which the shapes, functionality and behaviors of their predecessors come together in new combinations.

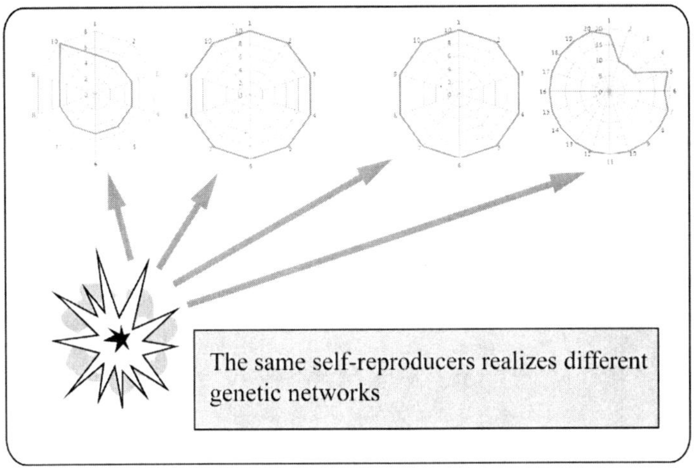

The development of the organism and its components involves several different replication, self-replication and developmental processes. Self-replication is only one of the possibilities. The origin of self-replication is accidental and random. However, once it is in place, it acts as if it had a goal. The goal is simple, namely to reproduce itself an infinite number of times. It recognizes specific parts of the system and replicates them as many times as is necessary. In this way, it acts like the cells of the skin or other self-replicating parts of the body such as nails, hair and liver cells.

Other evolutionary processes involve information processes connecting one network to another (Figure 12). These give rise to a number of changes such as mutations. Sometimes, these changes create genetically connected networks (genetically related networks); sometimes they create networks that are completely separate from one another. This process leads, via a series of variants, to the creation of new species of self-replicator (Figure 13).

At this point, the evolutionary process resumes, beginning with a new self-replicator and new evolutionary dynamics.

FUTURE TRENDS

All models of artificial life carry information. It is this information that determines the way the organism behaves and interacts with other components of the system.

Figure 12. Genetic creativity in two organisms with different genetic networks

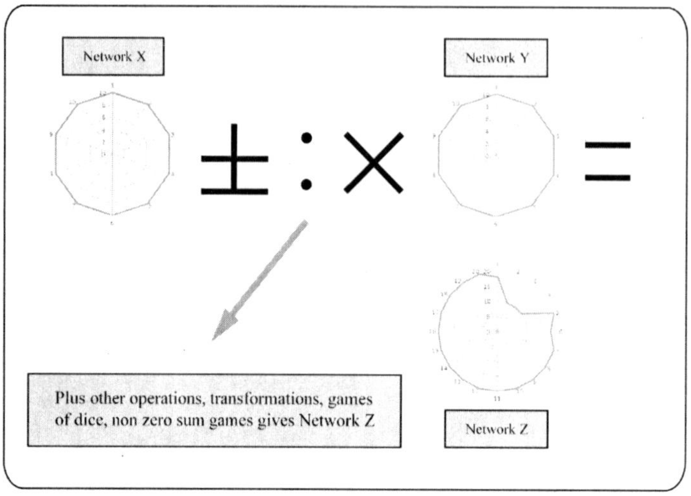

Figure 13. Evolution generates variants. Over long periods, these give rise to new species of artificial organism.

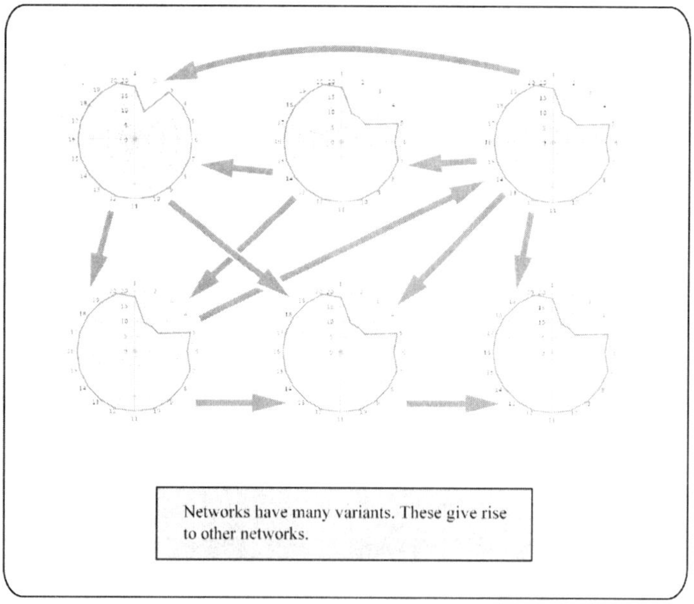

The meaningful parts of the information immediately attract the observer's attention. In the early stages of the evolutionary process, with random initial conditions, the system needs time to settle into its characteristic behavior. The transient may be chaotic or relatively orderly. Only afterwards does the system begin to display the real information it is carrying. It is as if the goal of the system were to reveal the meaning of this information. We observe a range of phenomena including spatio-temporal patterns and the growth of self-replicating structures.

One interesting issue for future work is the role of information. In what follows, we briefly present some of the open issues we will address in future research.

There are several elements in 1-D and 2-D CAs that we can consider as carriers of information:

- **Units.** These are relatively isolated or limited elements of information with their own distinctive character (for example, particles, gliders, self-replicating organisms or other kinds of elements that can be found in different ecosystems). In the case of 1-D CAs, for instance, the elementary units of information are the automata or other kinds of mathematical structure in which the genetic information is revealed. Every such unit of information has an "information content" and a "logical structure" (its algorithm) determining how it will behave in time and space. The problem we have to face in our modeling work is how to construct a taxonomy of minimal units of information that can provide a structural underpinning for our language or computational system. In other words, we have to distinguish between elements whose information content is different. Another problem is how to create a taxonomy of the rules (or algorithms) governing the way different information structures combine. These constitute the logical component in the modeling process.

There are many questions we would like to answer. Do there exist individual units–perhaps simple units–that contain the complete information content of the system? How is information distributed among the elements of the system? Is the distribution purely random? Or do some elements capture more information than others - playing a special role in the transmission of information, perhaps comparable to that of a dominant as opposed to a recessive gene? It is certain that the information depends on the role the element plays in the system. But what are these roles? Is it possible that some information could be neutral or dormant? Can we rely on our usual concepts of transmission, storage, management and retrieval of information or do we need new concepts, as yet unknown? What is the role of information itself? Is it purely semantic? Or is it meaningless to talk about the semantics of information? If we are going to talk about it, do we need an observer to attribute and to create conventions such as A (a specific kind of content with its own logic and evolutionary behavior) = B (the same element of information, this time associated with a conventional meaning to create a dictionary)?

- **Classes.** Set of elements identified by shared properties. Here the problem concerns the behavior of a particle's information algorithm when it is on its own or when it acts in combination with other particles. When it is together with other particles how does it behave? There are various possibilities.
  a. When a particle merges with another particle, the two particles may have the same or compatible properties. In this case, the two elements synchronize and the two algorithms fuse. We therefore say there is a correlation between the algorithms or that they have a lot of steps in common. What kinds of particles can fuse in this way? Is it their genetics which give them this ability or is it the environment? Are there different possible combinations for different rules?
  b. Sometimes, when two particles merge, their algorithms may be incompatible but collaborative. In this case, they can coexist, while remaining in different classes; the emergent behavior of these systems reflects the behavior of both particles.
  c. At times, the two algorithms create a symbiotic relationship. Sometimes the relationship is parasitic and only one of the algorithms benefits; sometimes it is truly symbiotic and both algorithms receive an advantage.
  d. At times, one particle suppresses the behavior of the other (this is the dominance of the strongest) or encourages it (as in social insects).

- **Relations and Relationships** among elements of information based on variables or algorithms describing relationships or on groups of algorithms sharing the same basis of information and somehow in contact.

- **Associations or links among elements** (in algorithms this kind of relationship can be implemented as a call to a subroutine.). Various kinds of association are possible:
  a. By quality
  b. By quantity
  c. By behavior (with the same evolutionary algorithm)
  d. By behavior (with different evolutionary algorithms)
  e. The same shape with different behaviors
  f. The same behavior with different shapes
  g. By spatial or temporal contiguity.

These elements of information can give rise to networks of relationships. In this case, the main problems concern the emergence of hierarchy in the network of relationship and the associated gain in meaning. Another problem is that the complexity of synthetic or biological life may requiring the nesting of algorithms inside other algorithms

- **Systems.** A system is an organized or structured set of linked or interacting elements of information. Each system represents a module with a specific organization and a specific function (analogous to biological modules such as the module responsible for visual perception). A system is characterized by a specific content of information and a group of interconnected algorithms, performing a specific function.

- **Transformations.** Transformations involve redefinitions, transpositions, revisions, updates, modifications, or evolutionary cycles in the content and function of information. Information evolves in steps. The way in which this occurs depends on the system's memory capability (or on genetically encoded innate knowledge). This implies that a specific system may allow some kinds of transformation and prohibit others. When considering transformations, we have to take account both of the diachronous aspects of the system (the way it evolves over time, assuming that the system exists and that it is not completely immutable) and its synchronous aspects (the way it may change at a specific moment in time or in a specific context).

- **Implications.** The extrapolation of information in the form of possibilities, predictions and known or supposed consequences.

Implications are knowledge about the system, in the form of an accurate description that can be stored in a catalog maintained by the system itself. An observer with access to this information - or even a small part of the information - can predict the behavior of the system, once certain basic elements or more complex units of information have been chosen.

Each element in each of the categories just described can be combined with other elements to forms larger sets or aggregates. Information can be structured hierarchically, creating a taxonomy of different levels of information. Each level may contain multiple elements, and different kinds of dynamic information process. Information may be transmitted in the following ways:

a. From one location to another, in all directions in space;
b. From one moment in time to another moment in time;
c. From one medium to another medium;
d. From one to many;
e. From one to one;
f. Inside the system;
g. From many to many;
h. From many to one.

As far as concerns the organization of information, we can consider the following entities:

- **Operations.** These are activities or essential processes that determine the way a system processes raw information. Operations include:
  a. The discovery of significant information (see above);
  b. The storage of information (not to be confused with existing memory repositories) - this could be the ongoing process that creates the repositories;
  c. The generation of new information from information received: here the focus is on the diversity and on the quantities of information that can be produced from a single piece of data (divergent production)
  d. The generation of information from information received when the latter entirely determines the former.
  e. The assessment of information and the matching of information against previously defined specifications.

- **Information content.** This can be divided by class and type of information. Information can be classified according to its nature.

- **Concrete information.** The way in which the information appears in the phenotype

- **Symbolic information,** present as signs that have no intrinsic meaning but whose meaning depends on the reality they represent (the conventional nature of the information and semiotic processes that give rise to an organized system of signs).

- **Behavioral information** describing the way information is structured, the way it can be transformed, and the way it can change, either spontaneously, or through the action of others.

- **Semantic information.** The information attributed to a specific informational behavior. Storage of information can occur:
  a. Using various kinds of memory;
  b. Through encoding schemes used to facilitate information retrieval;
  c. Through selection of meaningful information where the selection may privilege novelty, dynamic (as opposed to static) information, contrast, similarity, common fate, good form, closure, or various forms of illusory information (presence in absence, absence in presence).

- Finally **Information processes:** these are related to the fundamental arithmetic operations, namely:
  a. Combinatory processes capturing the combinatory intensity of the information through analysis of individual elements, structures and relations;
  b. Mutational processes: examining the diversity of an informational flow at different moments in time, exploiting characteristics such as diversity, originality, novelty, organization, and complexity perhaps measured by entropy metric).
  c. Transformational processes, using formal criteria (Hamming distance, different evolutionary runs, different ways of organizing information) to analyze the function of an information process.

CONCLUSION

In many biological systems, the only mode of reproduction is through processes of growth and separation. Other systems spread via mechanisms that project systems similar to themselves into new regions of time and space. In other words, any system that can generate new systems, in regions of time and space different from its own, is a self-replicator. Self-replicating systems are the result of the development of complex self-organizing systems with the ability to grow and spread in time and space. This process implies the transmission of information - an idea that derives in part from our vision of life as some kind of information processing machine. It is a concept with roots in neurobiology, logic, automata theory, cognitive psychology and even molecular biology. In its turn it has given birth to cybernetics, computer science and cognitive science: aided by McCullough and Pitt's use of mathematical logic to describe neural networks in the early 1940s and by Von Neumann's concept of self-replicating machines from the early 1950s. It was of course Von Neumann who first proposed the idea of a machine containing a complete description of itself and with the ability to self-replicate and assemble itself. Von Neumann saw his machine as a way of understanding the origins of life. (von Neumann 1966; Langton 1984; Derrida & Weisbuch, 1986). Since then, there have been many attempts to model biological organisms and functionality using the artificial languages of logic and mathematics. These models - that include the models of cognitive processes proposed by Artificial Intelligence - have often been implemented on computers, making it possible to simulate complex behaviors. It could be argued that the roots of cybernetics go back to the mechanistic tradition in biology. In the XVII and XVIII centuries, biology began to describe the functioning of the human body using analogies from the science of Galileo and Descartes (Canguilhem, 1966). In other words, the motion of the organism as information processing machine lies at the intersection between the metaphor of life as a machine and that of life as language.

This last line of thought introduced the idea that genetic material is organized as a code that acts as a vehicle for biological information on the hereditary characteristics of an organism. This is an idea we find for the first time in Weismann (Maynard Smith & Szathmàry, 1995). Once the structure of DNA become clear and researchers began to study the genetic code, they came to the idea that genetics controls organisms through biochemical mechanisms. However there was still very little knowledge on the epigenetic relationships between the phenotype and the genotype. In the meantime, Shannon and Weaver introduced their concept of information content (1948), as something analogous to instructions and algorithms in artificial programming languages. This concept exerted a powerful influence on biological thinking. Nonetheless many researchers criticized the application of information theory to this domain (Apter and Wolpert,1965; Waddington, 1968). In particular, Waddington (1968) argued:

"the 'information' is not merely being transcribed and translated but is operating as instructions: if you want to put it in fancy jargon, as "algorithms". The DNA makes RNA and the RNA then makes a protein and the protein then does something to its surroundings, which result in the production of more varieties of molecules than before. There is nothing very mysterious in this, unless you try to see it in terms of messages going down telephone wires."

Despite these critiques, the edifice of molecular biological was built on the metaphor of a genetic program: and this tendency was reinforced by simultaneous developments in cybernetics and computer science.

This is the complex background to all attempts to model biological organisms: regardless of whether the aim is to look at the essential properties of life or to investigate evolutionary processes, as in this chapter. In our own work, we have focused on the modeling of biological processes using 1-D and 2-D CAs. In our approach, we have treated self-replicators as organisms with life-like characteristics and the ability to adapt to different environmental conditions. Furthermore, we have returned to the metaphor of life based on a genetic code, treating the rules table of the CA as its genome. This approach has allowed us to describe CAs at a high level of abstraction. In this picture, space-time diagrams and representations of the structure of self-replicators are phenotypes.

In a series of experiments, we have shown that changes in the genetic code for 1-D CAs and k-totalistic CAs produce changes in the shape and function of the system. In both cases, we have classified genes in terms of their spatial positioning and their effects on the genetic information process. In this way, we discovered that changes in some parts of the genome have no influence over the shape and function of the system; other genes have a moderate influence; others have the ability to completely transform the spatial patterns and 3-dimensional structures produced by the CA. Using our computational tool, we found that the pattern produced by a 1-D CA, or a system of self-replicators in a 2-D CA can be associated with a genetic structure that we can represent with a glyph. In the case of 1-D CAs, we have developed a true "production language" which, if implemented algorithmically, would allow us to perform non-conventional forms of computation. Furthermore, we have found ways of representing the development of self-replicators with dynamic networks, showing the sequence of rules that the system uses during the various steps of the simulation. We have also simulated large scale evolution, using a multi-level model, displaying a range of information processes and language phenomena including homologies (self-replicators with identical genetic networks and different phenotypes and vice versa, self-replicators with identical phenotypes and different genetic networks) and families of self-replicators. In these families, as in semantic networks, information degrades slowly. As a result, we observe different degrees of genetic relatedness. In the end, these lead to bifurcations with the creation of new organisms and new classes of genetic network. These observations lead us to reflect on the role of genetic information and on how to organize this information in our CA models of biological phenomena.

REFERENCES

Apter, M. J., & Wolpert, L. (1965). Cybernetics and development, Part 1: Information theory. *Journal of Theoretical Biology*, *8*, 244–257. doi:10.1016/0022-5193(65)90075-5

Bak, P. (1996). *How Nature Works: The Science of Self-Organized Criticality*. New York: Copernicus Press.

Bilotta, E., Di Blasi, G., Giambò, S., & Pantano, P. (2008). Generating Multi State Cellular Automata by using Chua's "Universal Neuron . In Ruggeri, T., & Sammartino, M. (Eds.), *Asymptotic Methods in Non Linear Wave Phenomena* (pp. 12–24). Singapore: World Scientific Publishing Company.

Bilotta, E., Lafusa, A., & Pantano, P. (2003a). Life-like self-reproducers. *Complexity*, *9*(1), 38–55. doi:10.1002/cplx.10118

Bilotta, E., Lafusa, A., & Pantano, P. (2003b). Searching for complex CA rules with Gas. *Complexity*, *8*(3), 56–67. doi:10.1002/cplx.10063

Bilotta, E., & Pantano, P. (2001). Observations on complex multi-state CAs. Paper presented at Advances in Artificial Life: 6th European Conference ECAL 2001. In J. Kelemen, & P. Sosik (Eds.), *Lecture Notes in Computer Science* (pp. 226–235). Heidelberg, Germany: Springer-Verlag.

Bilotta, E., & Pantano, P. (2006). The Language of Chaos. *International Journal of Bifurcation and Chaos in Applied Sciences and Engineering, 16*(3), 523–557. doi:10.1142/S0218127406014988

Canguilhem, G. (1991). *The Normal and the Pathological.* New York: Zone Books.

Chomsky, N. (1965). *Aspects of the Theory of the Syntax*. Cambridge, MA: The MIT Press.

Crutchfield, J. P. (1994). The Calculi of Emergence: Computation, Dynamics, and Induction. *Physica D. Nonlinear Phenomena, 75*, 11–54. doi:10.1016/0167-2789(94)90273-9

Crutchfield, J. P., & Schuster, P. (2003). *Evolutionary Dynamics: Exploring the Interplay of Selection, Accident, Neutrality, and Function*. New York: Oxford University Press.

Crutchfield, J. P., & Young, K. (1989). Inferring Statistical Complexity. *Physical Review Letters, 63*, 105–108. doi:10.1103/PhysRevLett.63.105

de Jong, H. (2002). Modeling and Simulation of Genetic Regulatory Systems: A Literature Review. *Journal of Computational Biology, 9*(1), 67–103. doi:10.1089/10665270252833208

Derrida, B., & Weisbuch, G. (1986). Evolution of overlaps between configurations in random Boolean networks. *Journal of Physics, 47*(8), 1297–1303. doi:10.1051/jphys:019860047080129700

Eigen, M. (1971). Selforganization of Matter and the Evolution of Biological Macromolecules. *Naturwissenschaften, 58*, 465–523. doi:10.1007/BF00623322

Eigen, M., & Schuster, P. (1979). *The Hypercycle: A Principle of Natural Self-Organization*. Berlin: Springer.

Fogel, L. J., Owens, A. J., & Walsh, M. J. (1966). *Artificial Intelligence through Simulated Evolution*. New York: John Wiley.

Freitas, R. A. Jr, & Merkle, R. C. (2004). *Kinematic Self-Replicating Machines*. Georgetown, TX: Landes Bioscience.

Haldane, J.B.S. (1924) . A mathematical theory of natural and artificial selection. Part 1. *Transactions of the Camrbidge philosophical society*, 23, 19-41.

Hanson, J. E., & Crutchfield, J. P. (1997). Computational Mechanics of Cellular Automata: An Example. *Physica D. Nonlinear Phenomena, 103*(1-4), 169–189. doi:10.1016/S0167-2789(96)00259-X

Holland, J. (1975). Adaptation . In *Natural and Artificial Systems*. Ann Arbor, MI: The University of Michigan Press.

Holland, J. (1994). *Adaptation in Natural and Artificial Systems*. Cambridge, MA: The MIT Press.

Kauffman, S. A. (1993). *The Origins of Order: Self-Organization and Selection in Evolution*. New York: Oxford University Press.

Kauffman, S. A. (1995). *At Home in the Universe: The Search for Laws of Self-Organization and Complexity*. New York: Oxford University Press.

Kimura, M. (1968). Evolutionary rate at the molecular level. *Nature, 217*, 624–626. doi:10.1038/217624a0

Kimura, M. (1983). *The neutral theory of molecular evolution*. Cambridge, UK: Cambridge University Press. doi:10.1017/CBO9780511623486

Langton, C. G. (1984). Self-Reproduction in Cellular Automata. *Physica D. Nonlinear Phenomena, 10*, 135–144. doi:10.1016/0167-2789(84)90256-2

Langton, C. G. (1986). Studying artificial life with cellular automata. *Physica, 22D*, 20–149.

Langton, C. G. (1990). Computation at the edge of chaos. *Physica, 42D*, 2–37.

Langton, C. G. (1995). *Artificial life: an overview*. Cambridge, MA: The MIT Press.

Maynard Smith, J., & Szathmary, E. (1995). *The Major Transitions in Evolution*. New York: W. H. Freeman Press.

Mitchell, M. (1996). *An Introduction to Genetic Algorithms*. Cambridge, MA: The MIT Press.

Mitchell, M., Hraber, P., & Crutchfield, J. P. (1993). Revisiting the Edge of Chaos: Evolving Cellular Automata to Perform Computational Tasks. *Complex Systems, 7*, 89–130.

Murray, J. D. (1989). *Mathematical Biology*. Berlin: Springer-Verlag.

Ratner, V. A., & Shamin, V. V. (1980). The sysers: Mathematical modeling of fundamental properties of molecular biological organization, Part I, II, III . In *Mathematical models of evolutionary genetics* (pp. 66–110). Novosibirsk, Russia: ICG.

Schuster, P., & Swetina, J. (1988). Stationary Mutant Distributions and Evolutionary Optimization. *Bulletin of Mathematical Biology, 50*, 636–660.

Shannon, C., & Weaver, W. (1948). *A Mathematical theory of communication.* Urbana, IL: University of Illinois Press.

Turing, A. M. (1952). The chemical basis of morphogenesis. *Philosophical Transactions of the Royal Society of London, Series B, 237*, 37–72. doi:10.1098/rstb.1952.0012

von Neumann, J. (1966). *Theory of Self-Reproducing Automata (edited and completed by AW Burks).* Urbana, IL: University of Illinois Press.

Waddington, C. H. (1968). The Basic Ideas of Biology. In C.H. Waddington (Ed.), *Towards a Theoretical Biology: Prolegomena* (Vol. 1, pp. 1-32). International Union of Biological Sciences Symposium.

Wolfram, S. (1984). Universality and Complexity in Cellular Automata. *Physica, 10D*, 1–35.

Wuensche, A. (1999). Classifying Cellular Automata Automatically: Finding gliders, filtering and relating space-time patterns, attractors basins and the Z parameter. *Complexity, 4*(3), 47–66. doi:10.1002/(SICI)1099-0526(199901/02)4:3<47::AID-CPLX9>3.0.CO;2-V

Wuensche, A., & Lesser, M. (1992). *The Global Dynamics of Cellular Automata: An Atlas of Basin of Attraction Fields of One-Dimensional Cellular Automata.* Reading, MA: Addison-Wesley.

APPENDIX

Rule Codes

C3.1: $D = 1, k = 5, r = 1$

0300001230013000310140120031322313103000030321300400434034400
3000230313140300301031340203130011000200330201214030423334012030

C3.2: $D = 1, k = 5, r = 1$

1202331341420304230002200440434230140342343244000040040310032
0014023314420020010120222312100101101022404030032343300131134400

C3.3: $D = 1, k = 6, r = 1$

4000301022011455251020000114250221354505350030000040100200201 3
5441000451252002310053001015050051503204000033042005505342003 4020
4001435333102500302005542305250300030105232514202003401004040035 2
300534000314133334010004

C3.4: $D = 1, k = 4, r = 1$

2302322333023000101212333320230231103222301000201020210103213000

C3.5: $D = 1, k = 4, r = 1$

3031201131033113103202301200130233100122120203330311011230000031

C3.6: $D = 1, k = 4, r = 1$

3130031131300220011112232322020302113221102033023011330221230101

C3.7: $D = 2, k = 4, r = 1$

1033000033020003003200200330010000020300000000001020011000132002
0023003010000000000000001202000010000000000320001210022003200000000 0
0130001022001110000102300232321000000000002320010301310000000001301
020100000002000003223230 00

C3.8: $D = 2, k = 5, r = 1$

0100010301000000000000022000000300000000002000113300010000000000400
0004000000000000000000004000412230000000000000000000000330003100000002
4010120020022000000000001000000000300030102020002000000100021204300 0

00000000300133340004000000203000000010000000000000000000003401011000
40000000000000000004000000000000100030000000000010000000110203003000
00000021000030000030000004000000000200000100000000003400040000001000
00300000001000000200000000010000120000100000010004020000030000000000
00003440000030000000000000000100100000323010000000000000340000200001
30000000002000000000400440010200023320200240010303010040040004200 0
00200001000402040010000300000000000000220100001000000003000000000004
00020000000300000404000000000000000000000030010000000044000001003020

Chapter 4
Emergent Structures

INTRODUCTION

There are two classes of problem in the study of Cellular Automata. The *forward problem* is the problem of determining the properties of the system. Solutions often consist of finding quantities that are computable on a rules table and characterizing the behavior of the rule upon repeated iterations, starting from different initial conditions. Solutions to the *backwards* problem begin with the properties of a system and find a rule or a set of rules which have these properties. This is especially useful in the application of Cellular Automata to the natural sciences, when researchers deal with a large collection of phenomena (Gutowitz, 1989). Another approach is to *identify* the basic structures of a Cellular Automaton (Adamatzky, 1995). Once these are known it becomes possible to develop specific models for particular systems and to detect general principles applicable to a wide variety of systems (Wolfram, 1984; Lam, 1998). According to Adamatzky, the identification of a system consists of two related steps, namely *specification* and *estimation*. In *specification* we choose a useful and efficient description of the system: perhaps an equation and a set of parameters. The second step involves the *estimation* of parameter values for the equation: exploiting measures of similarity.

DOI: 10.4018/978-1-61520-787-9.ch004

In this chapter, we analyze the most important structures we observe in one dimensional CAs, more specifically gliders and regular domains. In this setting, we can view gliders as particles that travel inside these domains, interacting with each other in a non-linear fashion, on occasion annihilating each other or giving rise to new particles (Wolfram 2002; Langton1986; 1990). This analysis allows us to create a high level description of CAs as sets of interacting particles. This is the approach taken by Computational Mechanics which creates catalogs of formal descriptions for every system it studies (Crutchfield & K. Young, 1989; Crutchfield, 1994; Hanson & Crutchfield, 1997). An alternative approach is to study CAs' global dynamics identifying basins of attraction. In this way, we can study the tight relationship between attractors and the domains to which they give rise. However, computational mechanics and analysis of global dynamics are only practical for CAs with a small number of cells. Systems with large numbers of cells require new tools. In this chapter, therefore, we describe qualitative, largely visual, methods of analysis, and show examples of how they can be used to derive organizational laws for emerging structures. On the basis of these methods, we have created a computational tool that analyzes the pattern generated by a CA and used the tool to identify the basic elements of the system: objects and backgrounds which are subsequently decomposed into particles, emergent particles and uniform domains. In this way we can choose between a top-down and a bottom-up analysis. In this chapter, we also extend the Universal Neuron approach (Chua, 2006) to multi-state CAs, showing how this method can generate rules with a broad range of different characteristics (Bilotta et al., 2008). The final section of the chapter describes future developments and summarizes our conclusions.

BACKGROUND

One of the most spectacular characteristics of Cellular Automata is the emergence of complex structures, revealing their self-organizing capabilities. This kind of structure may emerge in a previously stationary domain against a range of different backgrounds. It can take many different forms. Some are stable; others decay immediately after they have come into being; others break down into substructures that reassemble after a certain lapse of time; others may generate new structures or absorb existing ones. When they collide, they may survive, they may be annihilated, or they may give rise to a cascade of new structures of increasing levels of complexity. The literature refers to some of these structures as "gliders".

Gliders

Gliders are local, sometimes periodic structures that remain stationary or move at a certain speed inside the pattern. Gliders are closely associated with so-called Glider Guns, and Rakes (Figure 1), which respectively, "shoot" and "absorb" gliders.

Gliders only appear in certain specific classes of complex CA. Several researchers, including Wolfram (1994; 2002) have produced proofs that the presence of gliders in these systems make them Universal Computers, in Turing's sense of the word. Whether or not this is correct, there can be no doubt that gliders allow the transport of information from one region of the system to another, even when the dynamics of the system depend entirely on local interactions. It is this possibility that gives rise to global patterns of behavior. Figure 1 shows examples of some of the best known complex CAs.

Figure 1. Spatio-temporal patterns in 1-D CAs. In the figures we can see Gliders, Glider-guns, Rakes, etc. The first two patterns were obtained using ECAs (54 and 110, respectively Figures a and b). The remaining 7 patterns (Figures c-i) were obtained with k=2,r=2. The relevant systems are rules C4.1-C4.7, listed in the Collection of Rules.

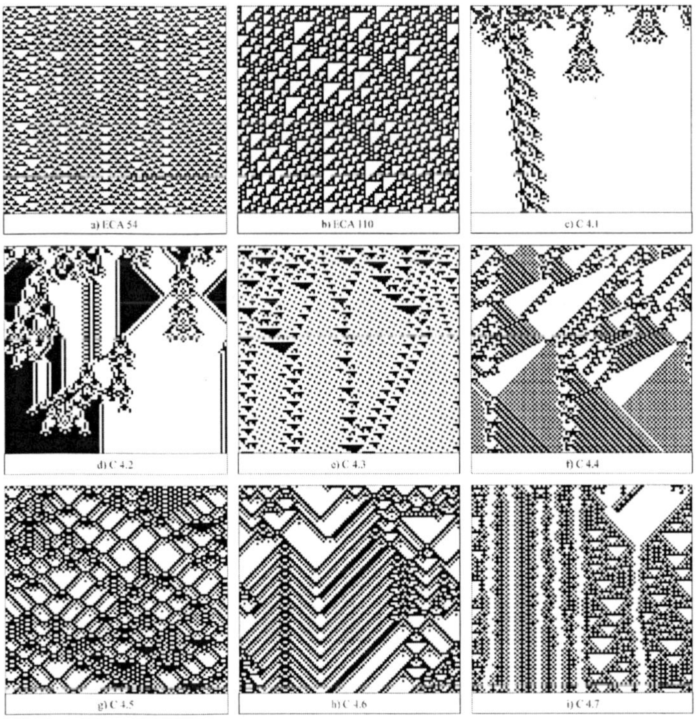

- Figures 1a and 1b shows two ECAs (54 and 110), discovered by Wolfram.

- Figures 1c and d describe two CAs with $k = 2$ and $r = 2$, discovered by Wolfram during his investigation of totalistic rules. These CAs use totalistic rules 20 and 52. The corresponding rules in the Collection of Rules of the book are systems coded with C4.1 and C4.2.

- Figures 1e and 1f show two CAs with $k = 2$ and $r = 2$, discovered by Wuensche. The corresponding rules are systems coded with C4.3 and C4.4.

- Figures 1g and 1h depict two CAs with $k = 2$ and $r = 2$, discovered by Aizawa et al. (1990). The corresponding rules codes are C4.5 and C4.6.

- Figure 1i shows the so-called "Raindrops" CA with $k = 2$ and $r = 2$. The corresponding rule is coded by C3.7.

If we carefully analyze the spatio-temporal patterns shown in the figures, we observe that gliders are coherent structures that stand out clearly against their backgrounds. The backgrounds themselves are uniform, periodic or display some other form of regularity. Gliders appear to travel at a well-defined speed. Their maximum speed is given by the radius of the neighborhood (r=1,r=2,r=3…). This plays a role, similar in some ways to that of the speed of light in relativistic physics. Gliders can be considered as emergent particles with well-defined characteristics, including the ability to absorb or emit other gliders (glider-guns). At times, we observe the emergence of composite gliders produced by periodic collisions between "sub-gliders".

The literature refers to gliders in many different ways according to the context in which they are observed. On different occasions, they have been called "solitons", "virtual automata", "information structures", "particle-like structures", "domain walls", and "domain faults". In many cases, they create a connection between two different types of background. In this sense, they can be considered as a wave or a disturbance that propagates within the system.

Computational Mechanics

One way of describing a CA would be to investigate the behavior of gliders and their collisions. This would a higher level of description than we could achieve through the study of individual cells. This can be interesting, but also quite *time-consuming*. Another way of obtaining a high level description of CAs is to apply so-called "Computational Mechanics". In this approach, structures ("particles") are treated as disturbances in a regular background. Computational Mechanics involves an in-depth study of the computational capabilities provided by these "particles" and their behavior in collisions (Crutchfield & K. Young, 1989; Crutchfield, 1994; Hanson & Crutchfield, 1997). However, this "particle-based" approach has several limitations. In some cases, gliders seem to behave like waves. In these instances they can be seen as disturbances in space-time, or as the boundary between regions with different characteristics. However, they also have internal structure and interact non-linearly with other structures, similar or dissimilar to themselves. A massive presence of gliders within a system leads to turbulence (CAs have been used to model turbulence and phase transitions). Gliders may even incorporate completely chaotic regions internal to the overall structure (Aizawa et al., 1990).

Figure 2 shows details of the gliders produced by totalistic rule 20 (rule code C4.1). The first two gliders move leftwards and rightwards respectively. The other two are stationary.

It is interesting to study the mechanisms underlying glider interactions. For instance, Figure 3 shows the interaction between gliders generated by Aizawa's rule (Aizawa et al., 1990) for type A solitons (rule C3.5). In this case, we have two types of glider, one moving leftwards (Period 2, code 1101, Speed 1) and one moving rightwards (Period 2, code 1011, Speed 1). If we separate the gliders with an

Figure 2. The top row shows the movement of the gliders over 100 time steps. The bottom row shows details of their internal structure. In Figure 2a, the glider travels leftwards with a period of 10, a minimum length of 8 characters -11011101–a maximum length of 15 characters–100101101010001–and a speed of 0.1; In Figure 2b the glider travels rightwards with a period of 10, a minimum length of 8 characters–10111011-, a maximum length of 15 characters–100010101101001-, and a speed of 0.1; Figure 2c shows a stationary type 1 Glider with a period of 2, a minimum and maximum length of 8 characters–11101001 and a speed of 0; Figure 2d shows a stationary type 1 Glider with a period of 1, a minimum and maximum length of 8 characters–10111101 and a speed of 0.

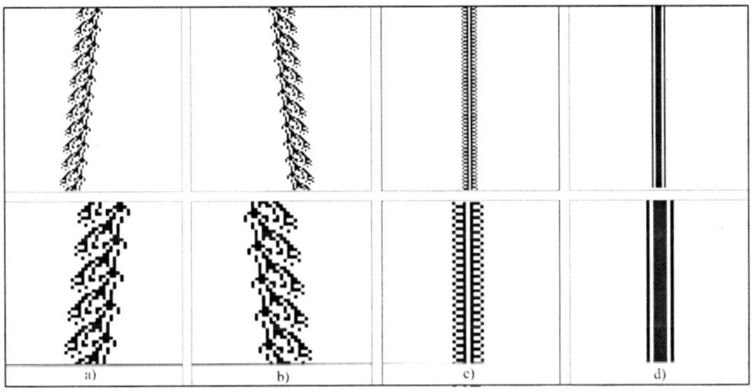

odd number of zeros, the two are superimposed. When the separation involves an even number of zeros, the result is a new structure, of higher complexity than that of the gliders entering the interaction. Where 3 or more gliders interact, we always obtain higher order structures.

Figure 4 shows the pattern generated by rule C2.1 with $d = 1$, $k = 3$, $r = 1$. In this pattern, we observe:

- 3 regular domains (A, B and C): A is at rest, with code 1; B is periodic 2-0; C is regular with code 1200;
- α: A first separator glider between (2) and (1); between (0) and (1), code 102 becomes code 021; the glider moves rightwards at the maximum possible speed;
- β: The glider moves rightwards, creating a separation between the regular domain and (1): 120022102111;
- γ: A separator glider between the regular domain and (1): 120012001111; this is the only glider that moves leftwards;
- δ: A separator glider between (1) and the regular domain:111112001200; the glider moves rightwards with a period of 6;
    - δ∋: A separator glider between (1) and the regular domain:120000002;
    - δ∋∋: A separator glider between (1) and the regular domain:12000002; in this case there are fewer zeros;
- ε: The glider moves in the uniform state (1): 0102;
- ϕ: The glider moves in a regular domain (1200221200) at a slower speed than in the previous examples;
- η: A regular, stationary separator glider between the two uniform regular domains (2222212001200);
- ι: A regular separator glider between two regular uniform domains (1200120022222).

Gliders of type δ can be distinguished by the number of zeros in their codes. There is a close correlation between gliders and domains, which we will discuss

Figure 3. Type A Solitons (Rule C4.5). Figure 3a show the superimposition of 2 gliders separated by an odd number of zeros. On a grid of 100 automata, we observe a period of 95. Figures 3b-d show the emergence of higher level structures when gliders are separated by an even number of zeros.

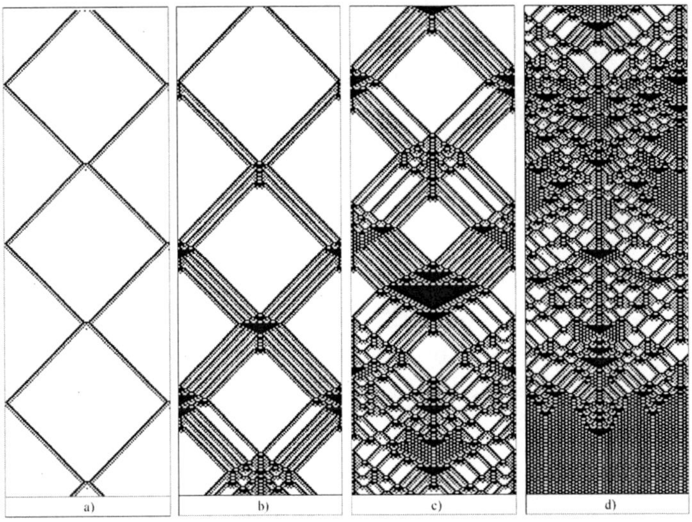

Figure 4. Spatio-temporal patterns generated by rule C4.1 with d = *1,* k = *3,* r = *1. The figure shows several domains and a series of gliders, together with the rules governing the interactions between the gliders.*

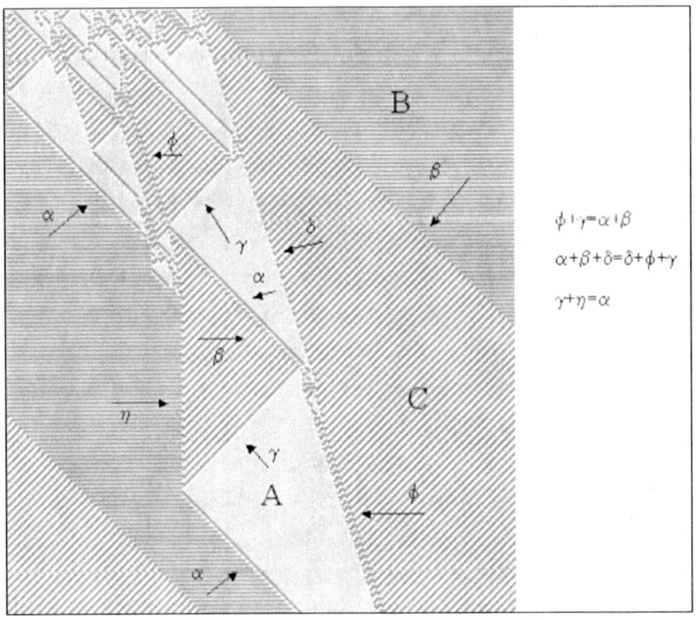

later in this chapter. Figure 4 shows interactions between gliders and the way these interactions can generate and annihilate other gliders.

The real limit of the Computational Mechanics approach is that it is impossible, in practice, to create a catalog of all possible emergent structures. The rules governing interactions between gliders and between gliders and domains generate a very large number of particles, closely tied to *M*, the number of cells considered in the analysis. An interesting possibility is that it may be possible to implement this kind of structure in hardware, creating non-standard computers with special computational capabilities (Adamatzky et al.,2007).

Basins of Attraction

Consider a CA composed of M cells, with k^M possible configurations. Table 1 shows the number of possible configurations for different values of k^k and *M*. As can be seen from the table, this grows rapidly with increases in the value of the parameters. The time the system takes to return to a previous configuration is always less than k^M.

In usual dynamic systems terminology (Hilborn, 2000; Strogatz, 2000), we say that systems that arrive at an equilibrium configuration and remain in that configuration have reached a *fixed point;* when they return to a configuration they have already visited, they enter a *limit cycle.* Given that they are deterministic systems, they will continue in this cycle indefinitely. Regardless of initial conditions, all systems eventually end up in a fixed point or a limit cycle. As in continuous dynamic systems, we will call these fixed points and limit cycles *attractors.* If a system starts from a given set of initial conditions, it will end up in an attractor, passing on the way through a sequence of *transient* configurations. The set of all configurations associated with an attractor is called its *basin of attraction.* If C is the set of all possible configurations, a *trajectory* is a sequence of configurations, $c\,0, c\,1, c\,2 \ldots\ldots c(n)$, where $c\,i \in C, \forall i$. Obviously, a CA can follow only a subset of the trajectories that are theoretically possible. Given that it is a deterministic system, a given set of initial conditions will always produce the same trajectory. The spatio-temporal diagram shown here represents one of these trajectories. Note that C contains configurations that are unreachable by any trajectory. The only way to produce these configurations is to set them as initial conditions. This kind of configuration is known as a "Garden of Eden."

C contains a finite number z of attractors αi, where $\alpha i \subseteq C, i=1,2 \ldots z$. In the case of a fixed point, αi contains a single element and the corresponding trajectory will consist of a finite sequence. In the case of a limit cycle, the sequence will be infinite and periodic. When we investigate the global dynamics of CA*s, what we are* doing is identifying the system's fixed points, limit cycles and basins of attraction.

Table 1. Number of possible configurations for different values of k and M

k	n. cells (M)	n. configurations (k^M)
2	8	256
2	10	1024
2	20	1048576
2	30	1073741824
2	50	1.1259*e^{15}
2	100	1.26765*e^{30}
3	8	6561
3	10	59049
3	30	2,05891*e^{14}
3	50	7,17898*e^{23}
4	8	65536
4	10	1048576
4	30	1,15292E+18
5	10	9765625
5	30	9,31323*e^{20}

Figure 5 shows the basins of attraction for two ECAs (rules 90 and 122), calculated for CAs with 5, 8 and 10 cells respectively. With M = 10 and rule 90, *we* can identify 46 attractors.

As shown in the Figure 5 (a and b), for ECA 90 with M=5, there are practically no transient states. The CA reaches the attractor in a single step. This result is closely dependent on the value of M (the number of cells). We see this in Figure 5c, where $M = 8$. In this case, we obtain a single basin of attraction and the attractor is a single fixed point. With M = 10 and rule *90*, we can identify 46 attractors:

- 1 of type 1 (see Figure 5d); this type of attractor consists of a fixed point and four elements;
- 40 of type 2 (see Figure 5d); these attractors have a limit cycle of 6 elements and a basin of attraction of 24 elements;

Figure 5. The first two images (a, b) show the basins of attraction for ECA 90. The first row shows the configurations corresponding to the various elements composing the basin of attraction. The second two figures (c, d) shows ECA 90 and ECA 122 for $M = 8$. *The last two rows show basins of attraction for the same ECAs with M = 10. The first three images (e, f, g) refer to ECA 90; the last 7 images (h, i, l, m, n, o, p) refer to ECA122.*

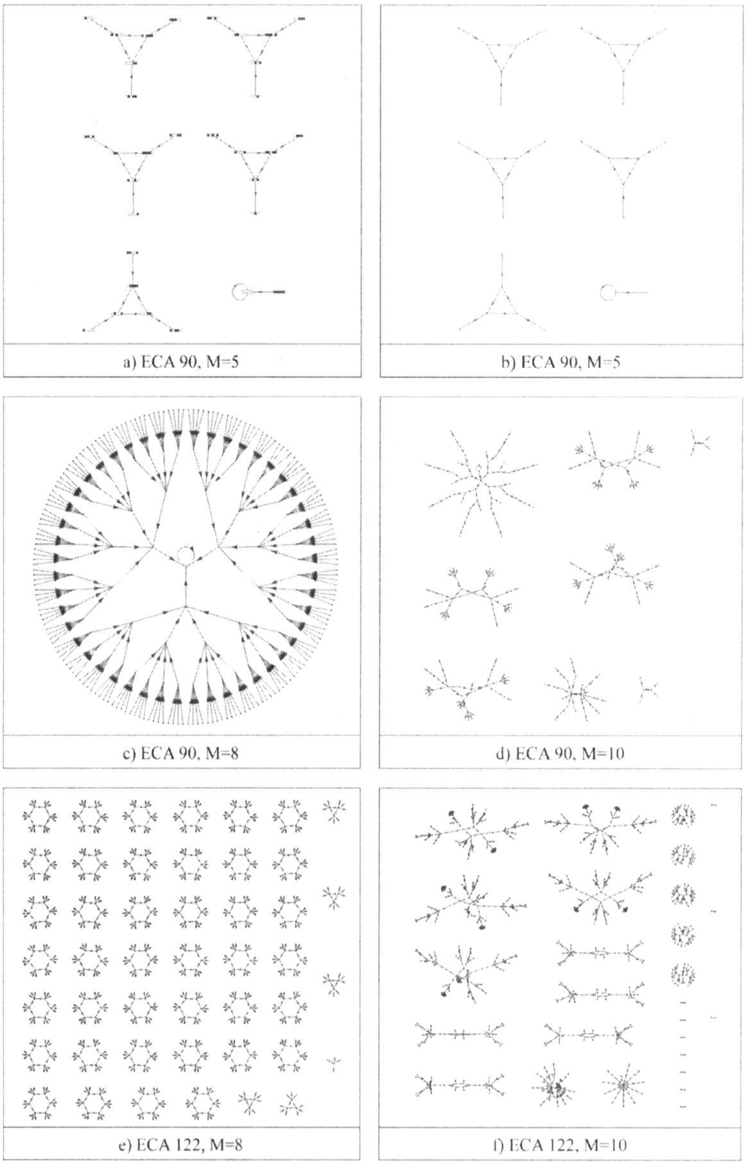

- 5 of type 3 (see Figure 5d); these attractors have a limit cycle of 3 elements and a basin of attraction of 12 configurations.

The last seven images show that ECA 122 is much more varied than ECA 90. The 27 basins of attraction belong to 6 different classes. The first, shown in the first image, has a single fixed point and a basin containing 42 configurations. Five are similar to the second image, with 1 6-element limit cycle and a basin of 48 configurations; 5 are like the third image, with 1 4-element limit cycle and a basin of 100 configurations; 5 are like the fourth image with 1 2-element limit cycle and a basin of 32 configurations. 10 are like the sixth image with 1 2-element limit cycle and a basin of two configurations, 1 is like the last image with 1 2-element limit cycle and a basin of 62 configurations. Two apparently different basins may nonetheless belong to the same class, as in the case of images 7 and 9, which are evidently symmetrical. These symmetrical basins of attraction generate symmetrical patterns - an issue we will discuss in greater detail in a later chapter. In general, the different basins of attraction have transients of differing lengths, ranging from 5 to 10. Studies of global dynamics attempt to produce a catalog of all possible configurations. This is possible only for CAs with a small number of cells. As will be shown in a later chapter, these analyses, with small M, can nonetheless yield interesting information about CAs with a larger number of cells.

Chua's Universal Neuron

Chua (2006; 2007; 2009) has published a series of papers in which he describes an innovative approach to the study of ECAs. From many points of view, Chua's work represents one of the most important steps forward since Wolfram's original pioneering work (Wolfram, 1994). In this section we will describe the main features of Chua's work.

We have already seen that the rules table for ECAs consists of 8 elementary rules. We can associate each rule with the vertex of a cube. Coloring the vertices in black and white, we obtain a visual representation of the rules table. If we place the origin in the center of the cube, we can associate each vertex $i = 0,1,...7$ with a vector, applying the expression:

$$\begin{cases} u_0 \equiv (-1,-1,-1), u_1 \equiv (-1,-1,1), u_2 \equiv (-1,1.-1), u_3 \equiv (-1,1,1), \\ u_4 \equiv (1,-1,-1), u_5 \equiv (1,-1,1), u_6 \equiv (1,1,-1), u_7 \equiv (1,1,1) \end{cases} \quad (4.1)$$

Given a vector $b \equiv (b_1, b_2, b_3)$, we project vectors u_i onto line *b*. In this way, we generate a sequence of 8 points which will be black or white according to the value (1 or 0) associated with vectors u_i (see Figure 6). It is possible, of course, that some of the points will be superimposed.

Following Chua, we indicate the scalar product of b and u_i with σ_i. In this way, the distance between a point on b and the origin, O (0,0,0) is given by:

$$\sigma_i = b \cdot u_i,\ i = 0,1,...7 \tag{4.2}$$

Given ***b***, every possible rules table is associated with a unique sequence of points along ***b***. Consider a function ***w(σ)***:

$$w(\sigma) = z_2 \pm \left|[z_1 \pm |z_0 + \sigma|]\right| \tag{4.3}$$

where z_0, z_1, z_2 are integers. This function, known as the *discrimant function,* will depend on six parameters $z_2, z_1, z_0, b_1, b_2, b_3,$ and two integers with value ± 1. Once these 8 parameters have been set, each elementary rule $i = 0, 1, 2..., 7$ can be associated with a specific value of the discriminant function, given by $w(\sigma_i)$.

Now consider the dynamic system:

$$\dot{x}_i = g(x_i) + w(\sigma_i) \tag{4.4}$$

Figure 6. The vertices of the Boolean cube projected onto vector b

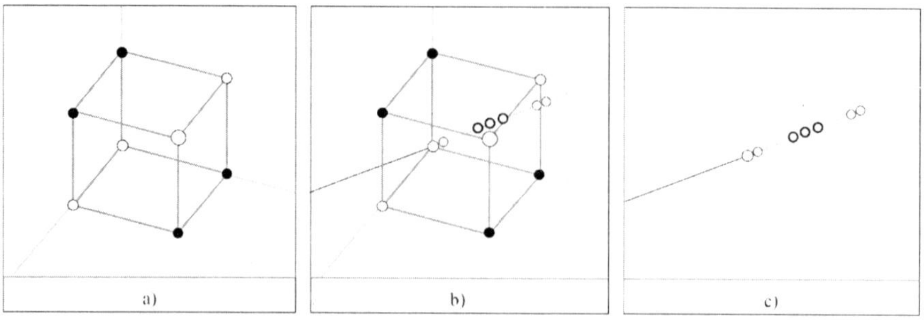

where $x = x(t)$ is a real function of the real variable, $\dot{x} = \frac{dy}{dt}$. Let us define a "guide function":

$$g(x_i) = -x_i + |x_i + 1| - |x_i - 1| \tag{4.5}$$

For all values of σ_i, (4.4) defines 8 distinct differential equations. With initial conditions $x(0) = 0$ (4.4) will yield a single positive attractor if $w(\sigma) > 0$, and a negative attractor in the opposite case. Figure 7 shows the system attractors (4.4) for different values of the discriminant function.

It follows that for any equation of form (4.4)-(4.5) with initial conditions x(0)=0, state x(t) converges monotonically towards attractor Q+ (positive) if w>0 and attractor Q- (negative) if w<0.

Let us introduce the 2-valued function (4.6). The output from this function is always -1 or +1:

$$y_i = \text{sgn}\left\{w(\sigma_i)\right\} \tag{4.6}$$

We consider this function as the *output* of the dynamic system (4.4). Thus, we have constructed a machine, a Universal Neuron, that takes as its input a neighborhood u_i and returns an output y_i. This allows us to create a sequence of outputs that we can interpret as elementary rules and use to build a rules table. The machine

Figure 7. Attractors of System (4.4) for different values of the discriminant function

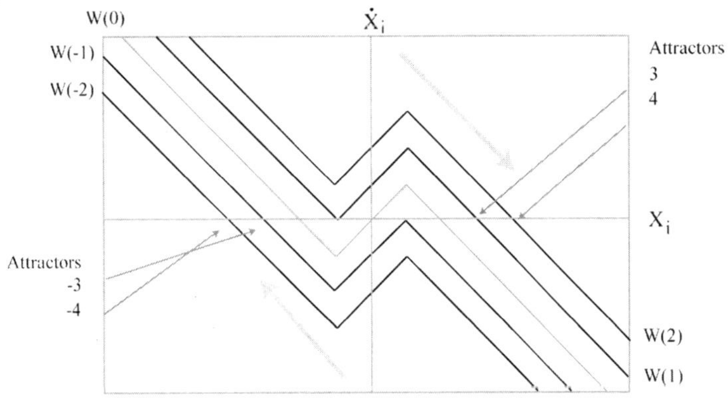

is controlled by 8 parameters $z_2, z_1, z_0, b_1, b_2, b_3, c_1$ and c_2, where $c_1$and c_2 are integers with value i ± 1 (Figure 8).

$$y_i = \text{sgn}\left\{z_2 + c_2 \left|[z_1 + c_1 \left|z_0 + b \cdot u_i\right|]\right|\right\} \tag{4.7}$$

Dynamic systems represented by (4.4) can be easily implemented on semiconductor chips, as CNN cells (Chua & Yang, 1988; Chua & Roska, 1993; Chua 1998). In this way we can create *analog systems* with the ability to simulate the behavior of the CA, and with much higher processing speed than current digital systems. (Chua et al., 2002; 2004).

The set of 8 parameters $z_2, z_1, z_0, b_1, b_2, b_3, c_1$e c_2 describes a *Universal Neuron.* By instantiating these parameters, we can generate every possible ECA rule. Chua examines the symmetries between cubes with different values assigned to the vertices and demonstrates corresponding symmetries in ECA behavior.

Given that parameters $z_2, z_1, z_0, b_1, b_2, b_3$ can take continuous values, the space of universal neurons is much larger than the rule space for ECAs. It follows that, for any given ECA, there may be more than one Universal Neuron and more than one dynamic system of type (4.4).

Figure 8. The Universal Neuron processes data from the neighborhood and returns the corresponding local rule

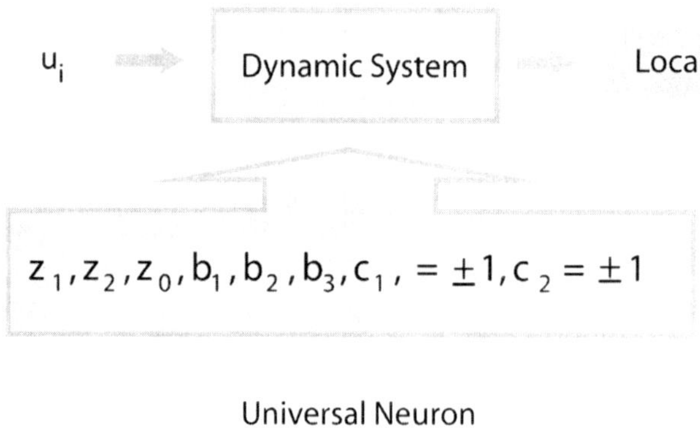

MAIN FOCUS OF THE CHAPTER

One of the most important problems with CAs is the identification of emergent structures in patterns. The problem is related to another, logically prior problem, namely, the identification of rules that generate emergent structures. We will discuss this problem in greater detail in Chapter 5. In this Chapter, we will present a number of techniques allowing the identification of structures inside patterns.

The first approach we present is a perception-based approach to pattern analysis. This will lead us to study the rules governing the visual organization of emergent structures and to discover a number of general tendencies, tied to the laws governing the organization of perceptual stimuli. Another approach is the use of automatic procedures to identify emergent structures such as regular domains and gliders. We go on to introduce constructs from linguistics and to show how they can be used to automatically generate new structures from pre-existing ones. Finally, with the same goal in mind, we extend Chua's treatment of the Universal Neuron to CA with $k > 2$.

Qualitative Analysis

Finding emergent particles in a CA pattern requires a detailed analysis of the pattern. Often the results of this analysis will be highly ambiguous. Particularly in chaotic patterns, it is very difficult to distinguish emerging structures. In the theoretical debate on the nature of complexity and emergence Chalmers (2002) and Crutchfield (1994) have claimed that the complexity and organization of a system are subjective properties that can be determined by the observer. Does this mean we use qualitative analysis to detect emergent structures in an intuitive way? Have we as observers the ability to detect structures in CAs patterns without a formal mathematical analysis? Are there other conceptual tools we can use? In this chapter we exploit the Gestalt theory of perception as the basis for a computational tool that can automatically detect emergent structures in a CA. The theory posits a small number of general laws (Koffka, 1935; Koehler, 1947; Werthcimer, 1912):

- **Proximity:** When the elements of a visual stimulus are close together we see them as part of the same object; when they are far apart we see them as part of different objects. This allows us to separate a vast set of elements into a much smaller set of objects (Figure 9a).

- **Similarity:** When the elements in a stimulus are physically similar, we see them as belonging to the same object; when they are dissimilar we see them as belonging to different objects. It is these visual differences that allow us to

Figure 9. Examples of perception laws of proximity (a), similarity (b), closure (c), good continuation (d), shared motion (e)

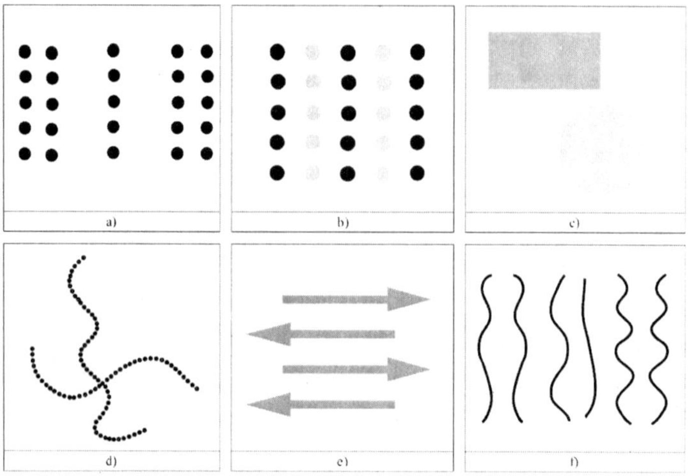

distinguish between objects. In the case shown in the Figure, we perceive two lines in two different colors, as distinct entities (Figure 9b).

- **Closure:** We perceive shapes as if they have a continuous border and ignore breaks in the border. This helps us to see them as if they were complete, even if they are partially hidden by other objects (Figure 9c).

- **Good continuation:** When the elements appear to be moving in the same direction, we see each element as a "good continuation" of its predecessors (this way of organizing the stimulus is typical of what we see with gliders) (Figure 9d).

- **Shared Motion:** When the elements move at the same time while other elements remain still or move in the same way, we perceive them as belonging to the same object (Figure 9e).

- **Symmetry:** When faced with objects that are symmetrical, the visual system tends to group them into a single object (Figure 9f).

Using these laws as tools, we detect emergent structures in CAs patterns.

Automatic Identification of Emergent Structures

Figure 10 provides a schematic representation of the tool just described. The tool allows us to analyze the patterns generated by a CA in terms of a syntax (a set of rules for combining elements into larger structures). We can then use this syntax to study the laws governing the way in which structures are created and the laws underlying this process. The patterns produced by a CA can be interpreted as a scene (created when the system comes into being) in which we can discriminate objects (particles and emergent particles) from their background. The backgrounds are what we call uniform domains (Crutchfield 1994; McTague & Crutchfield, 2006).

Using this approach, we have created a computational tool that represents the pattern generated by a CA as a hierarchy of components organized as a visual scene. The system performs a preliminary analysis to identify the basic elements of the system: objects and backgrounds which are subsequently decomposed into particles, emergent particles and uniform domains. In this way, if we move through the hierarchy starting at the top, we obtain a breakdown of the pattern into lower order elements. If we move in the opposite direction, starting from the bottom, we gradually compose elementary components into higher level structures. To make the approach feasible from a computational point of view, we need to consider the number of regular domains allowed by the system parameters.

Consider a CA with $k = 3$ and $n = 3$ (i.e. with a radius of 1), and with code 001220000100011010200211112. Figure 11a shows the pattern generated by the CA. Varying M between 3 and 10, we obtain the 33 regular domains described in

Figure 10. A representation of a CA, using the computational model

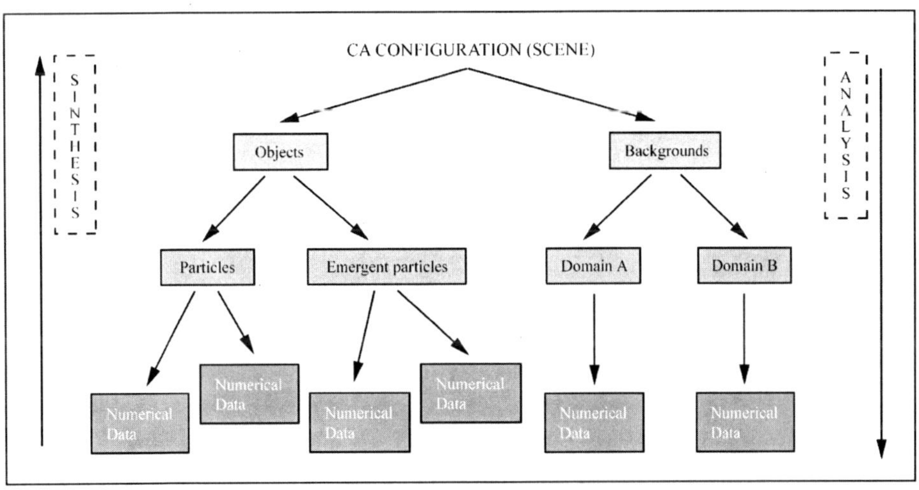

Figure 11. The first row shows the patterns produced by the 3 CAs. The second row shows the distribution of basins of attraction with period $\tau \leq 30$ as a function of M.

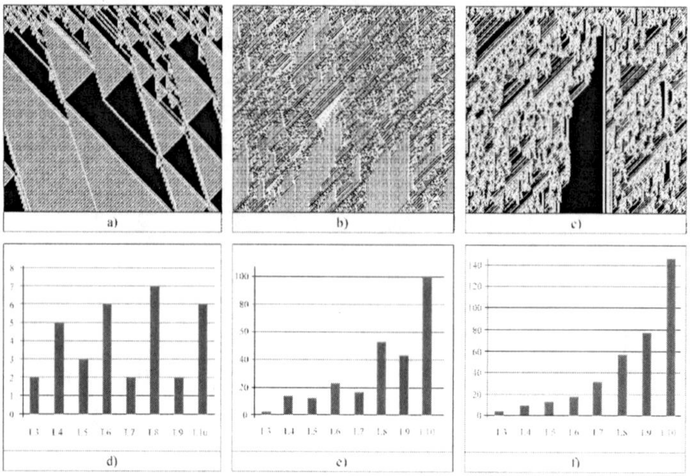

Table 2. Figure 11d shows the number of basins of attraction as a function of M, in this case 33.

For $M = 3$, the CA has $3^3 = 27$ states and 2 basins of attraction: one with a periodic 0-2 domain and one with a uniform domain with value 1. With $M = 4$, the CA has $3^4 = 81$ states and 5 basins of attraction. Figure 11d continues the analysis for strings of 5-10 characters. With $M = 5$, there are $3^5 = 243$ states and 3 basins of attraction; for $M = 6$, there are $3^6 = 729$ states and 6 basins of attraction; for $M = 7$, there are $3^7 = 2187$ states and 2 basins of attraction; for $M = 8$, there are $3^8 = 6561$ states and 7 basins of attraction; for $M = 9$ there are $3^9 = 19683$ states and 2 basins of attraction; for $M = 10$, there are $3^{10} = 59049$ states and 6 basins of attraction.

It is interesting to observe that all the CAs have the uniform basin of attraction 1 and the period 0-2 basin and that this behavior is independent of *M*. Other observations show correlations between domains in CAs with different values of *M*. For example domain d_{20} with $M = 8$. $M = 8$ is a composition of 2 domains previously observed with $M = 4$. In other words, $d_{20} = d_4 \,\&\, d_4$. As we will see in Chapter 9, the rules for composing domains can be very interesting. Once these rules are known, we can use analyses with *M* relatively small to learn about CAs in which *M* is very large indeed.

Table 2. Regular domains for a CA with d = 1, r = 1 and code 001220000100011010200211112, for different values of M. The last element in the phase is always the same as the first.

Number of cells M	Domain name	Phases
3	d_1	000, 222, 000
	d_2	111, 111
4	d_3	0000, 2222, 0000
	d_4	1000, 1021, 1002, 0010, 2110, 0210, 1000
	d_5	0100, 1102, 2100, 0001, 0211, 0021, 0100
	d_6	1200, 2001, 0012, 0120, 1200
	d_7	1111, 1111
5	d_8	00000, 22222, 00000
	d_9	21000, 00021, 02100, 10002, 00210, 21000
	d_{10}	11111, 11111
6	d_{11}	000000, 222222, 000000
	d_{12}	100000, 102221, 102002, 000010, 222110, 200210, 001000, 211022, 021020, 100000
	d_{13}	010000, 110222, 210200, 000001, 022211, 020021, 000100, 221102, 002102, 010000
	d_{14}	211100, 021101, 002111, 010211, 110021, 110102, 211100
	d_{15}	211010, 021110, 102110, 100211, 101021, 111002, 211010
	d_{16}	111111, 111111
7	d_{17}	0000000, 2222222, 0000000

continues on following page

Table 2. continued

Number of cells M	Domain name	Phases
	d_{18}	1111111, 1111111
8	d_{19}	00000000, 22222222, 00000000
	d_{20}	10001000, 10211021, 10021002, 00100010, 21102110, 02100210, 10001000
	d_{21}	01000100, 11021102, 21002100, 00010001, 02110211, 00210021, 01000100
	d_{22}	21111100, 02111101, 00211111, 01021111, 11002111, 11010211, 11110021, 11110102, 21111100
	d_{23}	12001200, 20012001, 00120012, 01200120, 12001200
	d_{24}	21111010, 02111110, 10211110, 10021111, 10102111, 11100211, 11101021, 11111002, 21111010
	d_{25}	11111111, 11111111
9	d_{26}	000000000, 222222222, 000000000
	d_{27}	111111111, 111111111
10	d_{28}	0000000000, 2222222222, 0000000000
	d_{29}	2212020000, 2020000221, 0000221202, 0221202000, 1202000022, 2000022120, 0022120200, 2120200002, 0200002212, 0002212020, 2212020000
	d_{30}	2100021000, 0002100021, 0210002100, 1000210002, 0021000210, 2100021000
	d_{31}	2111111100, 0211111101, 0021111111, 0102111111, 1100211111, 1101021111, 1111002111, 1111010211, 1111110021, 1111110102, 2111111100
	d_{32}	2111111010, 0211111110, 1021111110, 1002111111, 1010211111, 1110021111, 1110102111, 1111100211, 1111101021, 1111111002, 2111111010
	d_{33}	1111111111, 1111111111

As M increases, the number of basins of attraction varies, sometimes by quite large amounts.

As an example, consider a CA with $k = 4$ and $n = 3$ (i.e. with radius 1), and code

3000210133003300220031133320310310022102023030013301123202233203

Figure 11b shows the pattern produced by a CA with 300 cells ($m = 300$). The CA begins with a randomly generated pattern and evolves for 300 time steps. Here too, we observe a range of regular domains. Limiting our analysis to limit cycles with $\tau \leq 30$, we obtain a total of 268 regular domains. Figure 11e shows the distribution of these domains as a function of M.

For $M = 3$, the CA has $4^3 = 64$ states and 3 basins of attraction. With $M = 4$, there are $4^4 = 256$ possible configurations. Here we can identify 14 basins of attraction, with 13 limit cycles and 1 fixed point. One limit cycle has a period of 20 (domain 1), 4 have a period of 4 (domains 2, 4, 7 and 8) and 8 have a period of 2 (domains 3, 5, 6, 9, 10, 11, 12, 13). If we extend this analysis to strings of length 5-10, we obtain the results depicted in Figure 13e. With $M = 5$, there are $4^5 = 1024$ states and 13 basins of attraction. With $M = 6$, there appear to be $4^6 = 4096$ states and 23 basins of attraction. In reality, the number is higher. d_3 has phases 000032, 333331, 133303, 313030, 032313, 332032, 201332, 300101, 200300, 233102, 320002, 213321, 031003, 300131, 200003, 133333, 313330, 031303, 303231, 233203, 220133, 130010, 020030, 223310, 232000, 121332, 303100, 130013, 320000, 313333, 031333, 303130, 130323, 323320, 322013, 013001, 002003, 022331, 023200, 212133, 030310, 313001, 032000, 331333, 303133, 030313, 313032, 032332, 332201, 101300, 300200, 102233, 002320, 321213, 003031, 131300, 003200, 333133, 330313, 303031, 231303, 203233, 133220, 010130, 030020, 310223, 000232, 332121, 100303, 013130, 000320, 333313, 333031, 130303, 323130, 320323, 013322, 001013, 003002, 331022, 200023, 133212, 310030, 001313;

Figure 13. Different rules obtained from the same Neuron by varying the value of k

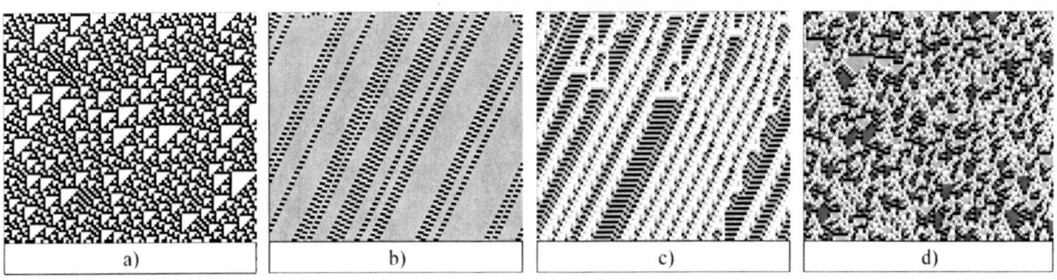

(Period $\tau = 84$). d_7 has phases:

002213, 022132, 121301, 203202, 333112, 230022, 300221, 202213, 112130, 220320, 233311, 223002, 130022, 320221, 011213, 022032, 123331, 222300, 213002, 132022, 301121, 202203, 112333, 022230, 221300, 213202, 130112, 320220, 311233, 002223, 022130, 221320, 213011, 032022, 331123, 300222;

(Period $\tau = 36$). This means the real number of basins of attraction is 25. For $M = 7$, there are $4^7 = 16384$ states, 17 basins of attraction with $\tau \leq 30$ and three basins with a longer period. With $M = 8$, there are $4^8 = 65536$ states, 54 basins of attraction with $\tau \leq 30$ and 1 with a longer period. With $M = 9$, there are $4^9 = 262144$ states, 43 basins of attraction with $\tau \leq 30$ and 4 with a longer cycle. With $M = 10$, there are $4^{10} = 1048576$ states, 100 basins of attraction with $\tau \leq 30$ and 3 with longer periods. As expected, the number of basins of attraction increases with the number of cells. However, in the majority of CAs, the transition from an even to an odd number of cells produces a slight reduction in this number.

In sum, automatic analysis can provide us with information about the number of emergent structures and their type. However, the analysis is limited by the computational capabilities of the system. The presence of limit cycles with very long periods makes it necessary to apply a fairly high time threshold. As we will see in the rest of this chapter and in Chapter 9, automated search can nonetheless produce a lot of information about the nature of these structures. Details of this analysis can be found in the Appendix of Chapter 9.

The automatic tool for the automatic investigation of a CA pattern can also identify gliders.

Let us now examine another CA, this time with $k = 4$, $n = 3$, and rule code

0310110121311121030032201110220030010012003110231103221202131223

Figure 11c shows the pattern produced by 300 cells ($M = 300$), starting with randomly generated initial conditions and allowing the system to evolve for 300 time steps. In this case too, we observe several regular domains. If we limit our analysis to limit cycles with $\tau \leq 30$, we find 363 domains. Figure 11f depicts the distribution of these domains as a function of M. The only uniform domain is the domain with code 2. Automated search detects a number of gliders. If we limit our attention to gliders with a maximum radius of 7 and evolve the system for a maximum of 30 steps, we obtain the gliders shown in Table 3. All belong to uniform domain (2).

Gliders with longer codes are more common than gliders whose codes are shorter. Both have the same behavior. In behavioral terms, we can classify all gliders into just two groups:

Table 3. Gliders in uniform domain (2) for rule 0310110121311121030032201110220030010012003110231103221202131220 The first column shows the radius of the glider, the second its code; the third shows when the glider is stationary, or in cases in which it moves, its direction of movement.

Radius of glider	Code	Direction	Radius of glider	Code	Direction
1	0	Left	7	0220220	Left
2	00	Left		0022220	Left
	30	Left		0222220	Left
3	110	Stationary		0022230	Left
	111	Stationary		0222230	Left
4	0220	Left	8	02202200	Left
	1113	Stationary		00222200	Left
5	02200	Left		02222200	Left
	00220	Left		13000110	Stationary
	02220	Left		11020110	Stationary
	02230	Left		13020110	Stationary
6	301200	Left		11022110	Stationary
	002200	Left		02200220	Left
	022200	Left		30120220	Left
	130010	Stationary		00220220	Left
	120110	Stationary		02220220	Left
	002220	Left		02202220	Left
	022220	Left		00222220	Left
	002230	Left		02222220	Left
	022230	Left		02222230	Left
7	0022200	Left		13000111	Stationary
7	0222200	Left		11022111	Stationary

a) Gliders that move leftwards
b) Stationary gliders

More complex gliders (those with longer strings) incorporate individual elements that are themselves similar to the smaller gliders.

Extending Chua's Universal Neuron to Multistate CA

The method used to generate rules tables from the Universal Neuron can be easily generalized for multistate CAs with $k > 2$. To achieve this goal, we choose specific points on the cube that are not necessarily vertices. For example, given a specific value of *k*, we can use a cube of size *k* - 1. We can then choose the points represented by vectors u_i with $i = 0,1,2......,(k^{(2r+1)} - 1)$. Each vector is defined by:

$$u = (p,q,r) \tag{4.8}$$

where $p \in S, q \in S, r \in S$ Thus

$$u_0 = (0,0,0), u_1 = (0,0,1), u_1 = (0,0,2)...............u_{k^{(2r+1)}-1} = (k-1,k-1,k-1)$$

Applying (4.8), we can now generate $k^{(2r+1)}$ vectors of type u and $k^{(2r+1)}$ points corresponding to these vectors. These points will have a one to one relationship with the rules in the rules table (see Figure 12).

As in the Boolean case (4.3), we can define the discriminant function:

$$w(\sigma) = z_2 + c_2 \left|[z_1 + c_1 \left|z_0 + \sigma\right|]\right|.$$

Figure 12. Representation of neighborhoods on the cube for rule C4.8 with k=3 and for rule C4.9 with k=4

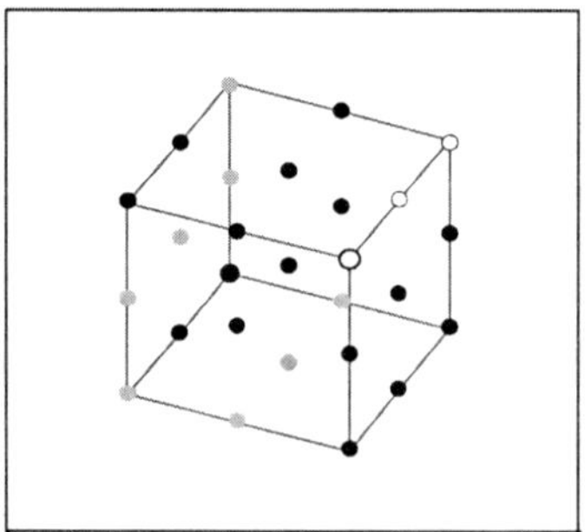

a)

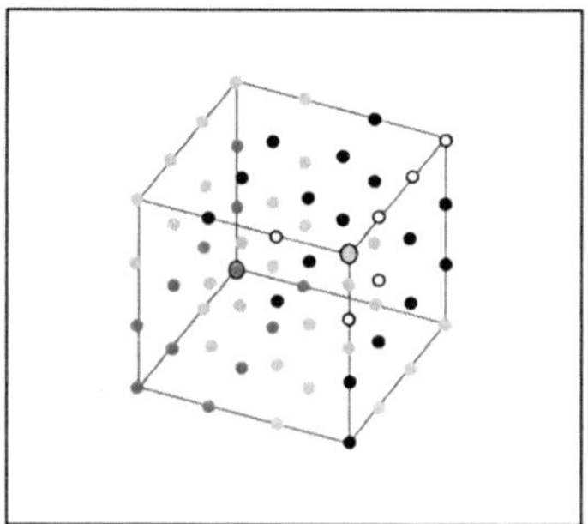

b)

Here too, the function has 8 parameters $z_2, z_1, z_0, b_1, b_2, b_3, c_2$e c$_1$ where c_2 and c_1 are integers with value ± 1. If we follow the same approach as in (4.4) and create a dynamic system with a guide function (4.5), the attractors generated by (4.4) will be the fixed points Q_i and the number of attractors will be equal to the number of neighborhoods defined in the rules table ($k^{(2r+1)}$). In this way, we extend the previous result, showing that for a neighborhood defined by a vector u(p,q,r), the state x(t) of a dynamic system with initial conditions x(0)=0 will converge monotonically towards attractor *Q(pqr)*. In this case, the output will no longer be Boolean, as in the previous case, but an integer in the range 0 - k-1. We thus obtain:

$$\overline{y(pqr)} = (k-1)\frac{Q(pqr) - \min}{(\max - \min)} \tag{4.9}$$

where max is the attractor with the highest value and min is the attractor with the lowest value. The output *y(pqr)* corresponding to neighborhood *u*(p,q,r) is the integer y_{pqr} closest to $\overline{y_{pqr}}$. We can thus write:

$$y_{pqr} = C\,\text{int}\left(\overline{y_{pqr}}\right) \tag{4.10}$$

As in the previous case, this method allows us to start from a Universal Neuron and to generate rules for a CA–this time a multistate CA. Figure 13 shows the patterns obtained by varying values of k for the Universal Neuron representing ECA 100.

The code for the Neuron allows us to obtain rule ECA 102. The other rules are obtained by varying the value of k (k = 3, 4, 5). Appendix lists the codes generated in this way as codes C4.8, C4.9 and C4.10.

The approach just described makes it possible to begin with a Universal Neuron and generate a very large number of multistate CAs. Given that these CAs are the output from a dynamic system, it is possible to implement then as parallel analog systems on semiconductor chips. These systems can perform the required computations much faster than traditional methods.

By using the Universal Neuron, we can operate at a higher level than the rules table. To investigate the possible behaviors of the CA, all we need to do is manipulate the parameters governing the behavior of the neuron. In Figure 14, we see the effects produced by varying the orientation (parameter b3) of the vector representing the points on the cube. Table 4 shows parameter values (column 1) and the corresponding rules (column 2) for a CA with k = 4 and r = 1.

It is interesting to observe the changes when parameters b_1 and b_2, take negative values. Changes in z have a less pronounced effect (Figure 14).

The generation of rules with specific characteristics, starting with Chua's Universal Neuron, can be a very useful tool in studies of the genetic behavior of the rules. It can also provide us with an alternative to Genetic Algorithms, as a way of generating families of rules.

Figure 14. Changes in pattern with changes in the value of parameter b3 for a CA with k = 4, r = 1

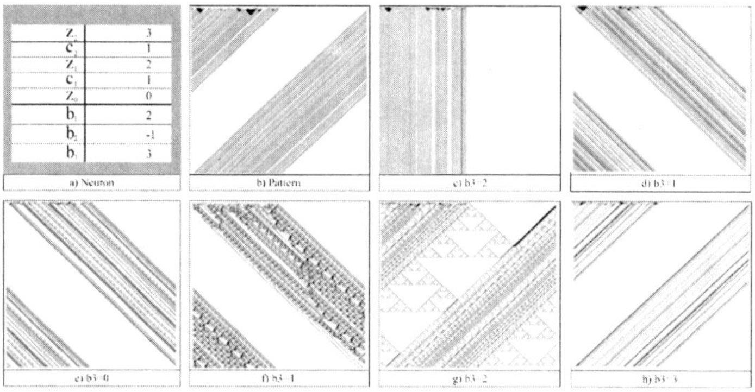

Table 4. Changes in neuron parameters and associated changes in rules strings in a CA with k = 4, r = 1

Neuron Parameters *z2,c2,z1,c1,z0,b1,b2,b3*	**Rule String**
3,1,2,1,0,2,-1,3	22113211322132221211022102211321121002110211022101101110021002110
3,1,2,1,0,2,-1,2	2211222132213222211022102211222111002100211022101001100011002100
3,1,2,1,0,2,-1,1	22112221322233222111021112211222110001100111021110011000110001100
3,1,2,1,0,2,-1,0	22222222222223333000011112222222220000000000000011112222111100000000
3,1,2,1,0,2,-1,-1	001201221222222231000000100120122221021001000000013222222122102100
3,1,2,1,0,2,-1,-2	1001101100120112210011011001101122102110210011013221321122102110
3,1,2,1,0,2,-1,-3	2101100110011012210021002101100122102210210021003221321022102210

FUTURE TRENDS

Cellular automata are important for complexity science because they are simple systems that nonetheless have the ability to produce emergent, self-organizing structures. In this chapter, we have described two classes of method used to identify these structures:

a) Quantitative methods
b) Qualitative methods

We have seen that classical quantitative methods can be extremely effective. However, they are time-consuming and difficult to apply. They are thus ill-suited for large-scale analysis of emergent structures. Our own qualitative approach, based on Gestalt laws, can be helpful for higher level analyses. To this end we have designed an automatic tool that has provided highly effective in detecting domains and gliders. The study of relations between gliders and between gliders and regular domains has allowed us to identify general rules we can explore with linguistic methods. These are themes we will discuss in Chapter 9.

In this chapter, we have also discussed our results for Chua's Universal Neuron, which we present as a quantitative tool for the generation of rules with specified complexity characteristics. We have extended the method to multistate CAs. Chua and his collaborators have demonstrated that certain ECAs share common properties, determined by symmetries in their respective rules tables. It may now to possible to extend this result to multistate CAs. It may also be possible to use Chua's neuron to generate rules, and to conduct genetic analyses of these rules. This approach can be seen as an alternative to Genetic Algorithms, as a way of generating rules with specified characteristics.

CONCLUSION

The overall aim of this chapter has been to illustrate quantitative and qualitative analysis, and other tools that make it possible to identify aspects of CAs that may be significant in the modeling of biological behavior and thus to extrapolate general laws of system behavior. We have seen how we can identify emergent structures generated by CAs. If we could identify all these structures and their interactions, we would be able to describe the CA at a higher level than computational mechanics. The Gestalt laws encapsulated in our computational tool can be helpful in this kind of analysis allowing a high level description of the CA. Automated tools make it possible to identify relations between different structures and suggest the possibility of rules governing the combination of simple structures into structures of gradually increasing complexity. This is a theme we will discuss further in Chapter 9. Another theme we have discussed is the generalization of Chua's Universal Neuron and the way it can be used to generate CAs with specified characteristics. This approach is an alternative to the Genetic Algorithms we will apply in the rest of this book. It is a theme that definitely deserves further investigation.

REFERENCES

Adamatzky, A., Bull, L., De Lacy Costello, B., Stepney, S., & Teuscher, C. (Eds.). (2007). Unconventional Computing. Bristol, UK: Luniver press.

Aizawa, Y., Nishikawa, I., & Kaneko, K. (1990). Soliton turbulence in one-dimensional cellular automata . In Gutowitz, H. A. (Ed.), *Cellular Automata: theory and experiment. Los Alamos: The MIT Press.*

Andrew, I., & Adamatzky, A. (1995). *Identification Of Cellular Automata*. London: Taylor & Francis.

Bilotta, E., Di Blasi, G., Giambò, S., & Pantano, P. (2008). Generating Multi State Cellular Automata by using Chua's "Universal Neuron . In Ruggeri, T., & Sammartino, M. (Eds.), *Asymptotic Methods in Non Linear Wave Phenomena* (pp. 12–24). Singapore: World Scientific Publishing Company.

Chalmers, D. (2006). Strong and weak emergence . In Clayton, P., & Davies, P. (Eds.), *The re-emergence of emergence: The emergentist hypothesis from science to religion* (pp. 244–254). Oxford: Oxford University Press.

Chua, L. O. (1998). *CNN: A Paradigm for Complexity*. Singapore: World Scientific Publishing Company.

Chua, L. O. (2006). *A Nonlinear Dynamics Perspective Wolfram's New Kind of Science* (*Vol. 1*). Singapore: World Scientific Publishing Company.

Chua, L. O. (2007). *A Nonlinear Dynamics Perspective of Wolfram's New Kind of Science* (*Vol. 2*). Singapore: World Scientific Publishing Company.

Chua, L. O. (Ed.). (2009). *A Nonlinear Dynamics Perspective of Wolfram's New Kind of Science* (*Vol. 3*). Hackensack, NJ: World Scientific Publishing Company.

Chua, L. O., & Roska, T. (1993). The CNN paradigm. *IEEE Transactions on Circuits and Systems*, *40*, 147–156. doi:10.1109/81.222795

Chua, L. O., Sbitnev, V. I., & Yoon, S. (2004). A nonlinear dynamics perspective of Wolfram's new kind of science, Part III: Predicting the unpredictable. *International Journal of Bifurcation and Chaos in Applied Sciences and Engineering, 14*, 3689–3820. doi:10.1142/S0218127404011764

Chua, L. O., & Yang, L. (1988). Cellular neural networks: Theory and Applications. *IEEE Transactions on Circuits and Systems, 35*, 1257–1290. doi:10.1109/31.7600

Chua, L. O., Yoon, S., & Dogaru, R. (2002). A nonlinear dynamics perspective of Wolfram's new kind of science, Part I: Threshold of complexity. *International Journal of Bifurcation and Chaos in Applied Sciences and Engineering, 12*, 2655–2766. doi:10.1142/S0218127402006333

Crutchfield, J. P. (1994). The Calculi of Emergence: Computation, Dynamics, and Induction. *Physica D. Nonlinear Phenomena, 75*, 11–54. doi:10.1016/0167-2789(94)90273-9

Crutchfield, J. P., & Young, K. (1989). Inferring Statistical Complexity. *Physical Review Letters, 63*, 105–108. doi:10.1103/PhysRevLett.63.105

Gutowitz, H. A. (Ed.). (1990). Cellular Automata: theory and experiment. Los Alamos: The MIT Press.

Hanson, J. E., & Crutchfield, J. P. (1997). Computational Mechanics of Cellular Automata: An Example. *Physica D. Nonlinear Phenomena, 103*(1-4), 169–189. doi:10.1016/S0167-2789(96)00259-X

Hilborn, R. C. (2000). *Chaos and Nonlinear Dynamics: An Introduction for Scientists and Engineers*. New York: Oxford University Press.

Koehler, W. (1947). *Gestalt Psychology: An Introduction to New Concepts in Modern Psychology*. New York: Liveright Publishing Corporation.

Koffka, K. (1935). *The Principles of Gestalt Psychology*. London: Routledge & Kegan Paul.

Lam, L. (1998). *Non linear physics for beginners*. Singapore: World Scientific Publishing Company.

Langton, C. G. (1986). Studying artificial life with cellular automata. *Physica, 22D*, 20–149.

Langton, C. G. (1990). Computation at the edge of chaos. *Physica, 42D*, 2–37.

McTague, C. S., & Crutchfield, J. P. (2006). Automated pattern detection: an algorithm for constructing optimally synchronizing multi-regular language filters. *Theoretical Computer Science*, *359*(1), 306–328. doi:10.1016/j.tcs.2006.05.002

Strogatz, S. H. (2000). *Nonlinear Dynamics and Chaos: With Applications to Physics, Biology, Chemistry, and Engineering*. New York: Perseus Books.

Wertheimer, M. (1912). Experimentelle Studien über das Sehen von Bewegung. *Zeitschrift fur Psychologie mit Zeitschrift fur Angewandte Psychologie*, *61*, 161–265.

Wolfram, S. (1984). Cellular Automata as Models of Complexity. *Nature*, *311*, 419–424. doi:10.1038/311419a0

Wolfram, S. (Ed.). (1994). *Cellular Automata and Complexity: collected papers*. Reading, MA: Addison-Wesley.

Wolfram, S. (2002). *A New Kind of Science*. Champaign, IL: Wolfram Media.

APPENDIX

Rule Codes

C4.1: $D = 1, k = 2, r = 2$ (Wolfram's Rule 20)

01101001100101101001011001101000

C4.2: $D = 1, k = 2, r = 2$ (Wolfram's rule 52)

11101001100101101001011001101000

C4.3: $D = 1, k = 2, r = 2$ (Wuensche's Triangular Gliders di)

11100000100010010111100000000001

C4.4: $D = 1, k = 2, r = 2$ (Wuensche's Bermuda Triangle)

10111100100000100010011100011100

C4.5: $D = 1, k = 2, r = 2$ (Aizawa's Solitons A)

10111111100010100001100011001000

C4.6: $D = 1, k = 2, r = 2$ (Aizawa's Solitons C1)

01011111001010101001110011001000

C4.7: $D = 1, k = 2, r = 2$ (Raindrops)

01000110011010001110110100010100

C4.8: $D = 1, k = 3, r = 1$

011112122011112121011112222

C4.9: $D = 1, k = 4, r = 1$

2011012212232233001211221223223301121122122322320112112222232333

C4.10: $D = 1, k = 5, r = 1$

3330133012301221123322344333013301201123122342234433301301120122
3122342234433011301220122312234233433301230122012231233423444

Chapter 5
Cellular Automata Metrics

INTRODUCTION

There have been many attempts to understand complexity and to represent it in terms of computable quantities. To date, however, these attempts have had little success. Although we find complexity in a broad range of scientific domains, precise definitions escape our grasp (Bak, 1996; Morin, 2001; Prigogine & Stengers, 1984).

One of the key models in complexity science is the Cellular Automaton (CA), a class of system in which small changes in the initial conditions or in local rules can provoke unpredictable behavior (Wolfram, 1984; Wolfram, 2002; Langton, 1986; 1990). The key issue, here as in other kinds of complex system, is to discover the rules governing the emergence of complex phenomena. If such rules were known we could use them to model and predict the behavior of complex physical and biological systems.

Taking it for granted that complex behavior is the result of interactions among multiple components of a larger system; we can ask a number of fundamental questions.

What are the necessary, qualitative and quantitative conditions that have to be met if a system is to produce complex behavior?

Is it possible to measure complex phenomena?

What are the laws of complexity?

DOI: 10.4018/978-1-61520-787-9.ch005

The first attempt to classify CAs on the basis of their behavior goes back to Wolfram (1984) whose quadripartite classification is still widely used today. The main limitation of his system is that it is based on phenomenological data (the development of the space-time diagram for the CA) and makes no attempt to identify the rules determining observed behavior. Since Wolfram, several authors have tried to overcome these limitations, offering more precise definitions.

Langton (1986) introduces a metric for measuring the complexity of a CA. Unlike Wolfram, who based his classification on CA phenotypes, Langton worked on their genotypes, introducing a metric for CA complexity, based not on pattern but on the rules generating patterns. According to Langton, complexity emerges on the *edge of chaos,* and requires complex rules to process, store and transmit information during the development of the CA. Metaphorically speaking, it is a fascinating idea. Unfortunately, however, there is little experimental evidence to back it up (Mitchell et al., 1993; 1994). Mitchell and Crutchfield have shown that there exist not one but many *edges of chaos.* The organization of CA parameter space (the ideal space in which we represent CA rules) is highly complex, and depends closely on the dimensionality of the system and its parameters. It has been shown, furthermore, that this is true not only for discrete dynamic systems (Bilotta et al., 2003) but also for continuous systems (Bilotta et al., 2007a, 2007b, 2007c, 2007d, 2007e, 2007f).

Alternative, more precise complexity metrics use information extracted directly from patterns (Li & Packard, 1990; Li et al., 1990; Gutowitz & Langton, 1995; Oliveira et al., 2001). Unlike rule-based metrics, this kind of measurement depends on the initial conditions and thus provides only a partial picture of the overall behavior of the CA. In this chapter, we will provide an exhaustive treatment of so-called Input-Entropy (IE), an example of this kind of measurement. IE provides us with clear evidence that complexity is indeed present in CAs. What we will see is that complexity is the result of interactions among a broad range of complex CA features such as regular domains, gliders and self-replicators and that the frequency of these phenomena varies with variations in IE. This kind of variability is, in fact, a hallmark of complex behavior in CAs. In this chapter, therefore, we will begin with a presentation of the original pioneering work by Wolfram and Langton - the historical backdrop for everything that follows. We will proceed to discuss qualitative and quantitative approaches and the various indexes developed by researchers in the field, including the authors' own work on IE in 1 and 2-D CAs. We will present experimental evidence showing that the presence of complex structures in a CA leads to large variations in IE and that the simultaneous presence of different complex phenomena in the same CA produces an additional increase in IE. We go on to describe how IE can be used to represent the development of 2-D CAs in phase space, a procedure we can apply both to the CA as a whole and to individual particles such as self-replicators.

BACKGROUND

Systematic studies of CAs (Culik & Yu, 1988; Baldwin & Shelah, 2000) and complex CA behaviors try to group the enormous variety of observed behaviors into categories, to count the number of categories, to identify the qualitative and quantitative characteristics that allow us to classify a CA as "complex" and above all to measure this complexity.

This poses a number of fundamental problems. The size of the rule space for a CA is k^{k^n}, where k is the number of states, and n the size of a local neighborhood in the CA. Is there any way in which we can represent the structure of this space in visual terms? How do the different categories of rule group inside this space? Do there exist clearly defined borders between regions of space containing rules in different categories? When we search the space, how should we look for rules with particular properties?

CA researchers have adopted a broad range of qualitative, quantitative, parametric, topological and computational approaches to these problems.

The qualitative approach is based on the analysis of the spatio-temporal patterns produced by the CA and on its asymptotic behavior - an approach that eliminates the transient. Unlike other approaches that study the characteristics of the rules table–the genotype of the CA–this is a "phenotypical" approach. The first author to examine CAs in this way was Wolfram in a wonderful series of papers written between 1983 and 1984 (Wolfram, 1984). In an analysis of ECAs and totalistic Boolean CAs, he identified four distinct classes of behavior:

- **Class I:** "Null rules": CAs which that evolve through a transient to a homogeneous state;
- **Class II:** Periodic rules: CAs that produce simple periodic structures (in a spatial as well as a temporal sense);
- **Class III:** Chaotic rules: CAs that produce completely random patterns with no periodicity or specific structure ;
- **Class IV:** Complex rules: CAs that evolve to produce complex emergent structures.

As we saw in the previous chapter, these structures usually emerge against a homogeneous or regularly patterned background. When they move inside the pattern and return in the same form after a certain period they are known as gliders. These structures evolve, self replicate, and spread, giving rise to new structures and interacting with other particles. On occasion they can disappear or generate new structures that are completely different (Hanson & Crutchfield, 1997) .

Wolfram's classification is based on the assumption that the patterns generated by the CA do not depend on the initial data but only on the rules that generate the patterns. Rules that belong to the same class give rise to behaviors that are qualitatively similar. Figure 1 shows examples of rules for each of Wolfram's four classes.

Like the equivalent continuous systems, discrete dynamic systems have attractors. In this sense, rules in class I specify attractors similar to fixed points; rules in class II produce fixed points and limit cycles; rules in classes III and class IV can also produce fixed points and limit cycles. However, these CAs include cells (sometimes a small number of cells) whose detection requires so much computation as to be impossible in practical terms. This means it is easy to determine the class to which a given CA belongs. Any CA that does not rapidly reach a fixed point or a limit cycle (after a brief transient) belongs either to class III or to class IV. The difference between the two classes depends on the presence or absence of emergent structures. The behavior of the third class is reminiscent of the chaotic behavior of certain continuous dynamic systems. By contrast, Class IV CAs continuously generate novel behaviors that differ from those we observe in continuous systems.

Figure 1. The CAs on the first row belong to Wolfram's first class (a and b) and second class II (c and d); the second row (e, f) shows CAs with complex behavior (Class IV); the second row (g, h) shows CAs with chaotic behavior (Class III); again the last row (i, j, k, l) shows CAs with complex behavior (Class IV).

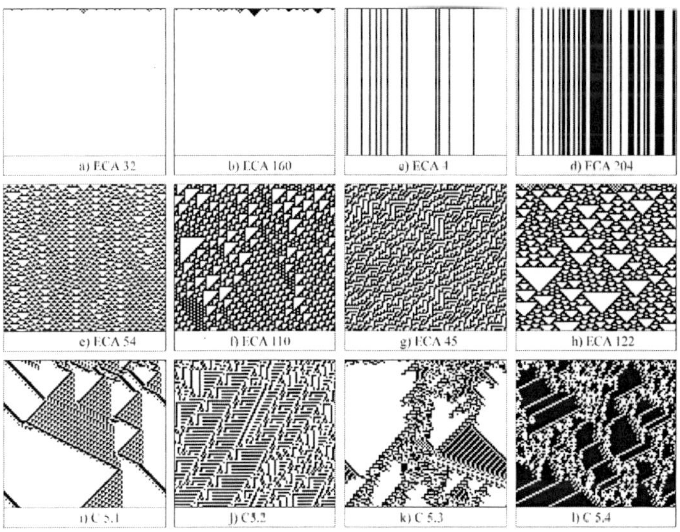

Wolfram's classification scheme is an interesting one. However it has obvious limitations. In purely qualitative approaches, there is often great doubt about the correct classification of specific CAs. Wolfram's Class IV includes CAs that produce very different behaviors. To make things worse, the classification is based entirely on the behavior of the phenotype and usually depends on a very limited set of experimental observations. Exploring the behavior of the CA with every possible set of initial values is impossible.

In the 1990s several authors extended Wolfram's pattern-based classification (Li, et al., 1990; Li & Packard, 1990; Li, 1992) and introduced new class definitions. One example is the scheme proposed by Oliveira et al. (Oliveira et al., 2001). This includes the classes described below.

- **Class I:** Null rules or homogeneous fixed points: CAs in this class move toward a fixed point in which all the cells in the automaton have the same value (Figures 1a and 1b).
- **Class II:** Heterogeneous fixed points: CAs in this class move towards a fixed point where different cells have different values. This class includes CAs whose fixed point changes over time but is invariant under translation (Figures 1c and 1-D).
- **Class III:** Two-cycle rules: these CAs move towards a limit cycle with two states. This category of CA includes CAs with a limit cycle of period 2 that undergoes translation in space.
- **Class IV:** Periodic rules: in this case the CA moves towards a limit cycle of period T>2.
- **Class V:** Complex rules: like members of Wolfram's class IV, CAs in this class produce complex structures that stand out against the background and interact (Figures 1e, 1f, 1i, 1j, 1k and 1l).
- **Class VI:** Chaotic rules; CAs in this class display completely chaotic behavior (Figures 1g and 1h).

The main difference compared to Wolfram's classification is that Class II is split in three. The limitations remain.

Pattern Metrics and Associated Classification Schemes

To overcome the obvious limitations of the methods described in the previous section, several authors have introduced new approaches based on metrics calculated over CA patterns. This makes it possible to quantify certain of their characteristics. In most cases, the approach is statistical. Many of the measures associate all possible patterns with a numerical value. They can thus be considered as metrics for pattern space. Adami (1998), Li & Packard (1990) and Wuensche (1999) provide examples of this strategy. Here too, as in the case of qualitative methods, the focus is on spatio-temporal patterns, the way the CA evolves, and the initial conditions, which directly influence the statistics taken.

In what follows we will focus on four specific metrics: Spatial and Vertical Entropy, IE, Mutual Information and the Difference Pattern Spreading Rate.

Spatial and Vertical Entropy

Given a spatio-temporal pattern, we can calculate, for each instant of time, the number of cells in any given state. Alternatively we can count the number of states expressed by a cell, over a given period of time. These are the simplest measurements it is possible to make on a pattern (see Figure 2). Two more interesting metrics are population or state entropy (also known as Horizontal Entropy or Spatial Entropy) and site entropy (Vertical Entropy or Temporal Entropy). These metrics, first introduced in Wolfram (1984), are closely linked to Shannon information theory (Adami,1998). Given a set of events $X = x_1, x_2, \ldots, x_n$, each of which occurs with probability $p_1, p_2, \ldots, p_n$, we can define a single monotonic function that measures the total uncertainty associated with X

$$H\left(X\right) = -c\sum_{i=1}^{n} p_i \log p_i \tag{5.1}$$

where c is a constant that depends on the base used for the logarithms. If all events have the same probability, $p_i = 1/n$. If we calculate logarithms to base 2, we obtain $c = log2 = 1$. Thus:

$$H\left(X\right) = -\sum_{1}^{n} p_i \log p_i <= \log n$$

Figure 2. Row 1: The first image (a) shows the spatio-temporal pattern produced by a complex CA with k=4 and r=1 (rule C5.5). The diagram shows a row and a column used to compute horizontal and vertical entropy. The second image, (b) shows the size of the population as a function of time (rule C5.5). Rows 1 and 3: patterns, populations and different kinds of entropy for rules with differing behaviors.

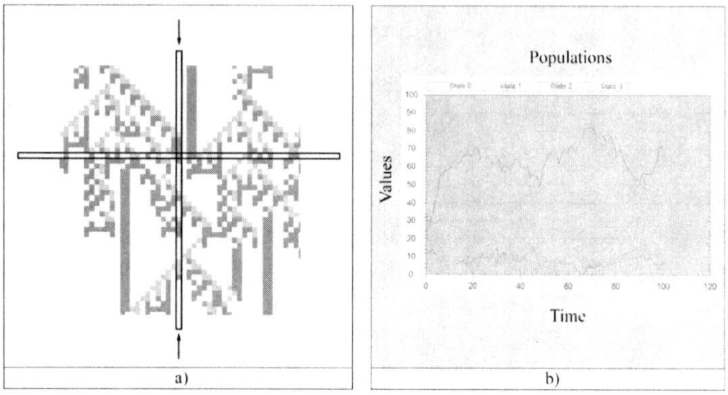

If the entropy of a series of events is lower than the maximum, we infer that the events are correlated.

There are several kinds of entropy that can be applied to a CA. Consider, for example, any one of the patterns that might be produced by a 1-D CA. In such a CA, we can count the number of states along a row or down a column (see Figure 2, Figure 3, Figure 4, Figure 5 and Figure 6, rule code C2.5). In the first case, we measure the frequency of events in time; in the second we measure their frequency in space (at a given site). In terms of our earlier discussion, events are the states taken on by specific cells. Thus, if we consider a CA whose cells can take on any integer value between 0 e k-1, we can compute the CA's *population entropy.*

$$H^t\left(X\right) = -\sum_{i=0}^{k-1} \frac{h_i}{M} \ln(\frac{h_i}{M}) \tag{5.2}$$

where $X = ai, i=1,2,\ldots M$ *ed* $ai \in S$ is a set of elementary events, hi is the number of cases of si at time t, with $si=0,1\ldots k-1$ and M is the total number of cells.

Site entropy *can be defin*ed using the same approach as for population en*tropy. Thus:*

$$H^j\left(X\right) = -\sum_{i=0}^{k-1} \frac{h_i}{T} \ln(\frac{h_i}{T}) \tag{5.3}$$

Figure 3. (rule C5.6) show CAs belonging to class II

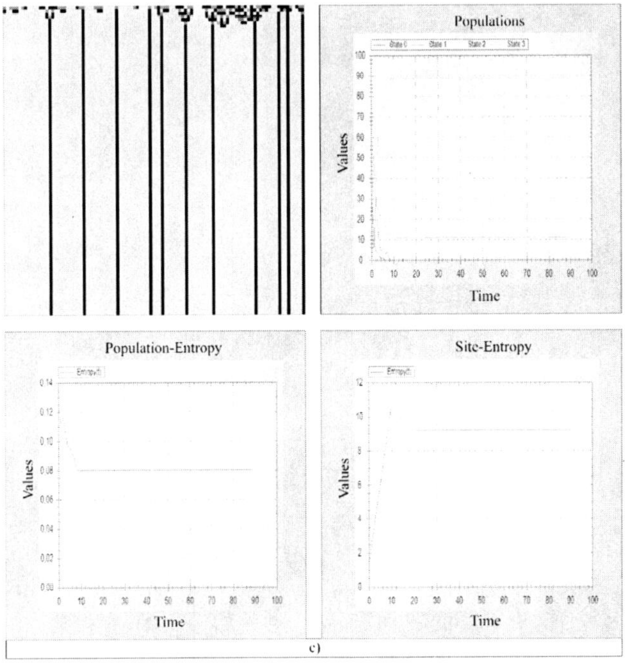

Figure 4. (rule C5.7) show CAs belonging to class II

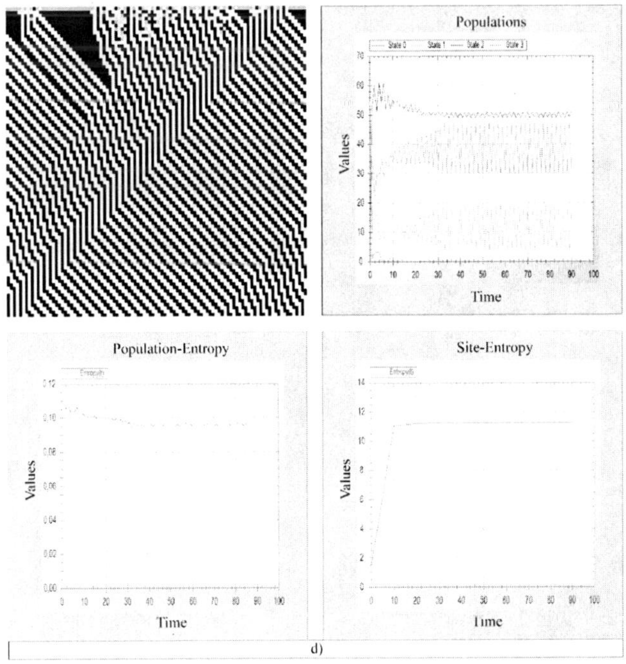

Figure 5. (rule C5.8) shows the behavior of a chaotic rule belonging to class III

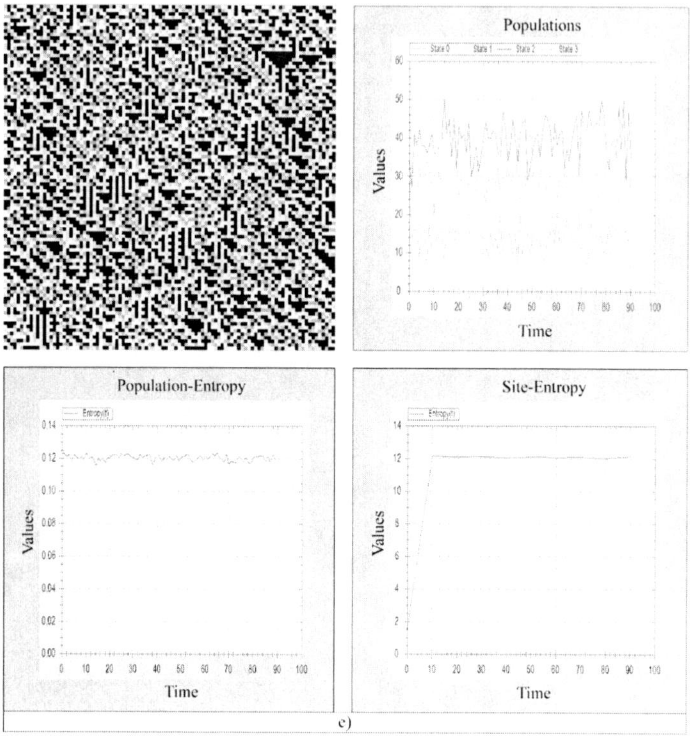

where *X*= *al,l*=1,2,...*T ed al*∈*S* is the set of elementary events, h*i* is the number of cases of *si* at site *j* and T is the total number of time steps.

Mutual Information

Another measure we can use to detect correlations between sites and to gain insights into CA complexity, is so-called Mutual Information (Li et al., 1990; Li, 1992). Let *pi* e *pj* be the probability that a given state, *s*=0,1...*k*−1, will be found in two distinct sites *i and j*. Let *pij* be the probability that the two sites are in the same state. We can then define Mutual Information *I*:

$$I = \sum_{i=0}^{k-1}\sum_{j=0}^{k-1} p_{ij} \ln \frac{p_{ij}}{p_i p_j}$$

When i and j refer to the same time but *are se*parated in space, we talk about Spatial Mutual Information; when t*he two sites are separated* in time, we talk about Temporal Mutual Information.

Figure 6. (rule C5.5) shows complex rules. Here we observe strong variations in entropy values t.

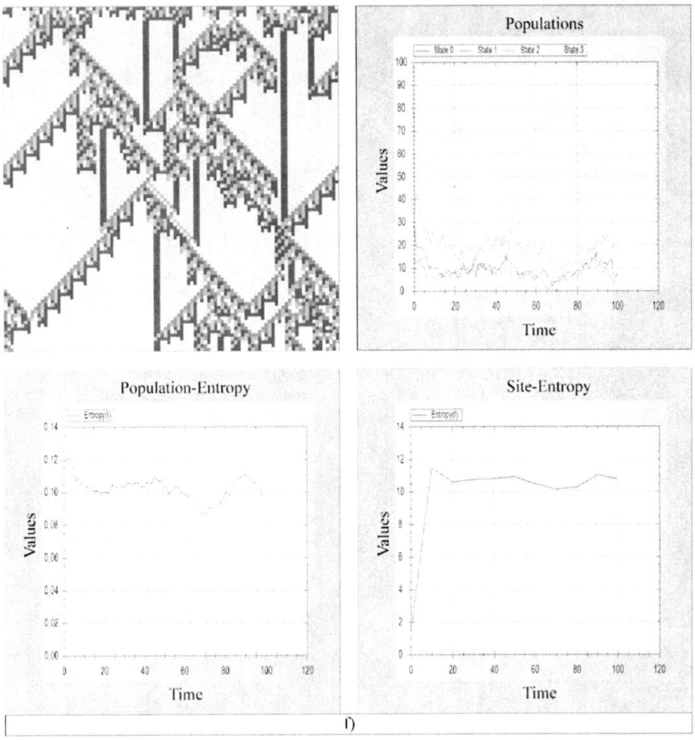

Difference Pattern Spreading Rate

If we compare different initial conditions for the same CA, the patterns the CA produces are sometimes similar and sometimes very different. The Difference Pattern Spreading Rate is a statistic based on comparisons between patterns, when the initial conditions for the pattern differ by only a single cell. The initial step is to detect cells with the same state in both cases and cells with different states. This data is used to generate a Difference Pattern whose cells a*re active if the c*ells in the original patterns are different and inactive if they are the same. The Difference Pattern Spreading Rate is defined as the sum of the speeds at which the right and left hand sides of the pattern move away from the center (Li et al., 1990; Wolfram, 1984). Observations on a large number of CAs show that "well-ordered" rules (null rules, fixed points, 2-cycle rules and periodic rules) have a Difference Pattern Spreading Rate of 0. By contrast, the rate for chaotic rules is almost as high as the theoretical maximum (given by the size of the neighborhood). Complex rules have highly variable rates.

Figure 7. The Figure shows the changes in pattern caused by changing the state of a single cell in the initial conditions. The two rows show, from left to right, an ordered (a), a chaotic (b), and a complex rule (c). The effects of the change in the initial data are shown in black. With ordered rules, the changes are limited; with chaotic rules the patterns change very fast; complex rules lie between these two extremes.

As can be seen in Figure 7, the Difference Pattern Spreading Rate allows a quantitative classification of CAs.

Ordered rules produce a low value, chaotic rules a high value; complex rules produce values between the two extremes.

Parametric Classification and Langton's Contribution

The main limitation of the classification methods discussed so far is that they measure the spatio-temporal patterns produced by the CA, which vary with variations in the initial data. This suggests it could be useful to make measurements directly on rules, investigating the properties of the strings (l strings) used to represen*t* them. If we could use these properties to classify CAs, we would not need to analyze the patterns they produce. In this way, we could avoid the variability introduced by changes in the initial data. We would also eliminate the problems caused by the high cardinality of rule space, making it possible to analyze the properties of the space without analyzing the way the CA evolves in time.

The simplest parameter for this purpose is the lambda (λ) parameter, first introd*u*ced by Langton (Langton, 1986; 1990). To compute the parameter, consider a rest state s. Then conside*r the numb*er of elementary rules leading to a state different from the rest state. For a given ru*le* $l=s1s2.............skn$, λ is defined by:

$$\lambda = \frac{k^n - q}{k^n} \tag{5.4}$$

where q is the number of elements in l that are in the rest state s. λ is easy to compute. Consider for example, ECA 91, in which string l expresses the local rule l=01011011. Given the rest state s=0,

$$\lambda = \frac{2^3 - 3}{2^3} = 0.625$$

(5.4) can be easily extended to any rule t expressed by a string l of length $Len(l)$.

$$\lambda = \frac{Len(l) - q}{Len(l)} \tag{5.5}$$

The values computed with (5.4) and (5.5) do not depend on the patterns generated by CA but only on the local rule. According to Langton (1986; 1990), λ can help us to identify complex systems. He hypothesizes that with the correct values of λ, the "average" behavior of the CA will be complex. In other words, rules are more likely to be complex when their λ value is close to these values. Patterns generated by rules with low λ rapidly develop towards a configuration in which all the cells are in the rest state. These CAs are what Wolfram calls Class I CAs. Higher values of λ produce behaviors of gradually increasing complexity, ranging from ordered (class II) through complex (class IV) to completely chaotic behavior (class III). Complex rules lie on the boundary between class II and class III automata. (see Figure 8).

Langton refers to this region as the edge of chaos. For Langton, the *importance of* complex rules lies in the fact that they transmit, store and process information. This means they can perform computations. The situation is similar to what we observe in biological systems when their survival calls for complex behavior. Evolution through natural selection tends to favor systems that are close to a phase transition, on the border between order and chaos (Holland, 1975). In the same way, the complex dynamics of a CA act as an evolutionary attractor in state space. From one point of view, λ can be considered as a measure of the CAs temperature or overall activity. *In this in*terpretation, order and chaos are two different phases corresponding to two different temperatures and the transition from order to chaos is a phase transition. Complex rules appear when activity is not so high as to destroy any kind of order but too high to leave the system in a stationary state. It goes without saying that in such a stationary state no kind of complexity or computation or life would be possible.

Langton's hypothesis that there exist well defined border regions separating order from chaos appears to be falsified by studies in density computation that we

Figure 8. A parameterized view of rule space according to Langton's classification

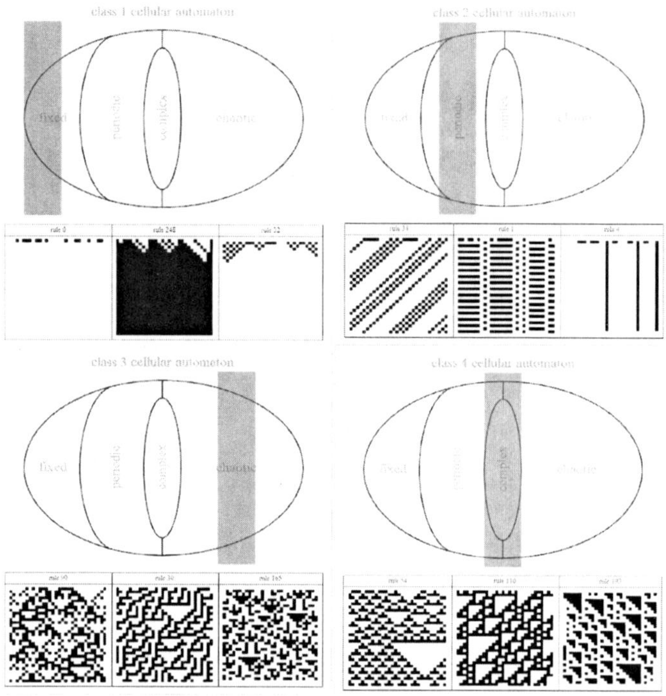

describe *in the next* chapter. Other negative evidence comes from our own work, (Bilotta et al., 2003), whose main results we will describe in Chapter 6.

Other Parametric Classification Methods

Quantitative measures of order and chaos can play an important role in studies of phase space and in the design of automatic techniques for exploring the vastness of rule space. Some of the most important of these measures are the Z parameter (Wuensche,1999), the Mean Field parameter (Li,1992), Sensitivity (Binder, 1991), Absolute Activity, Neighborhood Dominance and Activity Propagation (Oliveira, et al., 2001). In what follows we will examine these parameters in greater detail.

The Z Parameter

The Z paramete*r* is based on th*e* characteristics of the basins of attraction generated by a CA, as described in Chapter 4. To define the parameter, we identify the set of all pre-images for a specific configuration of the CA, where $c1$ is a pre-image of configuration *c* if a CA in configuration *c*1 develops to configuration c in the following time step. The search for pre-images can lead to various results.

- The configuration may have a single pre-image.
- It may have several pre-images.
- It may have no pre-images. In this case it is called a "Garden of Eden". Given that no configuration of the system leads to a Garden of Eden the only way to create it is to set it as a starting condition.

The Z parameter depends on the characteristics of the transient for these basins of attraction. It can be defined as the probability that a given configuration of the CA is uniquely determined. The parameter takes values in the range 0 to 1; low values of Z correspond to a high number of pre-images; high values of Z to a low number of pre-images. In Chapter 4, we have shown the basins of attraction for rule ECA 90 and rule ECA 122. Rule ECA 90 has a low number of pre-images. By contrast, rule ECA 122 has a high number. Wuensche hypothesizes that ordered rules have low values of Z and that chaotic rules have high values. He then uses the parameter to classify CAs and to find complex rules automatically.

Mean Field

As we will see later on, not all elementary rules are equally important in CA development. Mean Field theory (Gutowitz & Langton, 1995; Li, & Packard, 1990; Li,1992) is based on the hypothesis that some details of rules have little effect on the behavior of the CA and can thus be safely ignored. Mean field parameters are a generalization of the λ parameter as defined in (5.4) and (5.5). For CA with k = 2, r = 1, the mean *field parameters* ni for a state s are defined as the number of local rules that lead to state s when the neighborhood contains *i* cells whose state is also

Table 1. The rules table for ECA 88

111→0
110→1
101→0
100→1
011→1
010→0
001→0
000→0

s. Consider ECA 88, whose rules table is shown in Table 1. In the rules table, there is one condition (111) that generates a state of 0 when there are no 0s in the neighborhood; two conditions (010 and 001) that generate a state of zero when there is 1 0 in the neighborhood and just one condition (000) that generates a state of zero when there are 3 0s in the neighborhood. For this CA, therefore, the Mean field parameters for state 0 are:

$$n_0 = 1, n_1 = 1, n_2 = 2, n_3 = 1$$

The value of q is given by:

$$q = n_0 + n_1 + n_2 + n_3$$

In other words, the parameters are closely related to the λ parameter
To normalize the mean field parameters, we can apply the expression:

$$N_i = \frac{n_i}{\binom{m}{i}}$$

where m is the size of the neighborhood, i=0,1..m and $\binom{m}{i}$ is the number of neighboring cells that contain exactly i 1s.

Sensitivity

This parameter was first introduced by Binder in (Binder, 1991) and proposed again in Li (1992), where it is called Context D*ependence. The sen*sitivity μ of a CA measures the number of changes in the state of a cell, when the state of each neighboring cell is modified, one cell at a time. The parameter is measured taking account of all possible configurations of the neighborhood.

Other Approaches

The quantitative and parametric approaches, described in the previous paragraphs, are not the only way to classify CAs. The algebraic approach makes the classification in terms of algebraic properties of the CA such as additivity (Dow, 1997; Jen, 1990), invertibility (Chua, 2006) and self-replication (Bilotta & Pantano, 2006).

(Binder, 1991) proposes a topological classification. (Dubacq, et al., 2001) uses a computational approach based on Kolmogorov complexity. Given the high cardinality of rule space, an exhaustive analysis is impossible. It is possible that these methods will play an important role in future studies. For the moment, the results are unsatisfactory. However, we can already use qualitative and parametric methods in Genetic Algorithms (GAs). As we will show in later chapters, GAs can be a useful tool for exploring rule space.

Another way of investigating rule space is to use techniques from Gestalt theory, as described in Chapter 4, or to explore the properties of rules by using them to generate sounds and music (Bilotta et al., 2000; Bilotta & Pantano, 2001; Bilotta & Pantano, 2002; Bilotta et al., 2002; Miranda, 2001), as we explain in Chapter 15. Obviously what we are exploring here are phenotypes and specific behaviors. Expressing rules in sound and music makes it possible to detect tempo, rhythm, emergent structures etc. This kind of approach can play an important role in the study of dynamic systems, including continuous systems (Bilotta et al., 2005).

MAIN FOCUS

The main goal of this chapter is to provide an overview of the various methods currently used to analyze the behavior of discrete dynamic systems, to predict their future behavior and to capture the organizational laws underlying this behavior. These laws should enable us to answer a number of key questions.

a. Is it possible to create a classification of the various kinds of behavior produced by discrete dynamic systems such as CAs?
b. Do there exist general rules governing the development of these systems? Can we find these rules by applying quantitative metrics and systematically studying the set of all possible initial configurations?
c. Is it possible to create a description of the abstract spaces for these systems? Can we define these spaces in such a way that different classes of CA are assigned to different regions of space? And within these spaces, is it possible to identify regions of complexity, bearing in mind that for several authors, it is in these regions that we should find systems with the ability to store, transport and process information and thus to perform computations - a precondition for biological life?

In this Chapter we present a summary of these methods, and describe our own work in the area. We will then go on to present experimental data, showing how IE can provide us with information on the development of 1-D and 2-D CAs.

Input-Entropy

Input-Entropy occupies a special place among the various metrics that have been proposed for the classification of CAs - largely because of the way it has been used in the search for complex rules (Wuensche, 1999; Bilotta et al., 2003). This parameter, first introduced in Wuensche (1999) is based on the frequency with which the elementary rules in a CA rules table are actually used. As we have seen in Chapter 2, each rule is associated with a specific configuration of the neighborhood. If we consider a 1-D CA with a radius of *r*, each of whose cells can be in one of k states, there are $m = 2r + 1$ cells in each block and $k^{(2r+1)}$ possible configurations of the neighborhood. From (5.4) it follows that if Q_i is the frequency with which the system uses the ith configuration, and M is the total number of cells in the CA, the input-entropy of the CA at time t, is given by:

$$H^t\left(X\right) = -\sum_{i=1}^{k^{(2r+1)}} \frac{Q_i}{M} \ln(\frac{Q_i}{M}) \tag{5.8}$$

where *X* is the set of all possible configurations of the neighborhood and M the total number of cells in the CA. In this case, each block consulted by the CA has a size of n, the number of neighbors for a cell. At any given time, it is these blocks that determine the state of the cell, at the next time step.

The way IE is defined means that ordered rules that converge towards fixed points or periodic structures have low values. After the transient, these CAs use relatively few elementary rules. Chaotic rules use a large number of elementary rules, and thus have a high value. Complex rules generate highly variable values. This implies we can use the mean and the variance of the parameter to classify CAs as ordered, chaotic or complex (see Figure 9).

Wuensche (1999) observed that in ordered systems, input-entropy falls rapidly to a minimum value while in chaotic systems it always remains close to a maximum value. In complex systems, it varies significantly over time (see Figure 9.5). Variance in input-entropy can thus be considered as a good complexity metric for CAs. The same result holds true for population dynamics. Site entropy, on the other hand, appears to be less important.

As can be seen from Figure 9.5 and other examples to be presented later, IE measurements for complex rules display highly characteristic dynamics. In CAs with complex rules, unlike those with well-ordered or complex rules, IE varies enormously as the system develops from its initial, randomly determined state. We might expect that certain Class II (non-complex) rules with very long limit cycles,

Figure 9. Examples of an ordered rule (C 5.9, Figure a), chaotic (C5.10, Figure c) and complex rules (C5.11, Figure e).The images in the first column show the original patterns; the second column shows variations in IE as the CA develops. With ordered rules, IE begins with a high value, subsequently falling to a low constant value (Figure b). With chaotic rules, IE oscillates permanently around a very high value (Figure d). With complex rules, the value of the parameter undergoes major variations during the development of the system (Figure f).

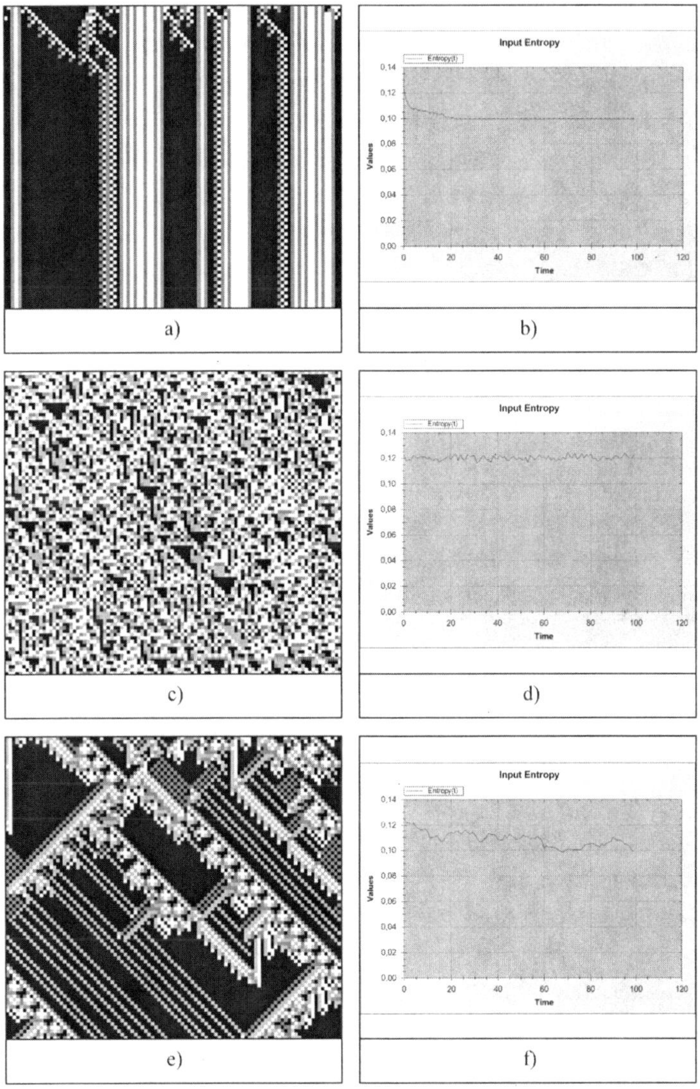

would also display very high variance in IE and. indeed, it may be possible to hand-design rules that behave in this way. However, we never found them in our experiments. To make matters worse, there are many cases in which it is hard to determine which class a rule belongs to. This is because the whole concept of CA complexity is very vague: if we had a universally accepted definition, we would not need so many metrics, and complexity measurement would not be such an important topic in CA studies. Be that as it may, IE captures many of the behaviors we would expect from a complex CA. It also has the added practical advantage that it is well suited for exploitation in the GAs we use to detect complex rules.

Input-entropy also has other applications. Given a pattern, generated by a CA, we can shift our attention from individual cells to pre-defined blocks of cells. In this way, we can define a more general entropy measure, in which we consider *Y* blocks of cells, each of length *m*. If the frequency with which particular blocks appear in the pattern is given by *p*(*Y*) we can define the 'Horizontal Entropy" of the CA using the expression:

$$H^t\left(X\right) = -\sum_{i=0}^{b-1} \frac{z_i}{M} \ln(\frac{z_i}{M}) \tag{5.6}$$

where m is the number of cells in the block, k is the number of possible states, $b=km$ is the number of all possible blocks, $X=yi,i=1,2,\ldots M$ is the set of elementary events, zi is the number of block i at time t and M is the total number of cells.

Similarly we can define Vertical Entropy, using the expression:

$$H^j\left(X\right) = -\sum_{i=0}^{b-1} \frac{z_i}{T} \ln(\frac{z_i}{T}) \tag{5.7}$$

where $X= yi,i=1,2,\ldots T$ represents the set of elementary events and T is, as usual, the total number of time steps.

These expressions maintain their validity for CAs with more than 1 dimension (d>1).

If we consider the mean values and variance of these entropy parameters for the total number of time steps (horizontal entropy) or the total number of cells in the CA (vertical entropy) we have a metric for the complexity of the CA. Trends in IE can also be used to evaluate the complexity of emergent particles. Consider, for example, the case of the gliders generated by totalistic rule 20 (k=2,r=2,d=1), as described in Chapter 4, Figure 6. We observe that the IE of the two gliders on the left varies periodically. Variations in the number of elementary rules used by the system have the same period as the glider itself. By contrast, the glider on the right uses a constant number of elementary rules. See Figure 10.

Complexity, Emergent Structures and Input-Entropy in 2D CAs

The information that IE can provide is not limited to 1-D CAs. IE can also provide insights into CAs with more than 1 dimension. For example, Figure 11 and Figure 12 show variation in IE for the 3-D emergent structures generated by rule C2.15 (see Chapter 2). The Figures show the initial structures and the way they have developed after 40 time steps. The last image compares trends in IE for each structure.

Let us examine the behavior of each structure. For simplicity, we renamed the structures from 1 to 7.

Structure 1: After a little while, this structure decays, producing two gliders, one of which moves towards the top of the image while the other moves leftwards. The creation of these gliders as separate structures requires only 4 steps. From this time on, the value of IE is perfectly periodical, oscillating between 0.4 and 0.38. In the last diagram, the oscillations are squashed close to the x axis and are hard to detect.

Structure 2: This structure includes a periodic glider gun that subsequently generates a self-replicator. The initial IE value, roughly 1.4, decays rapidly to 0.4, rises to 0.6 and after a long transient, stabilizes in a slow oscillatory movement. The presence of these oscillations shows the tight relationship between glider-guns and self-replicators.

Figure 10. The first row shows 4 gliders generated by totalistic rule 20 (d = 1, k = 2, r = 2). The second row shows changes in IE during the gliders' evolution. Variations in the entropy of the first two gliders follow the same periodic rhythm. The IE of the two gliders on the right remains constant.

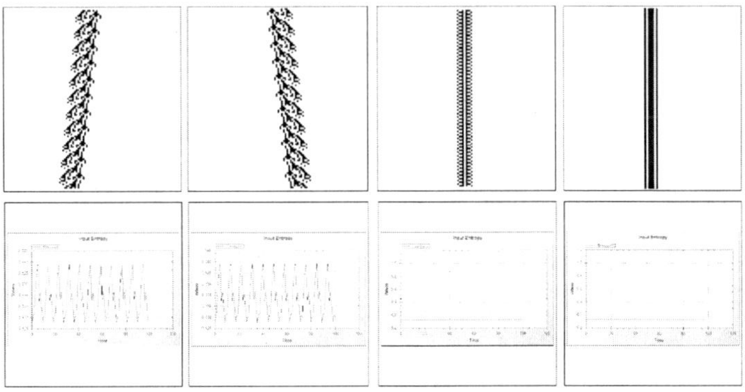

Structure 3: This structure, generated from structure 2, includes a complex self-replicator and a periodic glider gun. As we will see in Chapters 8-10-11, many self-replicators are much simpler than this. As the colony grows to fill the available space, the IE of the system increases slowly from 0.38 to 0.53, showing the growing complexity of the system as it uses an increasing number of rules.

Structure 4: The behavior of this very complex structure is similar to that of Structure 3. A glider-gun continuously generates self-replicators, giving rise to a beautiful cross-shaped colony, reminiscent of the initial symmetry of the system. IE increases slowly, from an initial low value of 0.4 up to a maximum of 1.2. It then begins to oscillate. The slow growth of the structure shows that the number of rules is also increasing slowly. Sometimes new symmetries emerge and the pattern becomes simpler. In these cases IE falls.

Structure 5: This structure incorporates a very simple self-replicator with a linear symmetry. After 40 steps, the structure evolves into 4 self-replicators whose shape differs from that of their common ancestor. In subsequent generations, the original shape returns. Although this is not clear from the image, which shows only a small number of time steps, the IE oscillates regularly. This behavior will be discussed in greater detail in Chapter 12. There a clear signs of the presence of higher

Figure 11. The rows show the structures describes by rule C2.15 reported in Chapter 2. The top images of panels a show 7 structures which evolve and depict the situation after 40 times steps. The bottom images of the same panel shows changes in IE during the structures' evolution.

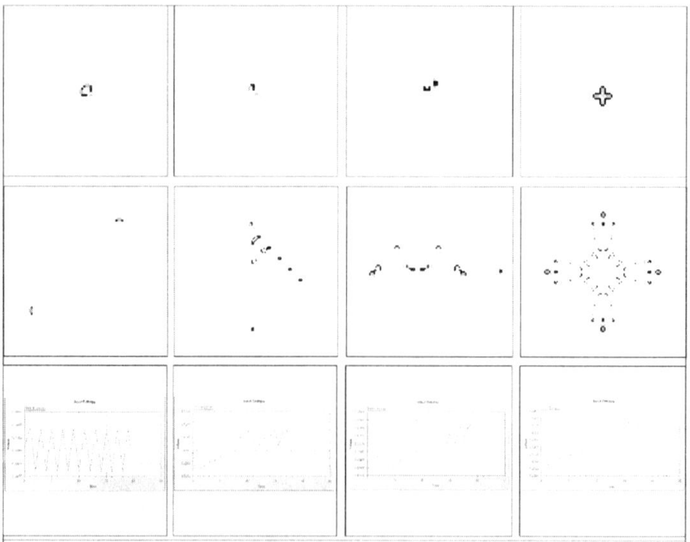

a)

Figure 12. The rows show the structures describes by rule C2.15 reported in Chapter 2. The top images of panel b show 7 structures which evolve and depict the situation after 40 times steps. The bottom images of the same panel shows changes in IE during the structures' evolution. The last row (panel c) compares trends in IE for all 7 structures.

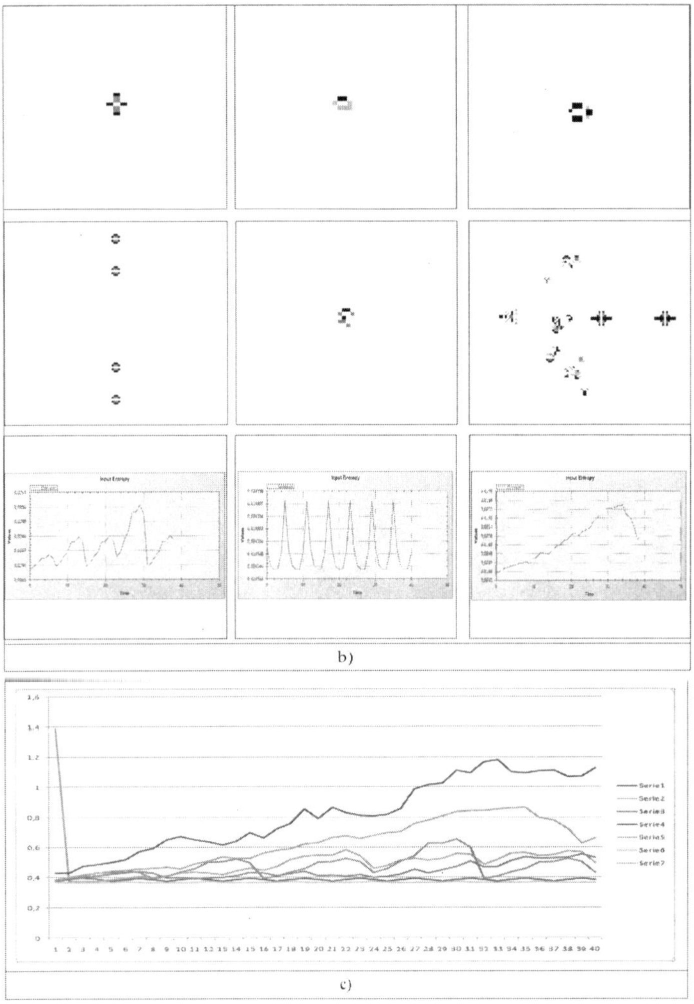

frequency harmonics. Values of IE oscillate from a minimum of 0.4 to a maximum of 0.65 - quite a large variation.

Structure 6: This structure consists of a glider with a period of 6. The same period is visible in the IE values which oscillate between a minimum of 3.66 and a maximum of 3.74: a very small gap. In the final graph, this curve is squashed making it hard to see the oscillation.

Structure 7: This structure reproduces the behaviors already observed in structures 3 and 4. Again we observe the simultaneous presence of glider-guns and self-replicators. The symmetry of the emerging colony is less marked than in the earlier structures. IE behaves in a very interesting way, first growing slowly, falling suddenly after 35 time steps and then returning to a pattern of slow growth.

Other Input-Entropy Metrics for 2-D CAs

IE provides a basis for other metrics. The first of these is the so-called I_1 index.

Let E_m be the minimum absolute value of IE in a given period of time.

Let E_M be the maximum value of IE after it has reached E_m.

Let E'_m be the minimum absolute value of IE after it has reached E_M.

We can thus define I_1 as follows:

$$I_1 = (E_M - E_m) + (E_M - E'_m) = 2E_M - E_m - E'_m \tag{5.9}$$

We can now define a second index I_2:

$$I_2 = \frac{I_1}{M} \tag{5.10}$$

where M is the mean value of IE during the time interval.

Unlike the variance of IE, which captures the variability of IE during the time period under consideration,I_{Im}akes it possible to identify the rules which create the largest gaps between the maximum and the two minima. In this way it focuses not on variations in general but on those that produce the largest effects. As we will see in greater detail in Chapters 7 and 8, these variations cause IE to oscillate but not to reach high values. I_2 captures this tendency, ensuring it is independent of the absolute value of IE.

Table 2 gives the values of I_1 for rule C5.3 and for the 7 structures generated by C2.15 and shown in Figure 11 and Figure 12. Figure 13 and Figure 14 show the way in which the index is calculated for the same rule and structures. For 1-D systems I_1 and I_2 do not seem to be particularly interesting. As we will see, however, they do provide significant insights into the behavior of 2-D systems. In general, they can be considered as a measure of the complexity of emergent structures. As we can see in the table, the structures with the lowest values for I_1 and I_2 are gliders (structures 1 and 5). Self-replicators and the structures from which they arise have significantly higher values.

Table 2. Value of the I_1 and I_2 complexity indexes for rule C5.3 and for the structures generated by rule C2.15. The values of the two indexes are visibly lower for gliders (structures 1 and 6) than for structures containing self-replicators.

Rules or Structures	I_1	I_1
Rule C5.3	I_1=1.94329420547592	I_1=0.676542113614535
Structure 1	I_1=2.83050085771665E-02	I_1=7.48486332991947E-02
Structure 2	I_1=0.317845826537709	I_1=0.656799990062873
Structure 3	I_1=0.184644898389437	I_1=0.423196746275121
Structure 4	I_1=0.791624130505913	I_1=0.951920028300739
Structure 5	I_1=0.521084985359368	I_1=1.10530886205907
Structure 6	I_1=1.45610911508269E-02	I_1=0.040646754554548
Structure 7	I_1=0.657604858551424	I_1=1.03557866695684

The Use of Elementary Rules

Input-Entropy, I_1 *and* I_2 are all based on the way the CA uses the elementary rules during its development: the more rules it uses, the higher the value of IE. Table 3 shows the number of rules used by the 7 structures generated by rule C2.15, measured after 40 and 60 time steps.

To show the rules used by the structure, we take the string that identifies the CA (2.18) and write an x in each position corresponding to a rule that is used. The other elements of the string remain unchanged. In this way, string *l*, (2.18) gives rise to a new string *l'*. Below we show values of *l'* for the structures generated by rule C2.15.

Structure 1

xxxxxxxxxxxxxxxxxxxxxxxxxxxxxxxxxxxxxxxxxxxxxxxxxxxxxxxxxxxxxx
xxxxxxxxxxxxxxxxxxxxxxxxxxxxxxxxxxxxxxxxxxxxxx0xxxxx0xxxxxxxxxxxx
xxxxxxx0xxx0xxxxxxxxxxxxx3xxxxxxxx002x0x3x0x0x3xxx0xxxx2x300x01x0
xxxxx00000x00003102000000

Figure 13. A graphical construction showing the computation of the I_1 and I_2 complexity indexes for rule C5.3

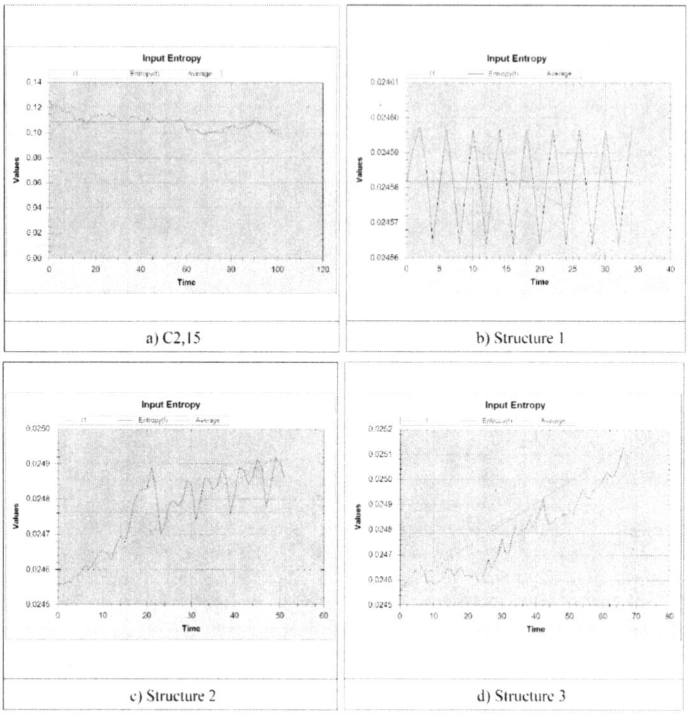

Structure 2

xxxxxxxxxxxxxxxxxxxxxxxxxxxxxxxxxxxxxxxxxxxxxxxx
xxxxxxxxxxxx0xxxxxxxxxxxxxxxxxxxxx0xxxxx0xxxxxxxxx
xxxxx130xxxxx0xxxx000xxxxx2xx1xxx2xxx0x00000xxxx00xx00300x30x000
0x000200330000030210220221300301200000000000000003102000000

Structure 3

xxxxxxxxxxxxxxxxxxxxxx0xxxxxx0xxxxxxxxxxxxxxxxxxxxxx
xxxxxxxxxxxxx0xxxxxxxx0xxxxxx0xxxxx0xxxxxxxxxxxxxxx13xxxxx0
0xxxxx000xxxx2xxxx2x2xx00x00000xx00000xx030x030x0x00x00020
0x300000302102202213003012000000000000000003102000000

Structure 4

2xxxxx3xxxxxxxxxxxxxxxxxxxxxxxxxxxxxxxxxxxxxxxxx
xxxxxxxx1x0x000xxxxxx3xxx2xxx0xxxx0xxx0xxxxx0xxxxx

Figure 14. A graphical construction showing the computation of the I_1 and I_2 complexity indexes for the structures generated by rule C2.15

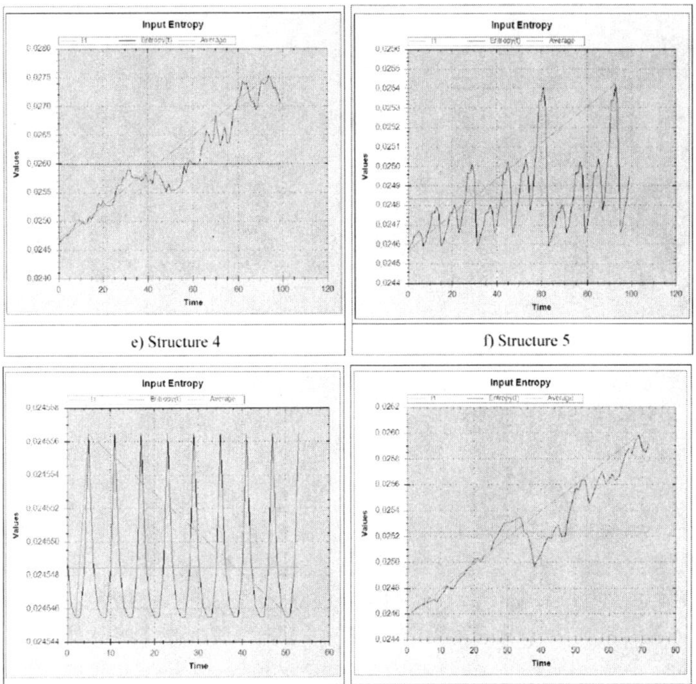

e) Structure 4 f) Structure 5 g) Structure 6 h) Structure 7

Table 3. Number of rules used by the structures generated by rule C2.15

Structure	Number of rules used after 40 time steps	Number of rules used after 60 time steps
Structure 1	40	40
Structure 2	89	89
Structure 3	93	94
Structure 4	114	128
Structure 5	51	51
Structure 6	28	28
Structure 7	132	142

xxx3313000xx00x2xx00003x002xx12xxx0xxxx000000x00000x0030x03030
0x00000200330000030210220221300301200000000000000003102000000

Structure 5

xxxxxxxxxxxxxxxxxxxxx2xxxxxxxxxxxxxxxxxxxxxxxxxxxxxxxxxx3xxxxxx
xxxxxxxxxxxxxxxxxxxxxxxxxxxxxxxxxxxx3xxxxxxxxxxxxxxxx0xxxxxxxxxxx
xxxxxxx00000xxxx0xxx003xxxxxxxxxxxxx02x0x3xxx0x30xxx2xxx213x0301x0
00xxx000000000003102000000

Structure 6

xxxxxxxxxxxxxxxxxxxxxxxxxxxxxxxxxxxxxxxxxxxxxxxxxxxxxxxxxxxxxxx
xxxxxxxxxxxxxxxxxxxxxxxxxxxxxxxxxxxxxxxxxxxxxxxxxxxxxxxxxxxxxxxxxx
xxxxxxxxxxxxxxxxxxxxxxxxxxxxx0xxxxxxxxxxxxxxxx0xxx1xxxxxx130x30xx0
xx0x0x00000x0x03102000000

Structure 7

xxxxxx3xxxxxxxxxxxxxx20xxxxxxx0xxxxxxxxxxxxxxxxxxxxxxxx13xx000x
xxxx33xxx20030xx010xxxxx0xxxxxx2xxxxx331300xx00002x000003x0021x122
x20x0x0000000000000x0030003030000000020033000003021022022130030120
0000000000000003102000000

CA Dynamics and Phase Space Diagrams

When investigating dynamic systems, it is very useful to study the way they behave in phase space or in 2- dimensional diagrams of phase space. If we consider a system with 2 degrees of freedom, we can represent its behavior as that of a point that starts in a particular state and follows a trajectory across the plane created by the two state variables (Hilborn, 2000; Strogatz, 2000). If the system has 3 degrees of freedom, as in the case of Chua's circuit (1993a; 1993b) or the Lorenz equations (Strogatz, 2000) the curve traced by the point will be in 3 dimensions. It is with this kind of curve that we discover the spectacular presence of strange attractors. (Bilotta & Pantano, 2008). In the case of non-linear, autonomous, continuous systems with two degrees of freedom, the attractors are fixed points or limit cycles. In these cases, Cauchy's problem has a single solution. Given the initial conditions, there is only one trajectory that the system can follow. In the case of non-autonomous systems, the situation is more complicated and trajectories tend to cross. In the case of non-linear systems, the dominant behavior is chaotic. We can observe the same behavior in discrete systems, where the presence of chaos is even more evident. Despite the complexity of these behaviors, phase diagrams provide us with useful insights into the behavior of the system.

CAs are dynamic systems with many degrees of freedom (as many as M - the number of cells in the CA). The phase space for a CA thus has M dimensions. Trajectories in such high-dimensional spaces provide us with little information of value. However, as we have already seen in the previous sections, we can still apply statistical time-dependent variables. Two examples are $pj(t)$ and $Q(t)$, where $pj(t)$, with j= 0,1,...,$k-1$, represents the number of cells in a given state j at time t, and $Q(t)$, the Input-Entropy of the system at time t. Using this kind of variable we can construct time series to which we can apply classical dynamic systems theory. This means we can use IE as a discrete state variable. By plotting the value of Q al time t against its value at t+1, we create a phase diagram, in which we trace the behavior of the system starting from a set of initial conditions. Figure 15 shows curves for some of the structures generated by rule C2.15, after 40 and 60 time steps

Figure 15. Trajectories in the phase space for the 7 structures generated by rule C2.15, captured after 40 and 60 time steps, top and bottom images respectively. Structures a and f, which are gliders, produce a limit cycle. Structures b and e are self-replicators and produce trajectories that are to some extent symmetrical. The trajectories generated by the other structures are more complex, with non symmetrical patterns.

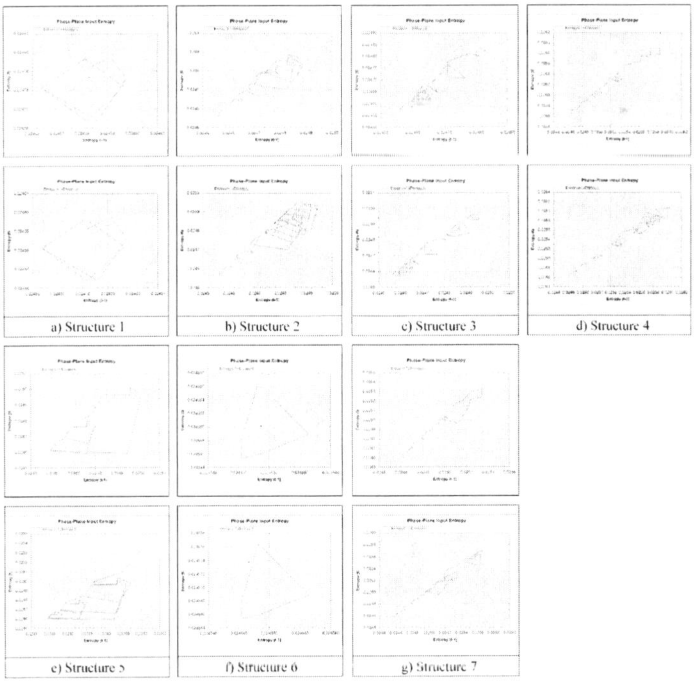

As expected, gliders generate limit cycles (see structures a and f in Figure 15). Self-replicators (structures b and e) generate more interesting behavior. The self-replication process is accompanied by the appearance of translational and scale symmetries and by the emergence of sub structures. The other structures produce more complex trajectories, due to interactions among elementary behaviors that are simultaneous present in the structure.

FUTURE TRENDS

In this chapter we have tried to answer three questions of crucial importance.

a) How does complexity manifest itself in a CA?
b) How can we classify CAs on the basis of their behavior?
c) How can we measure complexity in a CA?

As far as concerns the first question, we already offered a partial answer in Chapters 3 and 4, when we described gliders, regular domains, self-replicators etc. In this chapter, we saw how the dynamics of these structures affect the global dynamics of the CA. The exact mechanisms linking the two remain an open question. We will discuss this issue further in Chapter 8 when we examine self-replicators and in Chapter 9 when we subject these structures to a linguistic analysis. As we will see, a number of key problems remain unresolved - even though our understanding of the dynamics of these structures has greatly improved.

Wolfram's attempt to classify CAs on the basis of their behavior, first proposed in 1984, was a fascinating one. In reality, however, subsequent attempts to improve the classification have shown it to be weak and ambiguous. Its greatest limitation is that it is a purely qualitative approach. Yet in our opinion, it will be hard to improve until we have a clear answer to our first question.

We believe that a more useful strategy could be to investigate CA complexity metrics. Although Langton's parametric approach seems to have run out of steam, there is still much work to be done on pattern metrics. In (Ninagawa et al., 1998; Ninagawa, 2006) we discover that ECA 110 and Life - both complex rules - have power spectra that follow the 1/f pattern. This is very interesting. If this finding also applies to other complex rules, it could provide a basis for a new complexity metric we could use in Genetic Algorithms.

There exists a large literature in which complexity is identified with the "edge of chaos". In our opinion, however, complexity and complex behavior are far broader concepts. This, in fact, is what we will see, in the coming paragraphs, when we investigate specific regions of rules space. The idea of "an edge of chaos" is a narrower concept that presupposes a universally accepted metric of complexity or at least a set of metrics allowing us to characterize rules space. What is more, effective use of the concept would require a clear definition of the boundary between order and chaos, that so far escapes us. For these reasons we prefer to talk in terms of complexity. One possible topic for future research would "local activity", which is believed to be a necessary condition for complex behavior in a CNN (Chua, 1998). With suitable generalizations, it may be possible to link the two concepts.

What is certain is that for the moment, one of the most useful metrics is IE. IE-derived metrics are already a powerful tool for the investigation of specific CA behaviors. The development of metrics for the general behavior of CAs remains an open problem that requires further research. Nonetheless, IE-based metrics do capture certain aspects of CA complexity: which is why we refer to them as com*plexity indexes. Ch*apters 6 and 7 will describe how we have tried to evolve complex rules, using these indexes as fitness functions: and testing their usefulness in practice. At the same time, we can also use IE indexes to capture and measure the complexity of emergent structures and to show how this emerges from interactions between their elementary behaviors.

We can thus list the main questions that emerge from this chapter.

a) Is it possible to measure the complexity of a CA?
b) How can we ensure that this measure is independent of the initial data and thus that it represents a true characteristic of the CA?
c) Can we measure the complexity of emergent particles?
d) Are there any tools that allow us to study the behavior of these particles?
e) How does the behavior of individual particles influence the overall behavior of the system?

In Chapters 6 and 7, we will see that fitness functions based on IE and IE-derived indexes are very effective in evolving complex rules. This implies that IE can provide at least a partial solution to questions a) e b). As far as concerns the complexity of emergent particles, we observe that while the behavior of gliders is very simple, that of self-replicators is much more interesting and complex. This is a topic to which we will return in Chapter 8.

CONCLUSION

In this chapter, we have discussed several problems of critical importance for the investigation, classification and control of behavior in 1- and 2-D CAs. The goal has been to identify complex rules, and to characterize the structure of the ideal space where we can represent the parameters controlling CA behavior. We began by discussing the classical papers that set this important field of complexity science in motion. We then provided an almost exhaustive review of the tools available, leading up to our own work on IE. On the one hand, we have shown the problems inherent in the specific classification schemes we have discussed. On the other, we have presented an innovative approach in which we use IE to evaluate not only local structures generated by CAs (gliders and particles in 1-D systems; self-replicators in 2-D systems), but also the behavior of the system as a whole. This new method allows us to represent the development of 2-D CAs and their constituent particles (e.g. self-replicators) in 2 dimensional phase diagrams and to observe the geometrical correlations between them. It also allows us to infer fundamental characteristics of parameter space, a possibility we will discuss further in later chapters.

REFERENCES

Adami, C. (1998). *Introduction to Artificial Life*. New York: Springer-Verlag.

Bak, P. (1996). *How Nature Works: The Science of Self-Organized Criticality*. New York: Copernicus Press.

Baldwin, J. T., & Shelah, S. (2000). On the classifiability of cellular automata. *Theoretical Computer Science, 230*, 117–129. doi:10.1016/S0304-3975(99)00042-0

Bilotta, E., Di Blasi, G., Pantano, P., & Stranges, F. (2007d). A Gallery of Chua Attractors, Part IV. *International Journal of Bifurcation and Chaos in Applied Sciences and Engineering, 17*(4), 1017–1078. doi:10.1142/S0218127407017665

Bilotta, E., Di Blasi, G., Pantano, P., & Stranges, F. (2007e). A Gallery of Chua Attractors, Part V. *International Journal of Bifurcation and Chaos in Applied Sciences and Engineering, 17*(5), 1383–1511. doi:10.1142/S0218127407018099

Bilotta, E., Di Blasi, G., Pantano, P., & Stranges, F. (2007f). A Gallery of Chua Attractors, Part VI. *International Journal of Bifurcation and Chaos in Applied Sciences and Engineering, 17*(6), 1801–1910. doi:10.1142/S0218127407018105

Bilotta, E., Gervasi, S., & Pantano, P. (2005). Reading Complexity in Chua's Circuit by Music. Part I: A new way of understanding chaos. *International Journal of Bifurcation and Chaos in Applied Sciences and Engineering, 15*(2), 253–382. doi:10.1142/S0218127405012156

Bilotta, E., Lafusa, A., & Pantano, P. (2003b). Searching for complex CA rules with Gas. *Complexity, 8*(3), 56–67. doi:10.1002/cplx.10063

Bilotta, E., Lafusa, A., & Pantano, P. (2003b). Searching for complex CA rules with Gas. *Complexity*, *8*(3), 56–67. doi:10.1002/cplx.10063

Bilotta, E., & Pantano, P. (2001). *Artificial Life Music Tells Complexity.* Paper presented at Artificial Life Models for Musical Applications, 6th European Conference on Artificial Life, Prague.

Bilotta, E., & Pantano, P. (2002). Synthetic Harmonies: Recent results. *Leonardo*, *35*(2), 35–42.

Bilotta, E., & Pantano, P. (2006). Structural and Functional Growth in Self-Reproducers. *Complexity*, *11*(6), 12–29. doi:10.1002/cplx.20131

Bilotta, E., Pantano, P., & Stranges, F. (2007a). A Gallery of Chua Attractors, Part I. *International Journal of Bifurcation and Chaos in Applied Sciences and Engineering*, *17*(1), 1–60. doi:10.1142/S0218127407017161

Bilotta, E., Pantano, P., & Stranges, F. (2007b). A Gallery of Chua Attractors. Part II. *International Journal of Bifurcation and Chaos in Applied Sciences and Engineering*, *17*(2), 293–380. doi:10.1142/S0218127407017343

Bilotta, E., Pantano, P., & Stranges, F. (2007c). A Gallery of Chua Attractors, Part III. *International Journal of Bifurcation and Chaos in Applied Sciences and Engineering*, *17*(3), 657–734. doi:10.1142/S0218127407017537

Bilotta, E., Pantano, P., & Talarico, V. (2000). *Synthetic harmonies: an approach to musical semiosis by means of cellular automata.* Paper presented at Artificial Life VII. Cambridge, MA: The MIT Press.

Binder, P. M. (1991). Topological classification of cellular automata. *Journal of Physics. A. Mathematical Nuclear and General*, *24*, L31–L34. doi:10.1088/0305-4470/24/1/007

Chua, L. O. (1998). *CNN: A Paradigm for Complexity*. World Scientific Publishing Company.

Chua, L. O. (Ed.). (2006). A Nonlinear Dynamics Perspective of Wolfram's New Kind of Science: *Vol. 1. World Scientific Series on Nonlinear Science, Series A, 57*. Hackensack, NJ: World Scientific Publishing Company.

Chua, L. O., Wu, C. W., Huang, A., & Zhong, G. (1993a). A Universal Circuit for Studying and Generating Chaos. Part I: Routes to Chaos. *IEEE Transactions on Circuits and Systems*, *40*(10), 732–744. doi:10.1109/81.246149

Chua, L. O., Wu, C. W., Huang, A., & Zhong, G. (1993b). A Universal Circuit for Studying and Generating Chaos, Part II: Strange Attractors. *IEEE Transactions on Circuits and Systems*, *40*(10), 745–761. doi:10.1109/81.246150

Culik, K. II, & Sheng, Y. (1988). Undecidability of CA classification schemes. *Complex Systems*, *2*(2), 177–190.

Dow, R. A. (1997). Additive cellular automata and global injectivity. *Physica D. Nonlinear Phenomena*, *110*, 67–91. doi:10.1016/S0167-2789(97)00074-2

Dubacq, J. C., Durand, B., & Formenti, E. (2001). Kolmogorov complexity and cellular automata classification. *Theoretical Computer Science*, *259*(1-2), 271–285. doi:10.1016/S0304-3975(00)00012-8

Gutowitz, H., & Langton, C. G. (1995). Mean Field Theory of the Edge of Chaos. *Paper presented at ECAL*. Berlin: Springer-Verlag.

Hanson, J. E., & Crutchfield, J. P. (1997). Computational Mechanics of Cellular Automata: An Example. *Physica D. Nonlinear Phenomena*, *103*, 169–189. doi:10.1016/S0167-2789(96)00259-X

Hilborn, R. C. (2000). *Chaos and Nonlinear Dynamics: An Introduction for Scientists and Engineers*. New York: Oxford University Press.

Holland, J. (1975). Adaptation . In *Natural and Artificial Systems*. Ann Arbor: The University of Michigan Press.

Jen, E. (1990). Aperiodicity in one-dimensional cellular automata. *Physica D. Nonlinear Phenomena*, *45*, 3–18. doi:10.1016/0167-2789(90)90169-P

Langton, C. G. (1986). Studying artificial life with cellular automata. *Physica*, *22D*, 20–149.

Langton, C. G. (1990). Computation at the edge of chaos. *Physica*, *42D*, 2–37.

Li, W. (1992). Phenomenology of nonlocal cellular automata. *Journal of Statistical Physics*, *68*(5/6), 829–882. doi:10.1007/BF01048877

Li, W., & Packard, N. (1990). The structure of the elementary cellular automata rule space. *Complex Systems*, *4*, 281–297.

Li, W., Packard, N. H., & Langton, C. G. (1990). Transition phenomena in cellular automata rule space. *Physica D. Nonlinear Phenomena*, *45*, 77–94. doi:10.1016/0167-2789(90)90175-O

Miranda, E. R. (2001). *Evolving cellular automata music: from sound synthesis to composition,* Paper presented at Artificial Life Models for Musical Applications, 6th European Conference on Artificial Life, Prague.

Mitchell, M., Crutchfield, J. P., & Hraber, P. T. (1994). Dynamics, Computation, and the "Edge of Chaos": A Re-examination . In Cowan, G., Pines, D., & Melzner, D. (Eds.), *Complexity: Metaphors, Models, and Reality* (*Vol. 19*, pp. 497–513). Reading, MA: Addison-Wesley.

Mitchell, M., Hraber, P., & Crutchfield, J. P. (1993). Revisiting the Edge of Chaos: Evolving Cellular Automata to Perform Computational Tasks. *Complex Systems*, *7*, 89–130.

Morin, E. (2001). *La méthode* (*Vol. 5*). Paris: Seuil.

Ninagawa, S. (2006). 1/f Noise in Elementary Cellular Automaton Rule 110. In Unconventional Computation (LNCS 4135, pp. 207-216).

Ninagawa, S., Yoneda, M., & Hirose, S. (1998). 1/f fluctuation in the "Game of Life". *Physica D. Nonlinear Phenomena*, *118*, 49–52. doi:10.1016/S0167-2789(98)00025-6

Oliveira, G. M. B., De Oliveira, P. P. B., & Omar, N. (2001). Definition and Applications of a Five-Parameter Characterization of One-Dimensional Cellular Automata Rule Space. *Artificial Life Journal*, *7*(3), 277–301. doi:10.1162/106454601753238645

Prigogine, I., & Stengers, I. (1984). *Order out of Chaos: Man's new dialogue with nature*. London: Flamingo.

Strogatz, S. H. (2000). *Nonlinear Dynamics and Chaos: With Applications to Physics, Biology, Chemistry, and Engineering*. New York: Perseus Books.

Wolfram, S. (1984). Universality and Complexity in Cellular Automata. *Physica*, *10D*, 1–35.

Wolfram, S. (2002). *A New Kind of Science*. Champaign, IL: Wolfram Media.

Wuensche, A. (1999). Classifying Cellular Automata Automatically: Finding gliders, filtering and relating space-time patterns, attractors basins and the Z parameter. *Complexity*, *4*(3), 47–66. doi:10.1002/(SICI)1099-0526(199901/02)4:3<47::AID-CPLX9>3.0.CO;2-V

APPENDIX

Rule Codes

C5.1: $D = 1, k = 2, r = 2$

10010011011101100100001001001000

C5.2: $D = 1, k = 2, r = 2$

00000101100101000000010001010001

C5.3: $D = 1, k = 2, r = 3$

0110001111101010011010011100011000101101011001010000000000011111
00011001000101001010000011110110001101000100011101010101100110100

C5.4: $D = 1, k = 2, r = 3$

1110100101011110001100111111000010110010000101001001010101011001
011111010100110101010001111100101111110110001110101110111010110010

C5.5: $D = 1, k = 4, r = 1$

3313320003013200331223001102220013312121220130330113220223222322

C5.6: $D = 1, k = 4, r = 1$

1220123033221303232322232322332332333222333013303132032323322232

C5.7: $D = 1, k = 4, r = 1$

1322210333021122232322331223322322212303331023322112212323111332

C5.8: $D = 1, k = 4, r = 1$

2022230031231123213202233322132302323123132212323200311220222123

C5.9: $D = 1, k = 4, r = 1$

3132313331232221233332231113213232223223212211322320121233233222

C5.10: $D = 1, k = 4, r = 1$

3322112321221323031122323012213211212012310311132311233232121233

C5.11: $D = 1, k = 4, r = 1$

3330333031123203323022011203203112312301320333130310211320032011

Chapter 6
The Discovery of Complex Rules

INTRODUCTION

After Wolfram's proposal, other authors analyzed systems of CAs and proposed their own classifications (Gutowitz 1990; Li, Packard, and Langton 1990). Kurka (1997) proposed to classify them in 5 groups, based on the structure of their attractors. Walker (1990) based his characterization on connected Boolean networks. Many other researchers proposed classifications based on rule sets governing the CAs' behavior (Barbe 1990; Jen 1990; McIntosh,1990; Voorhees, 1990; Wootters & Langton 1990).

In previous chapters, we argued that the rules that best describe the behavior of biological systems are what Wolfram (1984, 2002) calls "complex rules". CAs with complex rules have the ability to process information. In other words, they are universal computers. If a system is capable of universal computation, the evolution of the system from a given set of starting conditions represents a finite computational process. A universal computer can mimic any other system generating a limitless range of complex behaviors. In these circumstances, it is not surprising that the search for complex rules and the analysis of their behavior has been one of the key themes in Complexity Science and Artificial Life. The proof that several CA systems are universal computers is based on Conway's Game of Life (Gardner 1983). The

DOI: 10.4018/978-1-61520-787-9.ch006

proof is based on the demonstration that CAs can generate structures corresponding to the components of an idealized digital computer. By connecting these components in different configurations it is possible to implement any algorithm, just as any known algorithm can be implemented on the simpler and better known Turing Machine. (Smith, 1971). Underlying Wolfram's characterization and classification of the behavior of Cellular Automata are two key questions (Wolfram 1994):

1. Is it possible to predict the global behavior of an automaton from knowledge of its local rules (the bottom up approach)?;
2. Is it possible to design local rules that generate complex behaviors (the inverse, top-down problem)?

Deducing complex behaviors from local rules is extremely difficult (Capcarrere 2002). The difficulty of the task has led many researchers to use techniques from evolutionary computing (Mitchell et a., 1993; Packard 1988; Richards et al., 1990). Starting with randomly generated rules, evolutionary algorithms have evolved methods to resolve the density classification and two cell synchronization tasks and to synchronize groups of cells (Crutchfield and Mitchell 1995; Crutchfield et al., 2003; Das et al., 1995; Umeo et al., 20021994).

Various metrics have been used to measure order and chaos in CA behavior and thus to characterize CAs' global dynamics. (Amoroso & Patt 1972; Kari 1990; Kari 1994; and recently Chua 2007; 2009). Other studies have used the "Lyapunov exponent" (applied to continuous dynamic systems in Bilotta et al., 2007a, b, c, d, e, f), a metric that measures how far the system depends on the initial conditions (Kaplan & Glass 1995) and "Hamming distance", which represents the number of cells whose states are different in two successive periods (Wolfram, 1984; Bilotta & Pantano, 2005). Geometric features such as spatial self-similarity or the fractal dimension of the patterns, generated during the development of the CA, can be analyzed using the *Hausdorff-Basicovitch* distance. Finally, there are a range of entropy measures that make it possible to measure the irreversibility of CA behavior (Wolfram 1994). Each of these measures counts the number of possible sequences of site values leading towards a specified pattern in time and space. For example, spatial entropy measures the size of the set of configurations that a CA can generate in a specified time, considering all possible initial conditions; temporal entropy counts the number of sequences in the time series of values for a specific site; topological entropy considers the number of possible configurations of the system. Chapter 5 provides a more detailed discussion of these metrics. However, none provides a clear response to our main question; none tells us how we can find local rules that produce complex global behavior. For binary CAs, Langton has proposed that predictions could be based on the value of what he has called the λ parameter. This is defined as the probability that a certain cell in a CA will take on a value of 1 in the next step in the CA's evolution. In other words, lambda measures the proportion of 1s in the binary rules table for the automaton. Langton (1990) proved that as λ increases, the overall behavior of the CA shifts from order to chaos. Other authors followed in Langton's footsteps proposing other local parameters (Wuensche & Lesser, 1992; Zwick & Shu, 1995).

To overcome these problems, we explore the use of Genetic Algorithms to search for complex rules with universal computational capabilities. By using Wuensche's entropy measures (Wuensche, 1999) as fitness functions in our Genetic Algorithms and combining it with Langton's approach, we have been able to discover complex rules for 1-D multistate CAs (Bilotta et al., 2003). This is the work we describe in this Chapter.

The Chapter is organized as follows. We begin by introducing Genetic Algorithms and their application to CAs, the theoretical framework for what follows. We then take a detailed look at the workings of GAs that evolve complex rules. In particular, we see how we can distinguish families of rules, examining common traits and traits that differ among individuals. This analysis allows us to introduce the concept of "genes" and "genetic networks" as applied to CAs. The latter depend, on the one hand, on the number of rules the system consults as it evolves in time and space; on the other on the evolution of populations of CAs. By using evolutionary techniques, we can monitor the genes for complex rules, as they pass from one generation to the next and see the privileged role of certain specific regions of the rule space, as characterized by λ. These are regions that display a global evolutionary dynamic in the sense that any initial population of rules is attracted to these regions, after a certain number of time steps. To find these regions, we repeat our experiments a number of times (multiruns). The results show that there are multiple "edges of chaos". This contradicts Langton's hypothesis that there is only one edge of chaos and casts doubt on the use of the parameter to characterize the boundary between order and chaos.

BACKGROUND

Genetic Algorithms (GAs), first introduced by Holland (Holland,1975), use evolutionary techniques to resolve specific classes of problem. GAs have given rise to a large body of research on optimization and problem solving - better known as evolutionary computing (Koza, 1992). Today GAs have many applications, ranging from electronic circuit design (Koza, 1994), through evolutionary robotics (Nolfi & Floreano, 2000; Bilotta & Pantano, 2008), to electronic art (Romero & Machado, 2008). Their flexibility and ease of use makes them a basic tool in modern science.

The basic idea is to evolve a population of possible solutions for a given problem and to select the best solutions in each generation, using a fitness function, reminiscent of natural selection. The transition from one generation to the next uses genetic operators, such as mutation and cross-over. In general terms, GAs operate as follows:

1. The system generates an initial population of individuals;
2. A fitness function measures the fitness of each individual;
3. The individuals with the highest fitness (the élite) are selected for reproduction;
4. The élite produces a new generation of individuals, applying genetic operators such as mutation and cross-over;
5. The process is repeated for the next generation.

Genetic Algorithms bring to mind the mechanisms underlying natural selection. The fitness function determines the gene pool (the traits), which will be conserved in the next generation, while mutations ensure that individuals become better adapted to their environment. In any given generation, some of the new individuals in the population will display mutations that were not present in the previous one, while others will be the result of cross-over (i.e. the crossing of the genome of two individuals). The evolutionary process continues until it finds a satisfactory solution to the original problem. Strangely, Genetic Algorithms are highly effective for a broad range of applications.

The genome of a Cellular Automaton is the string representing the rules for the system (2.10) or (2.18). The mutation operator, randomly changes some of the values in the string; the cross-over operator creates a new string joining an initial segment of one string to the final segment of another. In general terms, the GA works as follows.

- The system generates a population of randomly generated strings.
- Each CA is assessed on its IE value.
- The system generates a new population:
    1. It randomly selects two individuals with high fitness values.
    2. It uses the cross-over operator to generate two new individuals: one is encoded by the initial segment of the string for the first individual, joined to the final segment from the second individual; the second from the final segment for the first individual, joined to the initial segment from the second. The cross-over point is chosen at random.
    3. Randomly selected characters in the strings, for the new individuals, are subjected to random mutations.

- Individuals belonging to the élite from the previous generation (the individuals with the highest fitness) are carried over to the next generation without modification.

- The new population replaces the old population.

- The process is repeated starting with the first step.

The key problem is how to define a fitness function that selects CAs with specified characteristics or with the ability to perform a specific task.

As we saw in Chapter 2, the rule space for a cellular automaton is generally large. If rules are listed exhaustively, the number of rules for a CA, with a neighborhood of *n* cells and *k* possible states, is . Even with a more compact representation (see 2.17), the number of rules is equal to:

The high number of rules means it is not usually possible to find solutions to a problem without computational support. One example is the density classification task. The goal of the task is to identify a two state CA (a CA with states 1 and 0) whose output, after a certain number of steps, depends on the number of 1s in the initial configuration. If this number is higher than a given threshold, all the cells in the CA converge to a value of 1; in the opposite case, they converge to 0. The first author to address the problem with a GA was Packard (1988).

One important result from Packard's experiments was that all CAs capable of solving the problem had λ values close to a critical value. This seemed to confirm Langton's hypothesis about the "edge of chaos" (see Chapter 5). Mitchell et al. (1993; 1994) returned to the problem, showing that any solution requires coordination among cells, lying at arbitrary distances from each other. As we will see in the next chapter, the fact that complex CAs can solve the problem demonstrates the ability of gliders to transport information. The results obtained by these authors were satisfactory. However their complex rules had λ values a long way from the critical zone.

In synchronization problems, the goal is to design a CA which, after a certain number of steps, begins to oscillate from one state, in which all cells have a value of 0, to a second, in which all cells have a value of 1. In group (or Firing Squad) synchronization problems, the task is to design a 1-D CA, all of whose cells become active (fire) simultaneously after a given number of steps.

The first attempt to generate self-replicators with GAs was reported in (Lohn & Reggia, 1995; 1997) (see Chapter 7). However the authors were forced to make significant modifications to the mathematical model underlying the CA.

As reported in (Bilotta et. al. 2000; 2002), the authors have used GAs to find CAs that generate pleasing sequences of musical notes (see Chapter 15). In this

Chapter, we will concentrate on GAs that find complex rules (Bilotta et al. 2003). In other words, we will focus, not on CAs that resolve a specific task, but on CAs whose rules give rise to a specific behavior.

MAIN FOCUS

The main goal of this chapter is to show that GAs with a fitness function based on Input Entropy can find complex rules for 1-D, multistate CAs.

In a 1-D CA, with specified values for k and r, there are possible rules. As k and r increase, the rule space R becomes very large. From the observations in Chapter 5, we know that complex rules display large variations in Input-Entropy. By contrast, rules associated with orderly or chaotic behavior have very low and very high values respectively, and show no significant variation between rules. Knowing this, the authors used the variance in the entropy function as the basis for a fitness function. It was this fitness function they used to find complex rules for multistate CAs. (Bilotta et al., 2003).

Consider a CA, of radius r, each of whose cells can take on k distinct values. The local rule for the CA can be represented by string (2.10). The string contains characters. Altogether there are possible strings. The set of these strings L is a bijective mapping of the rule space R. Let Q_i be the frequency with which the system consults the ith configuration of the neighborhood; let M be the total number of cells in the CA and let X be the set of all possible configurations of the neighbor, Applying (5.8), the input-entropy of the system at time t is given by:

$$H^t\left(X\right) = -\sum_{i=1}^{k^{(2r+1)}} \frac{Q_i}{M} \ln(\frac{Q_i}{M})$$

The Genetic Algorithm operates as follows:

1. the system generates a random initial population of rules;
2. the rules with the highest fitness (the élite) are carried over to the next generation;
3. a certain proportion of the other rules is generated randomly; the remaining rules are created by cross-over between rules belonging to the élite followed by mutation.

We used the following control parameters:

- p, the number of rules per generation,

- *g*, the number of generations,
- *e*, the number of rules in the élite,
- *v*, the proportion of loci that undergo mutation
- *n*, the number of new rules that are generated randomly,
- *c*, cross over points,
- *w*, time window for IE calculations
- τ, number of initial time steps before beginning calculation of IE
- *i*, number of initial configurations for the simulation,
- *M*, number of cells in the CA,
- *T*, number of time steps for the simulation.

In our experiments, we discovered that, from the point of view of complexity, the only variations that count, are those that can be measured over a long period of time. To avoid rapid variations in entropy values, we introduced a time window *w*. In this way, we do not calculate entropy for each step in the simulation. Rather, we measure the mean value over the time window and compute the variance. We begin our computation of IE only after a pre-specified number of steps, τ. In this way, we allow the CA to settle into its normal pattern of behavior for the rule, moving away from its initial chaotic configuration. As is well known, the patterns whose fitness we calculate can vary greatly when we change the initial data. These changes lead to major changes in fitness. In these circumstances, it is important not to throw away rules with a high fitness in one generation and a low fitness in the next. To achieve this, we updated the fitness of a rule only when its fitness value was higher than the value for the previous generation. In all other cases, fitness was left unchanged. To counteract the influence of the initial data, we computed fitness as the mean fitness of the CA over 5 repetitions of the simulation, with 5 different sets of initial data and with different initial data in each generation. After these simple adjustments, the genetic algorithm generated a large set of complex rules. Some of these rules are described in Chapter 2. The patterns they produce are depicted in Figure 2.8.

The Appendix provides pseudo-code for a possible implementation of the Genetic Algorithm just described.

Families of Rules

To understand the results produced by the GA we analyze the individuals in the last generation of CAs. Rules C2.3 and rule C2.4 were generated by the same evolutionary process. Below we list the parameters.

- States number = 4
- Neighborhood = 3
- Generations = 40
- Populations = 40
- Length of CA = 75
- Steps of simulation = 430
- Ignore initial steps = 30
- Seeds = 1
- Elite (%) = 30.0
- Mean mutation (%) = 3.0

Generation mode: crossover + mutations

Rules C2.3 and C2.4 correspond to rules 1 and 26, in the last generation of CAs. As we can see the images produced by the two rules are very different. (Figure 1) depicts the patterns corresponding to each of the 40 rules in the last generation.

There are many ways of analyzing these rules in qualitative terms. One family of rules has 3 members (rules 1, 20 e 24); another very large family has 23 rules (rules 2-3-4-5-7-9-11-12 13-15-16-17-18-19-21–22-23-25-27-28-29-31-32); there are 6 families consisting of a single rule (rules 6-8-10-14-26-30); 7 chaotic rules (rules 33-34-35-36-37-38-39) and one Type II periodic rule (40). There is no doubt that all the rules in the first family and in the 6 single rule families are complex, in Wolfram's sense of the term.

We can gain further insights by analyzing λ values for the rules (see Table 1). The analysis shows fairly large differences (see Figure 2a). λ has a mean value of 0.401953 and a standard deviation of 0.066006. Maximum and minimum values

Figure 1. Patterns generated by the 40 rules in the last generation of CAs, produced by a GA using a fitness function based on variance in IE

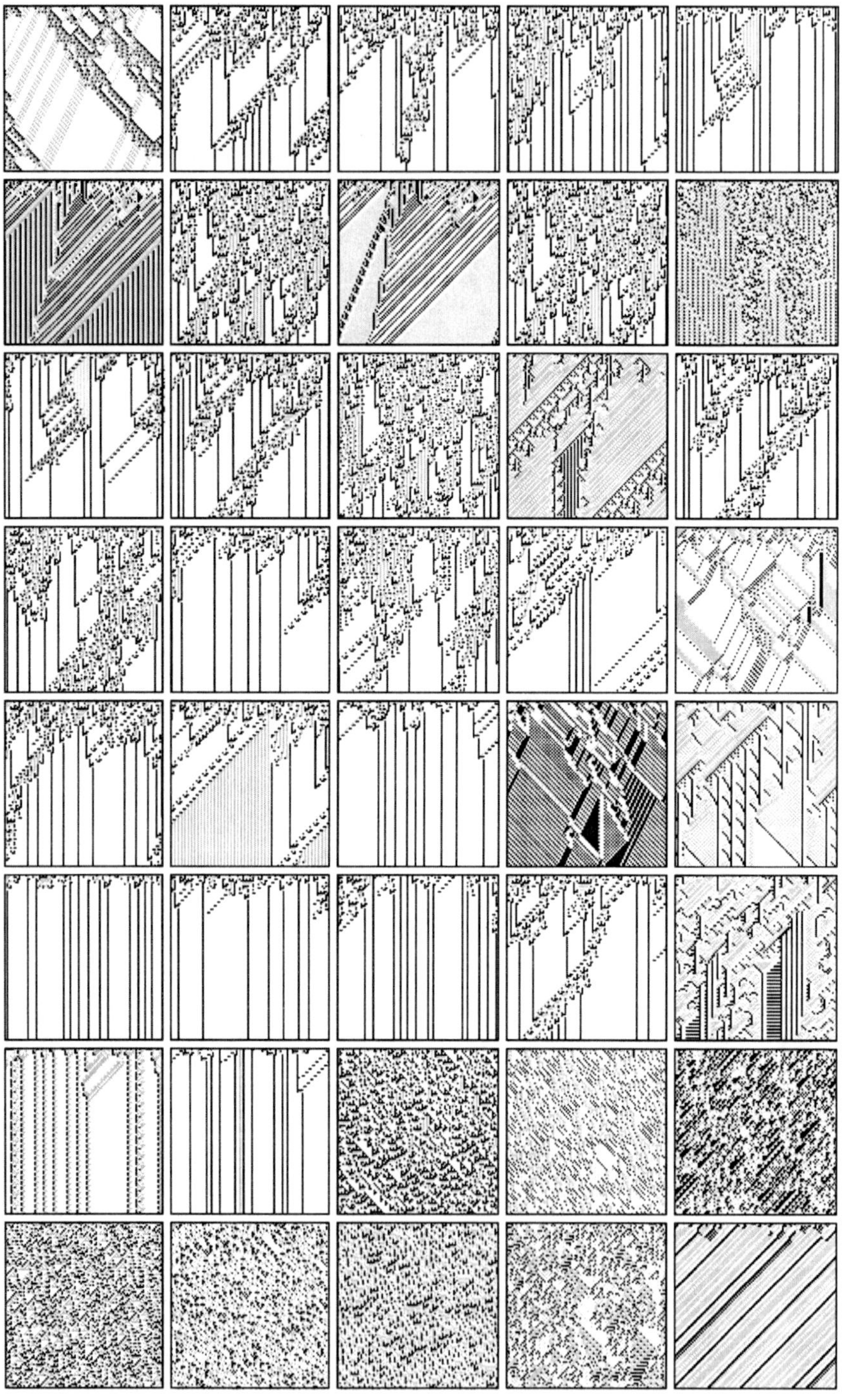

range from 0.265625 for rule 30 to 0.5 for rule 32. We observe that, although many complex and chaotic have lower λ values, rule 32, is in fact a periodic rule. This is more evidence against Langton's hypothesis that there exists a tightly defined range of λ values on the boundary between order and chaos and demonstrates that λ is an ineffective tool for structuring rule space. Results from other experiments provide further evidence pointing in the same direction.

It is interesting to analyze the distribution of λ values for complex rules, as shown in Figure 2b. The values range from a minimum of 0.28125 (rule 6) to a maximum of 0.390625 (rule 10). The analysis of the λ values shows a clear difference between the values for the three rules in the first family and identical values for rules 6 and 8. In these circumstances, an in-depth analysis can provide new insights into analogies and differences between families of rules. It is interesting to analyze λ, not only for state 0 but also for states 1, 2 and 3. In what follows, we will refer to these indicators as λ_1, λ_2 *e* λ_3. The mean values for the 40 rules are 0.20234375, 0.189453125 and 0.20625 respectively.

(Figure 2c) provides values of λ, λ_1, λ_2, *e* λ_3 for all rules in the last generation. (Figure 2d) provides the same data for complex rules only. The data in the first graph shows that 0s are more frequent than any other value and that 1s are more frequent than 2s and 3s. We observe that rules 6 e 8 have very different values for λ_2 and λ_3 but identical values for λ and λ_1.

Table 1. λ values for the 40 rules shown in (Figure 1) As it is possible to see from the data, many rules share the same λ values. As it is possible to see from the data, many rules share the same λ values.

Rule 1, $\lambda_=$ 0.359375	Rule 21, $\lambda_=$ 0.40625
Rule 2, $\lambda_=$ 0.46875	Rule 22, $\lambda_=$ 0.4375
Rule 3, $\lambda_=$ 0.40625	Rule 23, $\lambda_=$ 0.484375
Rule 4, $\lambda_=$ 0.421875	Rule 24, $\lambda_=$ 0.375
Rule 5, $\lambda_=$ 0.46875	Rule 25, $\lambda_=$ 0.40625
Rule 6, $\lambda_=$ 0.28125	Rule 26, $\lambda_=$ 0.328125
Rule 7, $\lambda_=$ 0.421875	Rule 27, $\lambda_=$ 0.484375
Rule 8, $\lambda_=$ 0.28125	Rule 28, $\lambda_=$ 0.484375
Rule 9, $\lambda_=$0.421875	Rule 29, $\lambda_=$ 0.484375
Rule 10, $\lambda_=$ 0.390625	Rule 30, $\lambda_=$ 0.265625
Rule 11, $\lambda_=$ 0.46875	Rule 31, $\lambda_=$ 0.375
Rule 12, $\lambda_=$ 0.421875	Rule 32, $\lambda_{= 0.5\ \text{maximum}}$ value of λ
Rule 13, $\lambda_=$ 0.421875	Rule 33, $\lambda_=$ 0.453125
Rule 14, $\lambda_=$ 0.296875, minimum value of λ	Rule 34, $\lambda_=$ 0.390625
Rule 15, $\lambda_=$ 0.421875	Rule 35, $\lambda_=$ 0.328125
Rule 16, $\lambda_=$ 0.421875	Rule 36, $\lambda_=$ 0.484375
Rule 17, $\lambda_=$ 0.46875	Rule 37, $\lambda_=$ 0.40625
Rule 18, $\lambda_=$ 0.421875	Rule 38, $\lambda_=$ 0.328125
Rule 19, $\lambda_=$ 0.4375	Rule 39, $\lambda_=$ 0.328125
Rule 20, $\lambda_=$ 0.3125	Rule 40, $\lambda_=$ 0.3125

Figure 2. First row: a) values of λ for rules generated in the last generation; b) values of λ for the 9 complex rules Second row: c)values of all λ s for rules in the last generation; d) value of all λ s for the 9 complex rules. Third line: e)Hamming distance of rules with respect to rule 1; f) Hamming Distance of rules with respect to rule 6.

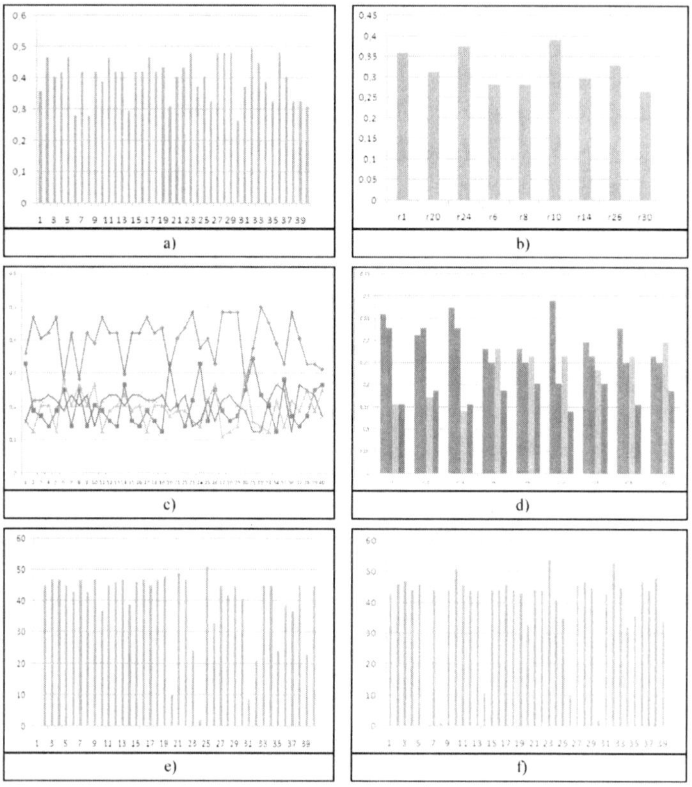

Genetic Issues and Dynamic Networks

In (Bilotta et al., 2003), we described a computational tool which allowed us to identify the rules used by a multistate CA with $k > 0$ or $k = 2$ and $r > 2$. We considered these rules as the CA's "gene pool". In this way, the phenotype of a CA is the pattern it generates; its genome consists of the rules it uses during simulations. In a system with specified values of *k* and *r*, we denote these rules by R1,.....Rn. This approach allows us to identify the genes for the whole pattern (considered as an artificial organism) and for individual cells (considered as a specific trait). Similarly we can identify the genes expressed during the evolutionary process as a whole, or for individual steps in the process.

During our experiments with GAs, we observed that certain genetic characteristics of the gene pool and of individual genes are transmitted from one generation to the next. The transmission of these elementary rules represents a form of genetic transmission of hereditary characteristics.

To measure changes in rules between generations, we used Hamming Distances to compare their respective gene pools. As mentioned earlier, the Hamming Distance between two strings of the same length measures the number of characters in the first string that differ from the corresponding characters in the second. Thus:

$$(6.1) \quad H : LxL \rightarrow N$$

$$(6.2) \quad v = H(l, l') \in N$$

with $v = \sum_{i=1}^{k^n} v_i$, where k^n is the length of the string and $v_i = \begin{cases} 0\, if\, a_i = a_i' \\ 1\, if\, a_i \neq a_i' \end{cases}$.

In the case we consider here, the length of the string is given by $Len(l) = 43 =$ ^{6}4. (Figures 2e) and (Figure 2f) show respectively the Hamming Distances between rule 1 and the other rules, and the distances with respect to rule 6. For rule 1, these distances oscillate between a maximum value of 51 for rule 25 and a minimum value of 2 for rule 20. Rule 20 has another low value (10) as does rule 31 (9). For rules 20 and 24, which appeared similar to rule 1, this result was expected. For rule 31, it was fairly surprising. When we measured Hamming Distances with respect to rule 6, we found other surprising results. In this case, the values varied between 1 (for rule 8) and 54 for rule 23. Rule 30 also gave a low value (2). In the case of rule 8, the low Hamming Distance was expected - we had already detected the similarity to rule 6 in our analysis of λ values. However, for rule 30 it was a complete surprise

Rules 1 and 24 are very similar. Let us analyze their behavior in greater detail.

If we compare λ values for the two rules, we obtain the results shown in (Table 2). Values of $\lambda 1$ and $\lambda 3$ are $_t$he same for both rules. However, rule 24 has an extra 0 and rule 1 has an extra 2. As a result, rule 24 has a higher value of $\lambda 0$ and r$_u$le 1 has a higher value of $\lambda 2$. The c$_h$aracters which make the difference are at position 32 in the string (neighborhood 200) where $a32 = 1$ an$_d$ $a32' = 0$ an$_d$ at position 48 (neighborhood 100) where $a48 = 2$ and $_{a4}$8′ = 1. Rul$_{es}$ 1 and 24 are very similar. Let us analyze their behavior in greater detail.

In reality, there are still several similarities between the two rules. In both cases, the rest states are 0 and 1. Both rules produce leftwards-moving gliders (01 and 103), all of whose cells have a state of 1. However in rule 24, the rightwards-moving glider (0002) is transformed into a glider-gun for glider (103) creating a regular

Table 2. λ values for rules 1 and 24

λ	Rule 1	Rule 24
λ0	0.359375	0.375
λ1	0.328125	0.328125
λ2	0.15625	0.140625
λ3	0.15625	0.15625

domain (103103). This difference is one of the most important factors determining the global behavior of the two systems, and the differences between the patterns they produce. The glider shown in (Figure 3a) has a period of 2. The system has 13 elementary rules: R8, R23, R27, R32, R34, R41, R43, R44, R48, R54, R55, R62, and R64. These include both of the elementary rules that distinguish rule 1 from rule 24. In the first step in the cycle, the glider uses rules R27, R43, R44, R48, R55, R62, R64; in the second step it uses rules R8, R23, R27, R32, R34, R41, R43, R54, R62 and produces the network shown in (Figure 3b). In rule 24, the glider becomes a glider gun (Figure 3c). The glider gun uses a network of 28 rules. After an initial transient, it produces the periodic pattern shown in (Figure 3d).

What we see here is the way in which a set of elementary rules can produce a pattern of genetic expression. The network can thus be seen as a gene pool, whose elements are expressed dynamically. If we change one of the elements in the gene, the result is a change in the way it is expressed. In some cases, this can lead to new patterns or drive the system towards chaos.

Figure 3. a) The glider (0002) generated by rule 1. b) the consultation network for the glider. c) the glider becomes a glider gun (in rule 24) creating a regular domain consisting of various gliders. d) the consultancy network for the glider-gun and its associated regular domain.

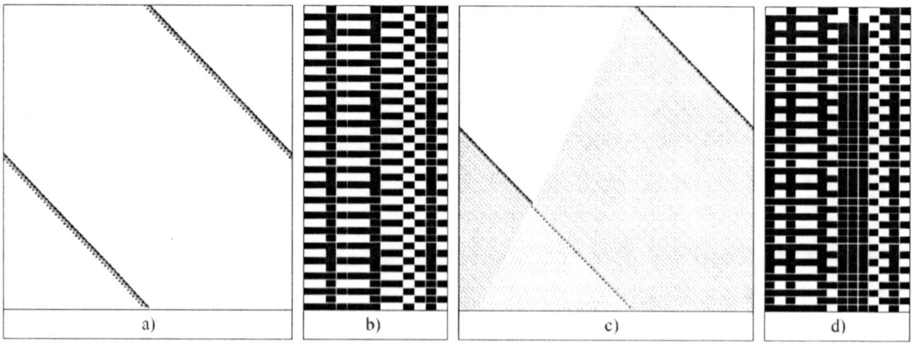

Let us now look at the behavior of rule 6 and of rules 8 and 30. These rules are very similar. (Table 3) provides their λ values. If we compare rule 6 and rule 8, we discover that the only difference between them concerns a1 (neighborhood 333), which has a value of 2 while a1' = 3. This difference produces a deep differentiation between the two patterns: rule 6 produces regular domains with states 0 and 1; rule 8 creates a new regular domain with state 3. As a result, rule 8 produces regions of uniform state, absent in rule 6. Rule 6 also generates cells whose state oscillates between 2 and 3. In rule 8, these cells take on a uniform value of 3. The result is that, although the two rules share many common elements such as gliders and regular domains, there are few obvious similarities between the large-scale patterns they produce.

If we compare rule 6 and rule 30, we find that the Hamming Distance between them is determined by the presence of different characters at position 32 (neighborhood 200), where a32 = 0 and $a32' = 1$, (neighborhood 100) and at position 48, where $a48 = 1$ and $a48' = 2$. This produces an excess of state 2 in rule 30 compared to rule 6. Here again some elements change, while others are conserved.

To conclude, it appears that the pattern a rule generates in time in space depends on a specific number of elements such as gliders, regular domains, etc. We denote this number by s. Each element corresponds to one of s genes *g1, g2,* ... *gs*, which together make up the CA's genome G = {g1, g2, ... *gs*}. Each gene corresponds to a single element in the spatio-temporal pattern. Each is also associated with a genetic *network,* σ_g *made* up of the elementary rules used by the element as it evolves in time and space. Each rule corresponds to one of the n bases $b_1^g, b_2^g \dots b_n^g$ defining the gene. The dynamics of the process-the process whereby different rules are consulted at different times-defines a dynamic network $\sigma_g(t)$. The dyna*mic behavior of the gene is* defined by the rules used during the generation of the corresponding pattern and the frequency with which they are used. It is important to note that several genes may use the same bases.

Table 3. λ values for rules 6, 8 and 30

λ	Rule 6	Rule 8	Rule 30
λ0	0.28125	0.28125	0.265625
λ1	0.25	0.25	0.25
λ2	0.28125	0.265625	0.296875
λ3	0.1875	0.203125	0.1875

Let us consider glider (103) in the uniform state 1, generated by rule 1, and denote its gene by g. The corresponding genetic network σ_g is composed of the following elementary rules:

$$\sigma_g = \{R11, R12, R35, R41, R43, R44, R45, R47, R51, R59\}$$

The corresponding dynamic networks are:

$$\sigma_g(1) = \{R11, R43, R44, R45, R51\}$$

$$\sigma_g(2) = \{R12, R35, R41, R43, R47, R59\}$$

This approach allows us to analyze the detailed behavior of the genes and the way they are expressed phenotypically. (Figure 4), for instance, shows that R43 is always active. By contrast, the other bases switch on and off as the system evolves.

Figure 4. The genetic network corresponding to the gene for glider (103) in the uniform state 1, generated by rule 1. The first panel (a) in the first column provides a graphical representation of the 10 elementary rules used compared to the complete set of elementary rules (220). The second column (b) shows the times at which different rules are used. The central column shows the dynamic network for time 1 and for time 2. The third column (c) provides an alternative representation of the dynamic networks showing only rules that are actually used.

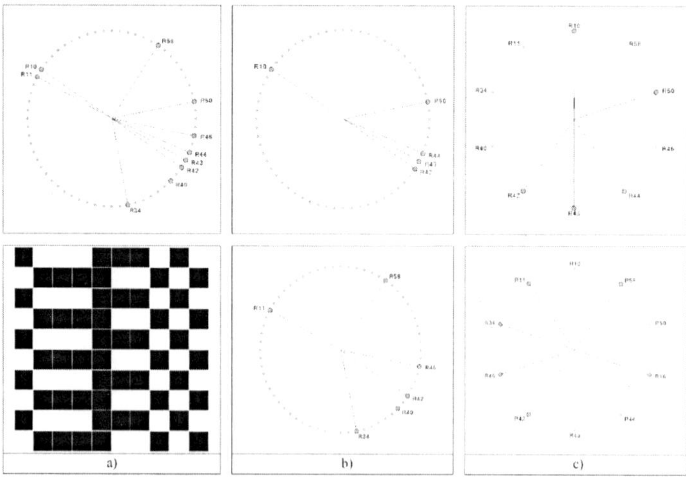

We can also see how the same gene is present in different CAs. For example, the gene that produces glider (103) in uniform domain 1 is present in rule 1, rule 20 and rule 24. By contrast, the gene for (0002) in the same domain is present in rule 1 and rule 20 but not in rule 24. In rule 24, the gene has undergone a single-based mutation, combining with other bases to produce new phenotypes. The gene corresponding to the regular domain (302302) can be found in rules 6, 8, 30; the gene for the oscillating domain (2-3) is present in a broad range of rules (for instance rules 4, 6, 7, 9, 12, 13, 14, 15, 16, 18, 19, 21, 29, 30, 34, 37, 38, and 39). The gene for uniform domain (3) is present in rules 8, 25 and 40.

Analysis of the Evolutionary Process

We will now examine the behavior of the genetic algorithm during the evolutionary process, using the same example already discussed in the previous section. (Figure 5a) shows trends in the mean and maximum fitness of CAs in the population. (Figure 5b) shows changes in the fitness of individual rules from generation to generation. As we proceed from right to left and from the bottom towards the top of the graph, fitness increases. This confirms the success of the Genetic Algorithm as the evolutionary process moves forward. The increased fitness of individual rules is confirmed by the increase in mean fitness. This is further evidence pointing in the same direction.

If we analyze the trend for maximum fitness, we observe a large increase in fitness after just 3 steps. This is evidence that the system has already discovered a rule with a high fitness. This rule remains the best until step 16 where we observe a small improvement in fitness. In step 18, there is another fairly large jump. All of this becomes clearer, if we examine Figure 5c. Here we see the CA's evolutionary tree. The fittest rules are on the left and evolution proceeds from the top down. In the tree, CAs with the same color are identical. At the end of the evolutionary process, the rule discovered in step 3 is still the second best in the population, even though for a short period (from step 30 to step 32) it is overtaken by another rule. This new rule maintains its third place until the end. These findings show how the genetic algorithm maintains fit rules once it has discovered them, and gradually improves their fitness over the evolutionary process. At times, rules that have disappeared are rediscovered. In this case, the system treats them as new rules. As shown in previous simulations, the elimination of these "repeated rules" does not improve the final result. Figure 5c shows the presence of a large number of rules belonging to the same family. This reflects the presence of many rules similar to the rule discovered in step 3. The tree also shows the trajectory of rule 6, discovered at step 25.

Another way of analyzing the evolutionary process is to analyze changes in the value of $\lambda_0 = 1 - \bar{\lambda}_0$. In (Figure 5d), it is clear that the parameter converges towards

Figure 5. a) trends in mean and maximum fitness during the evolutionary process; b) distribution of fitness for all rules for each generation; c) the genetic tree; on the left the fittest rules; evolution proceeds from the top down; d) distribution of $\bar{\lambda}_0 = 1 - \lambda_0$ *for all rules for all generations; e) mean* $\bar{\lambda}_0$ *for all generations; f) distribution of* $\bar{\lambda}_0$ *of all rules in the last generation of CAs*

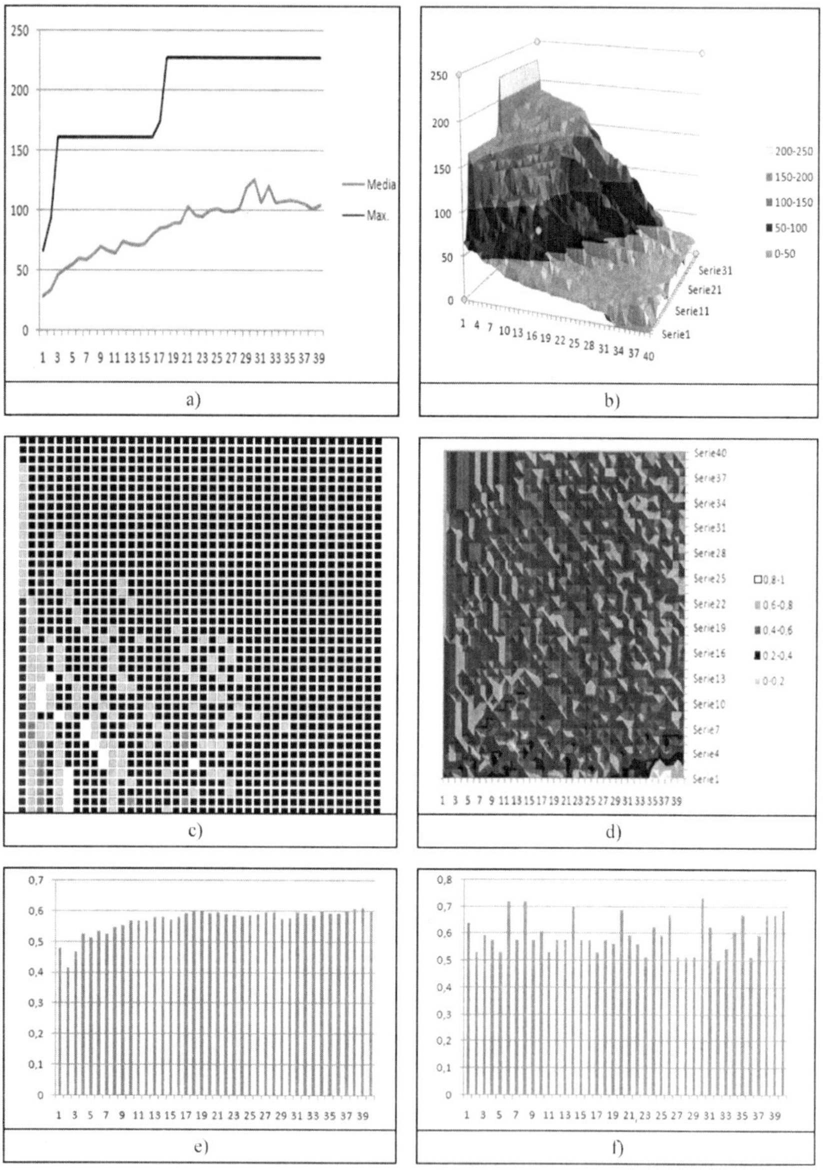

certain specific values. In Figure 5e, for instance, we see that the mean tends to a value of 0.598. Careful analysis of the value of $\overline{\lambda_0}$, for rules in the last generation, show a wide distribution around the mean, with values in the range 0.5 and 0.7 (see Figure 5f). This becomes clearer if we examine the evolution of $\overline{\lambda_0}$, as shown in Figure 6. In the early phases of evolution, $\overline{\lambda_0}$ has a fairly uniform distribution across the range of possible values. However, the evolutionary process is marked by a shift away from extreme values, with values on the left hand side of the distribution increasing past 0.5 and values on the right falling to below 0.7. The final result is shown in the next to last image in Figure 6 and the last image in (Figure 5f).

If we examine the distribution of $\overline{\lambda_0}$ on a fine scale, we discover that many of the values are concentrated in tightly restricted regions of the distribution. For example, 8 rules have values in the range $0,57 \leq \overline{\lambda_0} < 0,58$ and 5 in the range $0,51 \leq \overline{\lambda_0} < 0,52$. We will refer to these regions of the distribution as *evolutionary attractors*. As we will see, in the next section of this chapter, the presence of this kind of attractor is a general feature of many evolutionary processes.

At this point, it is interesting to monitor the presence of different genes during the evolutionary process. For example, the gene that expresses glider (103) in uniform domain 1, appears for the first time in generation 8, rule 7, is maintained in generation 9, rule 8 and 16, and then remains stable until the last generation (see Table 4). Figure 7a. provides another way of looking at this finding. Since each gene is associated with a rule, conserving the rule conserves the gene. As a result, genes, present in the elite, pass from one generation to the next. For the other genes, with rules subject to cross-over, the process is more complicated. Genes with bases at the far ends of rules have a better chance of conservation once they have joined the elite, and a higher probability of appearing in other rules. Sometimes recombination can lead to the reappearance of a gene that has previous disappeared. Obviously, genes with a relatively small number of bases have a better chance of being conserved than genes with a higher number. It is clear that the transmission of genes from one generation to the next is a highly complex process (see Figure 7).

On the basis of these results, we can say that genes correspond to specific features in the phenotype and that the way they are expressed depends on the way the CA uses genetic networks transmitted from generation to generation. From what we have seen, we know that we can investigate the underlying genetic mechanisms by analyzing the genetic tree and the evolutionary relationships between rules, seeing the forms of expression that the GA tends to select, and identifying networks that are highly prevalent or highly conserved during the evolutionary process. This analysis leads directly to an analysis of individual rules.

Figure 6. Distribution of $\bar{\lambda}_0$ for all rules for all generations. The next to last image shows the distribution in the last generation. The last image shows the same data on a finer scale. In this image, evolutionary attractors are clearly visible.

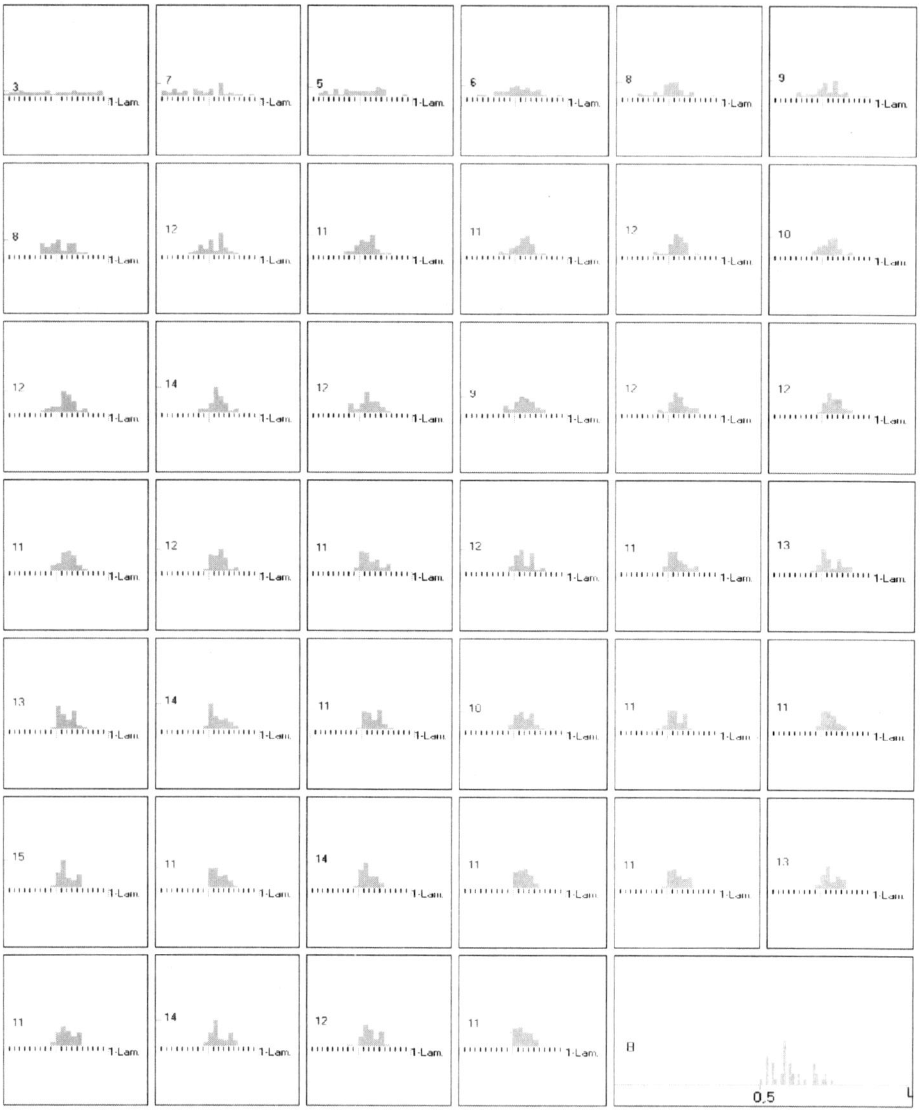

In reality, if we take a careful look at genetic networks and their composition in terms of bases (elementary rules) we discover that several elementary rules participate in different phenotypes. As in biology, there is no one to one correspondence of genes to function. Some phenotypical traits are mono-factorial, some are multi-

Table 4. Presence of the gene for glider (103) in uniform domain 1 in various rules. The gene appears for the first time in rule 7 in generation 8 and is conserved until the end of the evolutionary process.

Generation	Rules	Generation	Rules	Generation	Rules
8	7	9	8, 16	10	10, 17
11	10, 25	12	12, 16	13	3, 14, 29
14	3, 26	15	3	16	3
17	1, 20	18	1	19	1, 28, 37
20	1	21	1	22	1
23	1	24	1	25	1
26	1	27	1	28	1
29	1	30	1	31	1, 21
32	1, 38	33	1	34	1, 24
35	1, 37	36	1	37	1, 28
38	1	39	1, 20, 24		

Figure 7. This figure depicts the presence of different genes during the evolutionary process. a) the gene for glider (103), in uniform domain 1; b) the gene for glider (0002) in uniform domain 1; c) the gene of regular domain (302302); d) the gene of periodic domain (2-3); e) the gene of uniform domain (3); f) the gene for the rest state (0).

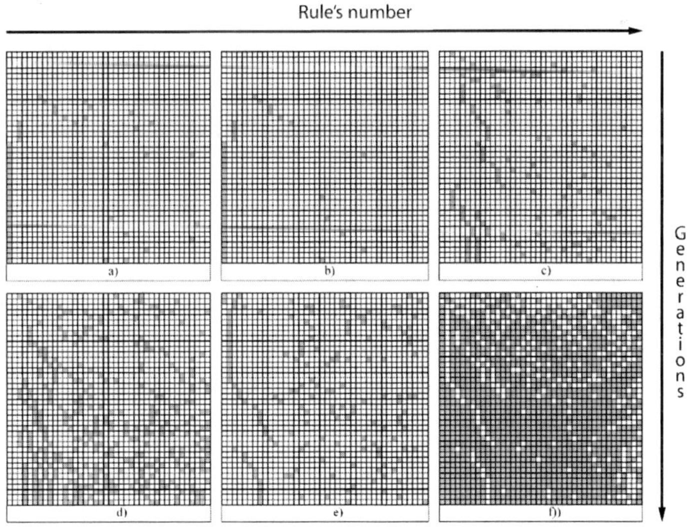

factorial. If we can identify the mono-factorial traits, we can identify the full set of genes in the genome. This allows us to analyze:

- Multi-factorial traits;
- Genetic networks;
- The complete phenotypes partially represented by patterns;
- The "genetic design" inherent in every genome.

Here we have an analogy with some of the most important problems in biology. In biology, however, they are too complicated for a direct attack. Furthermore, we have no model allowing us to approach them in a unitary manner. In this setting, mathematical models of CAs can provide useful ideas and insights.

One of the main limitations of the GA we have used in the work presented here, is that our evolutionary mechanisms have to take account of the way the genome is expressed phenotypically.

Multi Runs

In the previous section, we considered a single run of the evolutionary process. In what follows, we will analyze multiple runs. (Figure 8) shows the results of 6 evolutionary simulations with $k = 4$, $d = 1$, and $r = 1$. From (Figure 8a), we see that the basic trend is similar in each process. Each process begins with low levels of maximum fitness, followed by sudden jumps. The jumps are followed by periods of stability leading to new jumps. We also observe that several good rules are selected at a relatively early stage. The best result is for experiment 2: in which the initial maximum fitness was no higher than in the other experiments. This result suggests that the characteristics of the initial population have relatively little influence on its evolutionary future. (Figure 8b) shows the mean fitness trend for each experiment. We observe how the initial fitness is low and grows over time; in the case of experiment 6 we actually observe a four-fold increase. The growth in mean fitness is tightly correlated with maximum fitness. However, the experiments with the highest mean do not necessarily have the highest maximum value.

It is very interesting to observe the distribution of the $\overline{\lambda_0}$ s in the individual experiments. The three central rows of Figure 8 show the patterns produced by the best rule in each experiment and the distribution of the $\overline{\lambda_0}$ s in the last generation. (Table 5) shows the λ values for the best rules for each experiment. Although the

Figure 8. The first two rows (a and b) shows trends in maximum and mean fitness for six applications of GAs to the development of CAs with k = 4, d = 1, and r = 1. The following 3 rows (c-n) show the spatio-temporal patterns generated by the fittest rule in each experiment and the distribution of $\bar{\lambda}_0$ s for all rules in the last generation of each experiment (c, d, e, f, g, h). The last row shows the distribution of $\bar{\lambda}_0$ s for sets of experiments. From the left: 6 experiments with k = 4, r = 1 (o); 5 experiments with k = 4, r = 2 (p); 5 experiments with k = 5, r = 1 (q); 6 experiments with k = 6, r = 1 (r).

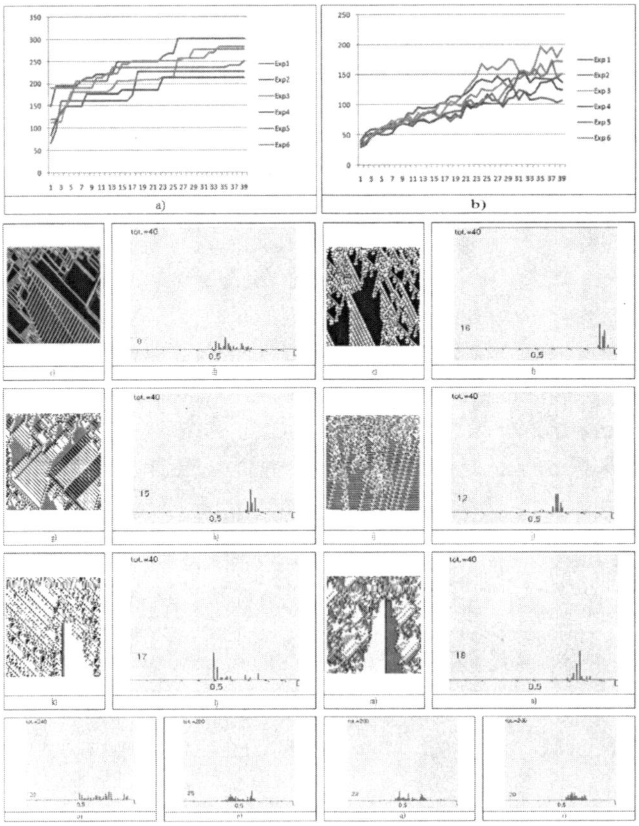

values are distributed between a minimum value of 0.0625 in experiment 1 and a maximum of 0.546875 in experiment 5, it is clear from the patterns that these are complex rules. This is further evidence that there is no such thing as an edge of chaos where all complex rules reside.

What is more interesting is an analysis of the distribution of $\bar{\lambda}_0$ s in the last generation. Here too, as in the previous example, we observe fine structure. In experiment 2, for example, the distribution is concentrated to the right of the diagram where λ values are low. In this case too, the rules are concentrated in certain restricted regions. 16 rules have values of $\bar{\lambda}_0$ s in the range 0.9-0.91; 8 rules lie in the interval 0.92-0.93, 12 in the interval 0.93-0.94. In experiment 3, the $\bar{\lambda}_0$ s are all distributed in the interval 0.7-0.8 and many lie in smaller regions. 15 rules lie in the interval 0.73-0.74, 9 in the interval 0.76-0.77, and 7 in the interval 0.71-0.72. In experiment 4, the $\bar{\lambda}_0$ s are concentrated in the range 0.6-0.7 even though a few isolated rules have values in other regions. The main attractors are the intervals 0.64-0.65 which contains 12 rules, 0.65-0.66 which has 12 rules and 0.67-0.68 which has 6. In experiment 5, the $\bar{\lambda}_0$ s are mainly concentrated around a value of 0.5, through a few are distributed across a wide region. The main attractors are the intervals 0.49-0.5 with 17 rules and 0.51-0.52 with 8 rules. In the last experiment, the $\bar{\lambda}_0$ s are distributed across the interval 0.65-0.85, with a strong concentration in the intervals 0.78-0.79, which contains18 rules and 0.76-0.77 which has10. The presence of these concentrations in specific regions demonstrates the presence of what we have called *evolutionary attractors.*.

(Figure 8) presents data from various experiments using different values of k and r. The last row clearly shows the presence of large numbers of evolutionary attractors: sometimes more than one in a single experiment. The left most image (o) shows 6 experiments, described earlier in this chapter, with $k = 4$, $r = 1$. The highest number of rules (23) corresponds to values of $\bar{\lambda}_0$ in the range 0.76-0.77. The next image (p) shows 5 experiments with $k = 4$, $r = 2$. Here the highest number of rules (25) is for values of $\bar{\lambda}_0$ in the range 0.62-0.63. The following image (q) shows 5 experiments with $k = 5$, $r = 1$; the highest number of rules (22) is for $\bar{\lambda}_0$ s between 0.47 and 0.48. The last image (r) depicts 5 experiments with $k = 6$, $r = 1$: the highest number of rules (20) is for $\bar{\lambda}_0$ between 0.59 and 0.6.

Table 5. λ values for the fittest rule from each experiment

λ	Exp 1	Exp 2	Exp 3	Exp4	Exp5	Exp6
λ0	0.359375	0.0625	0.265625	0.546875	0.484375	0.21875
λ1	0.328125	0.296875	0.265625	0.140625	0.234375	0.25
λ2	0.15625	0.328125	0.3125	0.125	0.125	0.28125
λ3	0.15625	0.3125	0.15625	0.1875	0.15625	0.25

The concept of evolutionary attractors is useful for understanding the way the Genetic Algorithm works. It also allows us to look for genetic attractors in experiments. The existence of these attractors shows that GAs tend to select specific genetic traits that always win out over the long period. However, there is no evidence for "global" attractors present in all experiments. In other words, our experiments provide no support for the hypothesis of an "edge of chaos": even in a modified form. It is thus unlikely that that we can use the concept to structure the rule space.

The concept of evolutionary attractors raises a number of interesting questions:

- What are the manifestations of an evolutionary attractor?
- Is it possible to characterize them, quantitatively and qualitatively?
- What is the relationship between evolutionary attractors and the parameters of the rule space?
- Is it possible to identify universal traits that consistently drive evolution in a pre-determined direction?

Comparing Different Evolutionary Algorithms

It is possible to conceive GAs whose fitness functions are based on some of the other metrics presented in Chapter 5. (Figure 9) shows rules obtained using fitness functions based on site-entropy and population entropy. As can be seen from the images on the first row of (Figure 9), fitness functions based on site-entropy find few complex rules. Although the patterns they produce may be very interesting, and include complex regular domains, these fitness functions mainly produce rules belonging to Class II. They also produce much less variety in rules than fitness functions based on Input Entropy.

By contrast, fitness functions based on population entropy are a useful tool for finding complex rules - as we can see in the second row of (Figure 9). Nonetheless they are less effective than functions based on Input Entropy and produce less varied rules. Input Entropy provides the best basis for an effective fitness function.

FUTURE TRENDS

In this chapter, we have seen that GAs can be an effective tool for finding CAs with complex behavior. In the next chapters, we will examine the use of GAs in greater detail. In chapter 7, we will describe the use of GAs to find complex rules for 2-dimensional CAs. In future work we plan to:

1. Further develop the concept of an "edge of chaos", applying the same genetic approach adopted in this chapter;
2. Test other metrics for the evolution of complex CAs;
3. Make comparative assessments of the effectiveness of different GAs;
4. Manipulate genetic operators to guide the GA in specific directions, avoiding the destruction of emergent structures by cross-over and mutation, without preventing the emergence of new structures;
5. Investigate the possibility that the rule space for multistate CAs contains multiple edges of chaos; seek to visualize these multidimensional spaces.

CONCLUSION

In this chapter, we have used GAs to evolve complex rules for multistate CAs. The most important component of the GA is the fitness function. It is this function that describes the level of adaptation of individuals during the evolutionary process. We have seen that the variance of IE-one of the indexes we introduced in Chapter 5-provides an effective measure of fitness. The study of the evolutionary processes

Figure 9. Rules from a GA using site-entropy (row 1, a, b, c, d, e) and population entropy (row 2, f, g, h, i, j). The patterns were obtained from CAs with k = 5, r = 1, (a and b), k = 6, r = 1 (c and d), and k = 7, r = 1 (e); r = 1 and k = 5, 6, 7, 8 (f,g,h and i). The last image (j) on the second row was obtained from CA with k = 3 and r = 2. The rules are described in the Appendix of this chapter with reference codes C6.1-C612

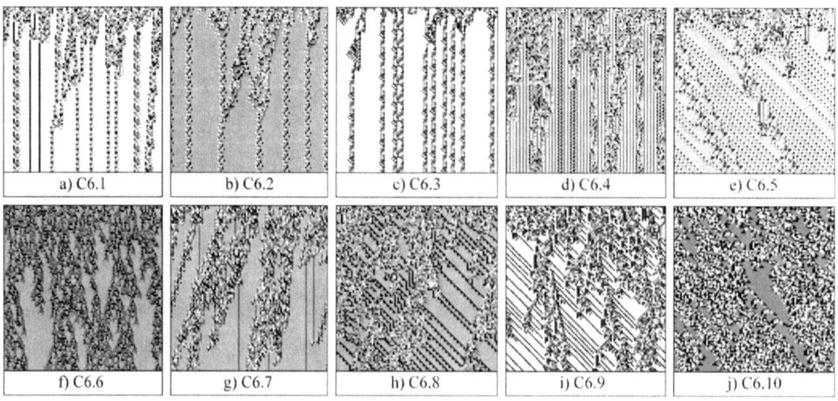

generated by these GAs has allowed us to identify zones in rule space that represent complex rules. We found these rules, which Bilotta et al (2003) have called *evolutionary attractors*, in all our experiments. However their characteristics are far from homogeneous. In each individual evolutionary process, complex rules occupy different regions of rule space. At the phenotypical level, this lack of homogeneity is reflected in the extraordinary complexity of the patterns we discovered. This reveals the multidimensional nature of rule space. If we consider the rules of a CA as its genotype, we can treat families of rules as a gene pool and examine the evolution of the gene pool from generation to generation. This approach allows us to investigate how complexity emerges from the conservation and destruction of genes, where each gene corresponds to an elementary particle such as a glider or a regular domain. We will use this method again in Chapter 11, when we investigate problems related to pattern formation.

REFERENCES

Amoroso, S., & Patt, Y. N. (1972). Decision Procedures for Surjectivity and Injectivity of Parallel Maps for Tessellation Structures. *Journal of Computer and System Sciences*, *6*(5), 448–464.

Barbe, A. M. (1990). A cellular automaton ruled by an eccentric conservation law . In Gutowitz, H. A. (Ed.), *Cellular Automata: theory and experiment*. Los Alamos, NM: MIT Press.

Bilotta, E., Di Blasi, G., Pantano, P., & Stranges, F. (2007d). A Gallery of Chua Attractors, Part IV. *International Journal of Bifurcation and Chaos in Applied Sciences and Engineering*, *17*(4), 1017–1078. doi:10.1142/S0218127407017665

Bilotta, E., Di Blasi, G., Pantano, P., & Stranges, F. (2007e). A Gallery of Chua Attractors, Part V. *International Journal of Bifurcation and Chaos in Applied Sciences and Engineering*, *17*(5), 1383–1511. doi:10.1142/S0218127407018099

Bilotta, E., Di Blasi, G., Pantano, P., & Stranges, F. (2007f). A Gallery of Chua Attractors, Part VI. *International Journal of Bifurcation and Chaos in Applied Sciences and Engineering*, *17*(6), 1801–1910. doi:10.1142/S0218127407018105

Bilotta, E., Lafusa, A., & Pantano, P. (2003b). Searching for complex CA rules with Gas. *Complexity*, *8*(3), 56–67. doi:10.1002/cplx.10063

Bilotta, E., & Pantano, P. (2005). Emergent Patterning phenomena in 2D Cellular automata. *Artificial Life Journal*, *11*(3), 339–362. doi:10.1162/1064546054407167

Bilotta, E., & Pantano, P. (2008). Cellular Non-Linear Networks as a new paradigm for Evolutionary Robotics . In Iba, H. (Ed.), *Frontiers in Evolutionary Robotics*. Wien, Austria: I-Tech Education & Publishing.

Bilotta, E., Pantano, P., & Stranges, F. (2007a). A Gallery of Chua Attractors, Part I. *International Journal of Bifurcation and Chaos in Applied Sciences and Engineering, 17*(1), 1–60. doi:10.1142/S0218127407017161

Bilotta, E., Pantano, P., & Stranges, F. (2007b). A Gallery of Chua Attractors, Part II. *International Journal of Bifurcation and Chaos in Applied Sciences and Engineering, 17*(2), 293–380. doi:10.1142/S0218127407017343

Bilotta, E., Pantano, P., & Stranges, F. (2007c). A Gallery of Chua Attractors, Part III. *International Journal of Bifurcation and Chaos in Applied Sciences and Engineering, 17*(3), 657–734. doi:10.1142/S0218127407017537

Bilotta, E., Pantano, P., & Talarico, V. (2000). *Synthetic harmonies: an approach to musical semiosis by means of cellular automata.* Paper presented at Artificial Life VII. Cambridge, MA: The MIT Press.

Bilotta, E., Pantano, P., & Talarico, V. (2002). Evolutionary Music and Fitness Function . In Anile, A. M., Capasso, V., & Greco, A. (Eds.), *Progress in Industrial Mathematics at ECMI 2000* (*Vol. 1*, pp. 118–125). Heidelberg, Germany: Springer-Verlag.

Capcarrere, M. (2002). *Cellular automata and other cellular systems.* Lausanne: Ph.D. thesis, EPF.

Chua, L. O. (Ed.). (2007). *A Nonlinear Dynamics Perspective of Wolfram's New Kind of Science* (*Vol. 2*). Hackensack, NJ: World Scientific Publishing Company.

Chua, L. O. (Ed.). (2009). *A Nonlinear Dynamics Perspective of Wolfram's New Kind of Science* (*Vol. 3*). Hackensack, NJ: World Scientific Publishing Company.

Crutchfield, J. P., & Mitchell, M. (1995). The evolution of emergent computation. *National Academy of Sciences USA, 92*(23), 10742–10746. doi:10.1073/pnas.92.23.10742

Crutchfield, J. P., Mitchell, M., & Das, R. (2003). The evolutionary design of collective computation in cellular automata . In Crutchfield, J. P., & Schuster, P. K. (Eds.), *Evolutionary Dynamics-Exploring the Interplay of Selection, Neutrality, Accident, and Function* (pp. 361–411). New York: Oxford University Press.

Das, R., Crutchfield, J. P., Mitchell, M., & Hanson, J. E. (1995). Evolving globally synchronized cellular automata. In L. J. Eshelman (Ed.), *Proceedings of the Sixth International Conference on Genetic Algorithms*. San Mateo, CA: Morgan Kaufmann.

Das, R., Mitchell, M., & Crutchfield, J. P. (1994). A genetic algorithm discovers particle-based computation in cellular automata . In Davidor, Y., Schwefel, H. P., & Manner, R. (Eds.), *Parallel Problem Solving from Nature-PPSN III* (pp. 344–353). Berlin: Springer-Verlag.

Gardner, M. (1983). *Wheels, Life and other Mathematical Amusements*. San Francisco: W. H. Freeman.

Gutowitz, H. A. (Ed.). (1990). *Cellular Automata: theory and experiment*. Los Alamos, NM: MIT Press.

Holland, J. (1975). Adaptation . In *Natural and Artificial Systems*. Ann Arbor, MI: The University of Michigan Press.

Jen, E. (1990). Aperiodicity in one-dimensional cellular automata. *Physica D. Nonlinear Phenomena, 45*, 3–18. doi:10.1016/0167-2789(90)90169-P

Kaplan, D. T., & Glass, L. (1995). *Understanding Nonlinear Dynamics*. New York: Springer-Verlag.

Kari, J. (1990). Reversibility of 2D cellular automata is undecidable. *Physica D. Nonlinear Phenomena, 45*(1-3), 386–395. doi:10.1016/0167-2789(90)90195-U

Kari, J. (1994). Reversibility and surjectivity problems of cellular automata. *Journal of Computer and System Sciences, 48*(1), 149–182. doi:10.1016/S0022-0000(05)80025-X

Koza, J. (1992). *Genetic Programming: On the Programming of Computers by Means of Natural Selection*. Cambridge, MA: The MIT Press.

Koza, J. (1994). *Genetic Programming II: Automatic Discovery of Reusable Programs*. Cambridge, MA: The MIT Press.

Kurka, P. (1997). Languages, Equicontinuity and Attractors in Cellular Automata. *Ergodic Theory and Dynamical Systems, 17*, 417–433. doi:10.1017/S014338579706985X

Langton, C. G. (1990). Computation at the edge of chaos. *Physica, 42D*, 2–37.

Li, W., Packard, N. H., & Langton, C. G. (1990). Transition phenomena in cellular automata rule space. *Physica D. Nonlinear Phenomena, 45*, 77–94. doi:10.1016/0167-2789(90)90175-O

Lohn, J. D., & Reggia, J. A. (1995). Discovery of self-replicating structures using a genetic algorithm. *Paper presented at IEEE International Conference on Evolutionary Computation (ICEC'95)*, Perth, Australia.

Lohn, J. D., & Reggia, J. A. (1997). Automatic discovery of self-replicating structures in cellular automata. *IEEE Transactions on Evolutionary Computation, 1*(3), 165–178. doi:10.1109/4235.661547

McIntosh, H. V. (1990). Wolfram's class IV automata and a good Life . In Gutowitz, H. A. (Ed.), *Cellular Automata: theory and experiment*. Los Alamos, NM: MIT Press.

Mitchell, M., Crutchfield, J. P., & Hraber, P. T. (1994). Dynamics, Computation, and the "Edge of Chaos": A Re-examination . In Cowan, G., Pines, D., & Melzner, D. (Eds.), *Complexity: Metaphors, Models, and Reality* (*Vol. 19*, pp. 497–513). Reading, MA: Addison-Wesley.

Mitchell, M., Hraber, P., & Crutchfield, J. P. (1993). Revisiting the Edge of Chaos: Evolving Cellular Automata to Perform Computational Tasks. *Complex Systems*, *7*, 89–130.

Nolfi, S., & Floreano, D. (2000). *Evolutionary Robotics – The Biology, Intelligence, and Technology of Self-Organizing Machines*. Cambridge, MA: The MIT Press.

Packard, N. H. (1988). Adaptation toward the edge of chaos . In Kelso, J. A. S., Mandell, A. J., & Shlesinger, M. F. (Eds.), *Dynamic Patterns in Complex Systems* (pp. 293–301). Singapore: World Scientific Publishing Company.

Richards, F. C., Meyer, T. P., & Packard, N. H. (1990). Extracting Cellular Automaton Rules Directly from Experimental Data. *Physica D. Nonlinear Phenomena, 45*, 189–202. doi:10.1016/0167-2789(90)90182-O

Romero, J., & Machado, P. (2008). *The Art of Artificial Evolution: A Handbook on Evolutionary Art and Music*. Berlin: Springer Verlag.

Smith, A. R. III. (1971). Simple Computation-Universal Cellular Spaces. [JACM]. *Journal of the Association for Computing Machinery, 18*(3), 339–353.

Umeo, H., Maeda, M., & Fujiwara, N. (2002). An Efficient Mapping Scheme for Embedding Any One-Dimensional Firing Squad Synchronization Algorithm onto Two-Dimensional Arrays . In *Lecture Notes in Computer Science* (*Vol. 2493*, pp. 69–81). Berlin, Heidelberg: Springer Verlag.

Voorhees, B. (1990). Nearest neighbor cellular automata over Z_2 with periodic boundary conditions . In Gutowitz, H. A. (Ed.), *Cellular Automata: theory and experiment*. Los Alamos, NM: MIT Press.

Walker, C. C. (1990). Attractor dominance patterns in sparsely connected boolean nets. *Physica D. Nonlinear Phenomena*, *45*, 441–451. doi:10.1016/0167-2789(90)90201-Y

Wolfram, S. (1984). Universality and Complexity in Cellular Automata. *Physica*, *10D*, 1–35.

Wolfram, S. (Ed.). (1994). *Cellular Automata and Complexity: collected papers*. Reading, MA: Addison-Wesley.

Wolfram, S. (2002). *A New Kind of Science*. Champaign, IL: Wolfram Media.

Wootters, W. K., & Langton, C. G. (1990). Is there a sharp phase transition for deterministic cellular automata? In Gutowitz, H. A. (Ed.), *Cellular Automata: theory and experiment*. Los Alamos, NM: MIT Press.

Wuensche, A. (1999). Classifying Cellular Automata Automatically: Finding gliders, filtering and relating space-time patterns, attractors basins and the Z parameter. *Complexity*, *4*(3), 47–66. doi:10.1002/(SICI)1099-0526(199901/02)4:3<47::AID-CPLX9>3.0.CO;2-V

Wuensche, A., & Lesser, M. (1992). *The Global Dynamics of Cellular Automata: An Atlas of Basin of Attraction Fields of One-Dimensional Cellular Automata*. Reading, MA: Addison-Wesley.

Zwick, M., & Shu, H. (1995). Set-Theoretic Reconstructability of Elementary Cellular Automata. *Advances in Systems Science and Applications*, *1*, 31–36.

APPENDIX

Pseudo Code of GA

```
/* Pseudo code of a genetic algorithm applied to CA */
/* method that evaluates fitness function.
 * the parameters are three:
 * the Cellular Automata (ca);
 * the number of CA's simulation steps (steps);
 * the number of CA's simulation steps to skip before
start evaluating the fitness value (ignore);
 */
 double evaluateFitness( CA ca, int ignore, int
steps);
 /* this method return a random ruled CA.*/
 CA getRandomCA(int states, int radius, int width, int
height);

 /*this method extract a CA from the list randomly.
   CA are selected from the elite ( first "eliteCount"
elements in the list)
 */
 CA extractRandomCA( List<CA> list, int eliteCount );

 /* Crossover parent1 and parent2 and returns child1
as child2*/
 void crossover( CA parent1, CA parent2, CA child1, CA
child2);

 /* mutate the ca with "rate" rate.*/
 void mutate ( CA ca, double rate );


 void main(){
//CA setup
int statesNumber  = 4;
int neighborhoodRadius = 1;
int sizeX = 40;
int sizeY = 40;
```

```
int ignoreSteps = 50;
int simulationSteps = 350;
//GA setup
int generations = 20;
int population = 40;
int seeds = 1;
double elite  = 50.0; //% ratedouble mutationRate =
2.0;//% ratedouble crossoverRate = 10.0;//% rateint
eliteCount = elite * population / 100.0;
List<CA>[ generations ] populations;
for( int i = 0; i < population; i++) {
  CA ca = getRandomCA(statesNumber,neighborhoodRadius,
sizeX,sizeY);
  populations[0].add( ca );
}

 //main loop
 for( int generation = 0; generation < generations-1;
generation++ ){
  //Step 1. Evaluate Fitness
  foreach( CA ca in populations[generation])
   ca.Fitness = evaluateFitness( ca, ignore, steps);
  //Step 2. Select Elite.
  //sort current population in descending order
  populations[generation].DescentSort();

  //Step 3. Create next generation
  //we need an even population.
  int count = (population - seeds) - (population -
seeds) % 2;
  while( population[generation+1].Count < count ){
    CA parent1 =  extractRandomCA(
populations[generation], eliteCount );
    CA parent2 =  extractRandomCA(
populations[generation], eliteCount );
    CA child1,child2;

    if( random(1, 100) < crossoverRate ){
      crossover( parent1, parent2, child1, child2 );
    }else{
```

```
        child1 = parent1;
        child2 = parent2;
    }
    mutate(child1, mutationRate);
    mutate(child2, mutationRate);

    population[generation+1].add( child1 );
    population[generation+1].add( child2 );
  }
  //Step 4. Insert seeds
  for( int i = 0; i <seeds; i++)     {
    CA ca = getRandomCA(statesNumber,neighborhoodRadiu
s,sizeX,sizeY);
    population[generation+1].add( ca );
  }

  //Step 5. Add other needed CA.
  while( population[generation+1].Count < population )
{
    CA ca =  extractRandomCA( populations[generation],
eliteCount );
    mutate(ca, mutationRate);
  }

   //New population is ready!
 }
}
```

Rule Codes

C6.1: $D = 1, k = 5, r = 1$

1024000021332203413444221242121422232133142144441403442213421 2223341331003303012000440420110040202000023023430204122143011404

C6.2: $D = 1, k = 5, r = 1$

2111232342042121221341213244401011041221240144004231400231413121 2333141031124033424102302401011412122432334314032213014030331

C6.3 $D = 1, k = 6, r = 1$

2211301314243003434441501433510340030534400444342441532042351140445114420435504435431043332110443001302331445244045441035343243450500433355450420021504413103434132453425121052231420104343445004045323444301013013413 04

C6.4: $D = 1, k = 6, r = 1$

0105215311322403022102111302321033220250505121204334241532531031522200434520314241302251150211001051411222112253521505535424554440101553151353351055555115035255550234505424231504551323311044132323312024541523000302 13

C6.5: $D = 1, k = 6, r = 1$

2301406555663602360205132205104313033555210000533125106652603161324024351430216500214022101552104215553252412416065003240106002423225042441335611021252501340613341066625025003553305111161402300233621302662201324525221116321122262442522244432020030565201410452330322113622400211022212322221122242405121103050055224226350420051112662152040626343

C6.6: $D = 1, k = 5, r = 1$

331044021410021100322244020313234324010432302013001400202424233411020034103200034121110040404204041343300044311312322302044 00

C6.7: $D = 1, k = 6, r = 1$

3104010003323000121300532325454133301413002132510150143150240010254223204022153244530533155214210315044402301525004504205351510402400331052252354512114314341231550103031152541013005220352100405323205404115515500353 00

C6.8: $D = 1, k = 7, r = 1$

556252056663356216540434550036660550431652423402556565405535043236246063145560445524023556465034144244354514654553535525504446554322604401162502336216546635335300031235360004016666445554434324151106341504602466330050466351455640152263462066631554206142650242340600000634024203154662503655662166063665151662354654415533004454253140055555062431 0

C6.9: $D = 1, k = 8, r = 1$

5700512523630000055341406556461051644603004334301036565200321674073024527326761272232720054445260014060573020622645676010402020022636216014137731640075133724640756532132501453257730671042010460111127246540501600352111176410755156236056130561014521203104002046234634100262210021624415223500031662735176436622417070405514005043171406226403500531046522252770461065323700710611120003120023201465650545120101740221412502310111317517031220070620054670642225607226753561036201771666214330506110272200610426476026053005

C6.10: $D = 1, k = 3, r = 2$

0102112112000122010221211201020012211201120221100221202202101202220201221020122101102112020111200221201122121021201011211110121011212101021121221120120210011021202001110211100000020112120020212212110010212021211001101001101120212112112122111102

Chapter 7
Searching for Self-Replicating Systems

INTRODUCTION

In Complexity Science (Bak, 1996; Morin, 2001; Gell-Mann, 1994; Prigogine & Stengers, 1984) and Artificial Life (Langton, 1995; Adami, 1998), almost all attempts to simulate or synthesize living systems in new media are somehow related to the influential work of John von Neumann (1966).These studies can be grouped into four basic categories (Sipper et al., 1998):

1. Attempts to build universal constructors based on von Neumann's self-replicating automaton. This work mainly dates to the 1950s and 1960s. (von Neumann 1966; Codd 1968; Vitànyi, 1973);
2. Attempts to create a minimal system capable of non-trivial self replication. This line of study began with Langton (1984) with later contributions from (Byl 1989; Reggia et al.1993; Morita and Imai 1996).
3. The enhancement of self-replicators with additional computational capabilities. This was a key topic of research in the 1990s (Tempesti 1995; Perrier, Sipper and Zahnd 1996; Chou and Reggia 1998);
4. Implementation of evolving, emergent self-replicators. This work began in the 1990s and is still in progress (Lohn and Reggia 1995 and 1997; Chou and Reggia 1997; Sayama 1998; 2000; 2004; 2006).

DOI: 10.4018/978-1-61520-787-9.ch007

The basic idea underlying all this work is that an artificial self-replicating system can be described by a logical sequence of steps or an algorithm. Until a few years ago, we were aware of only a relatively small number of such systems. Today, however, many researchers have developed highly effective methods for generating these systems. One of these methods is to use Genetic Algorithms (Mitchell et al., 1994; Mitchell, 1996). Particularly important is Wuensche's method (1999), which measures order and chaos in CA rules by measuring Shannon Entropy in the frequency of use of specific elementary rules.

The values for this distribution vary over time, following the evolutionary process. Wuensche identified characteristic features in the dynamics of his indicator and used them as the basis for an empirical classification of CA behavior. Ordered systems converge rapidly to equilibrium, after which they generate no further structural change apart from short period oscillations. As a rule, they use relatively few rules, some frequently and others very rarely. In chaotic systems, the frequency distribution is uniform. These systems have no preference for specific rules or for rules with a special ability to process information. Finally, complex systems use just a few rules, characterized by a high variance in Input Entropy. Frequencies of use of other rules are uniform.

Starting our own work from the method devised by Wuensche, we began with 2-D cellular automata, as described in Chapter 2, and used evolutionary techniques to generate a wide range of self-replicating systems. Later we came to see self replicators as living organisms that adapt to the environments they inhabit (Bilotta et al., 2003). These changes involve changes in their genomes. In our work, we discovered that some of these changes change the morphological structure of patterns and their behavior. Speaking in biological terms, we can say that some change is adaptive, while some has less adaptive value and some leads to the death of the organisms concerned. The environment selects specific traits in organisms. So individual agents, with their own specific traits, all adapt in their own specific ways, following their own evolutionary trajectories. *A priori*, these trajectories are unpredictable. So, we used Genetic Algorithms to simulate long term evolutionary processes and to analyze the relationship between organisms in the initial population and those evolved by the algorithm. To this end, we conducted a phylogenetic analysis that generated explicit, verifiable data. Our initial idea was that a group of agents with a common behavior is likely to have a common evolutionary history and that organisms belonging to the group would thus be more closely related to each other than to other organisms. In our work, we were guided by three basic assumptions. First, any group of artificial organisms must necessarily share a common ancestor. Second, evolution involves specific sequence of bifurcations. Third, changes in traits take place over time. So, once we had found a self-replicator we began to modify its genetic networks, trying to identify universals and fundamental schemas and looking at possible mutations.

Later we performed simulations in which we evolved self-replicators for many generations and analyzed their emergent evolutionary dynamics.

BACKGROUND

Chapter 6 described the application of GAs to 1-D CAs. A GA generates an initial population of strings, which is evaluated on its ability to perform a given task. Then the algorithm generates a new population, from which two individuals with high fitness values are randomly selected and used to generate, new individuals, applying cross-over and mutation. Individuals belonging to the élite from the previous generation (the individuals with the highest fitness) are carried over to the next generation without modification. The new population replaces the old population. The process is repeated many times starting with the first step. In the case of a 2-D CA, CAs belonging to a population are defined by their strings (2.10):

$$l = s_1 s_2 \ldots\ldots\ldots\ldots s_{k^{(2r+1)}}$$

The total number of possible strings is given by $k^{(k^{(2r+1)})}$. If the GA is to function, it is necessary to define the fitness function used to select the best individuals. In Chapter 5 we introduced various measures of complexity. The most suitable measure appears to be variance in Input Entropy. This result was confirmed by the experiments described in Chapter 6.

For non-Boolean CAs with $d = 2$, we need to make a number of modifications with respect to the 1-D case. As is clear from Table 8 in Chapter 2, the total number of rules is very high. The high cardinality of the rule space and the length of the strings, makes it computationally very expensive if not impossible, to apply the GA. In other words, we need a way of reducing the number of rules. One way of doing this is would be to restrict the GA to the totalistic rules or the outer totalistic rules introduced in Wolfram [1983, 1984, 2002] (see Chapter 2). We therefore adopt what seems like a reasonable compromise, and use k-totalistic rules. In this way, we significantly reduce the cardinality of the rule space, without imposing undue restrictions on its diversity. In this scheme, each individual is described by a string in the form (2.18)

$$l = \left(s_1, s_2, \ldots\ldots\ldots, s_{(k)^{N_T}} \right)$$

where

$$N_T = \binom{k+n-1}{n} = \frac{(k+n-1)!}{(n)!(k-1)!}$$

and n is the number of cells in the neighborhood.

Having resolved the problem of how to encode individual genomes, we apply genetic operators in the same way as in the 1-D case. This leaves the choice of which complexity measure to use in the fitness function open. We decided that as well as using IE, as in the experiments reported in Chapter 6, we would also use I_1 and I_2 - two additional IE- derived measures introduced in Chapter 5. Later experiments showed that these measures are as effective as IE in evolving complex 2-D k-totalistic rules.

MAIN FOCUS

The main goal of this chapter is to show that by applying evolutionary methods to a particular formalization of 2-D CAs, we can evolve systems similar to Conway's Game of Life (Gardner, 1970), and that these simulations can generate a very wide range of self-replicators. We have presented an initial taxonomy of these systems in (Bilotta & Pantano, 2006; 2008). In what follows we describe how we applied GAs to these systems. We go on to describe initial simulations of families of rules in 2D CAs, demonstrating the presence of numerous self-replicators. We then present a detailed analysis of alternative IE-based fitness functions. We describe a taxonomy of rules for self-replicators and a technique of genetic analysis which allows us to monitor specific genetic traits over a period of time. In this, as in the previous cases, we apply the same computational methods used to analyze 1-D multistate CAs. Here too we use multirun experiments to track the evolution of families of attractors.

Genetic Algorithms for 2D CAs

In the GA, we use the same parameters used in the 1-D case.

As in the earlier case, we introduce a time window *w*, which eliminates rapid variations in IE. We begin our computation of IE only after a pre-specified number of steps. In this way, we allow the CA to move beyond the transient and settle into its normal pattern of behavior for the rule.

Complex Rules and Emergent Structures

The GA just described is very effective in evolving complex rules. Figure 1 shows a number of rules evolved using variance in IE (the first and the last image a and f), the I_1 index (the 3rd image e) and the I_2 - index (the 2nd b) . Beneath each image, we provide a graph showing trends in IE, I_1 and I_2 for 60 time steps.

If we examine the patterns carefully, we observe that they all display the presence of emergent structures. We also see that the patterns corresponding to rules generated by I_1 and I_2 are less "full" than those based on IE. If examine the graphs below these patterns, we observe that IE falls during the initial phase of the evolutionary process and then slowly increases, with many oscillations around the central trend. With these rules, it is the emergent structures that determine the large-scale dynamics of the system. Given that the space is less full, they replicate with little or no interference.

If we compare the first and the second rule, we observe that the rule generated using I_2 has low maximum and mean values for IE. We also observe differences between I_1 and I_2. However, the increase is far faster for the second rule than for the first. This suggests that I_2 is more effective in isolating structures from interference. In our discussion of self-replicators in the next chapter, we will see that they are more easily generated by the second index than with other measures of complexity.

Let us now analyze the evolutionary processes for the first 3 rules.

Evolution with IE

The first rule in (Figure 2) was obtained using the following evolutionary parameters:

- States number = 4
- Neighborhood radius = 1
- Generations = 20
- Population = 40
- Size x = 40
- Size y = 40
- Steps of simulation = 350

Figure 1. A primordial soup of complex 2-dimensional CA rules generated by the GA. Each image refers to the state of the CA after 20 steps, beginning with random initial data. The graph beneath each pattern shows values for IE, I1 and I2 for 60 steps. Row 1: a) rule C7.1, b) rule C7.2. Row 2: c) values of the IE for C7.1; d) values of the IE for C7.2. Row 3: e) rule C7.3, f) rule C7.4. Row 4: g) values of the IE for C7.3, h) values of the IE for C7.4. Rules C7.1-C7.4 are reported in the Appendix of the chapter.

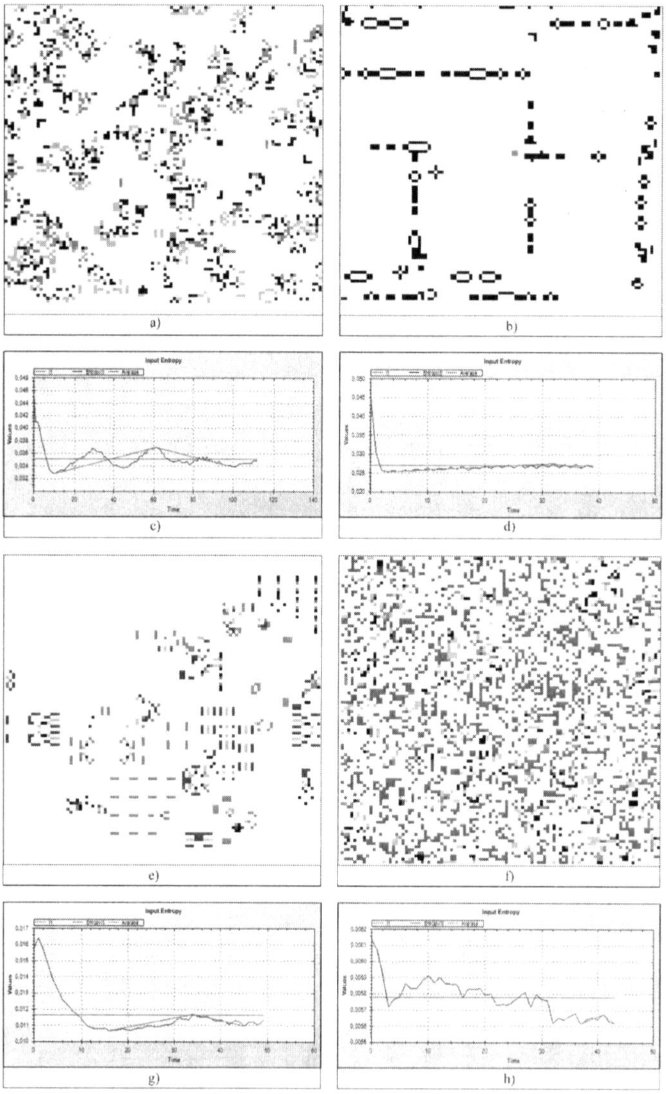

Figure 2. The first image (a) shows the dynamics of maximum, mean and elite fitness for 20 steps. The second image (b) shows a phylogenetic tree in which we observe the discovery of the best, second best and third best rules. The third image (c) shows the dynamics of lambda during the evolutionary process. The last image (d) displays the distribution of $\bar{\lambda}$ in the last generation.

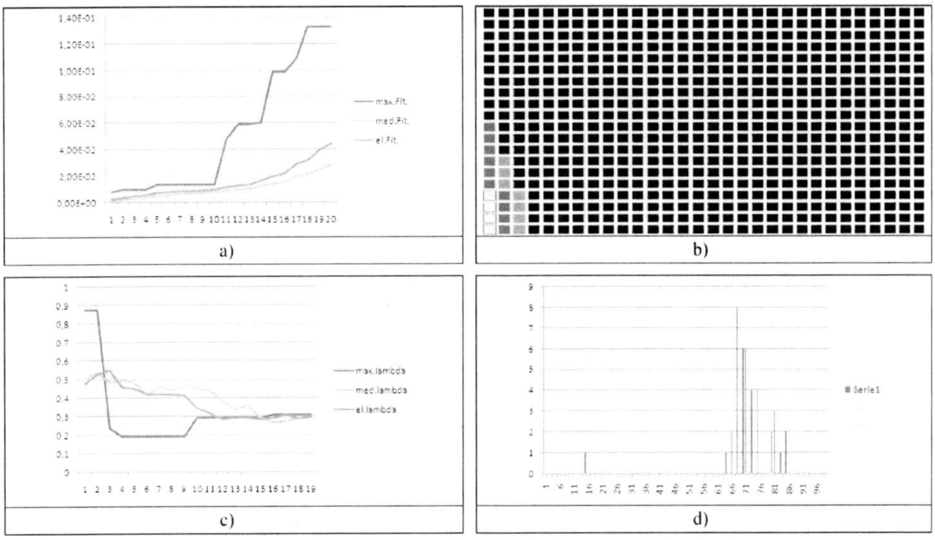

- Ignore initial steps = 50
- Seeds = 1
- Elite (%) = 50.0
- Mean mutation (%) = 2.0
- Generation mode: mutations

The image shows the fittest rule in the last generation of the evolutionary process. (Figure 2) shows the dynamics of the various fitness indicators during this process. We observe a gradual increase both in elite and mean fitness. However the discovery of the best rules obviously involves jumps. The rule with the highest fitness of all is discovered in generation 17 and is conserved until the end of the simulation (see Figure 2b). The second best rule is discovered in generation 11 and improved in generation 14. In generation 17 it is replaced by a better rule.

The dynamics of λ are also very interesting. In the third image in (Figure 2c), we observe that the mean value of λ for the whole population and for the elite both converge towards 0.3. The λ value for the best rule converges to the same value. As in the 1-dimensional case, we observe the presence of fine structure in the distribution of . This reveals the presence of evolutionary attractors (Figure 2d).

As in Chapter 6, we can investigate specific structures. For example, one self-replicator has code (00220202200). As it develops it maintains a uniform state of (0). The gene for the structure has 9 bases which combine to form the string below:

xxxxxxxxxxxxxxxxxxxxxxxxxxxxxxxxxxxxxxxxxxxxxxxxxxxxxxxxxxxxxxx
xxxxxxxxxxxxxxxxxxxxxxxxxxxxxxxxxxxxxxxxxxxxxxxxxxxxxxxxxxxxxxxxxx
xxxxxxxxxxxxxxxxxxxxxxxxxxxxxxxxxxx0xxxxxxxxxxxxxxxxxxxxxxxxxxxxxx
xxxx0xxx1xxxxx2xx0xx0x100

(Figures 3a) and (Figure 3b) show the rules used by the self-replicator and their relative positions in the dynamic network. We observe that the self-replicator is present in very many rules: 314 out of 800, 39.25% of the total (see Figure 3c).

Below we see the code for another self-replicator.

0011100
0011100
0011100

As can be seen from (Figure 3d) and (Figure 3e), this structure uses 13 elementary rules. The reference string for the rule is:

Figure 3. The first two columns show the dynamic networks for the two structures. The last column identifies rules in which both structures are present.

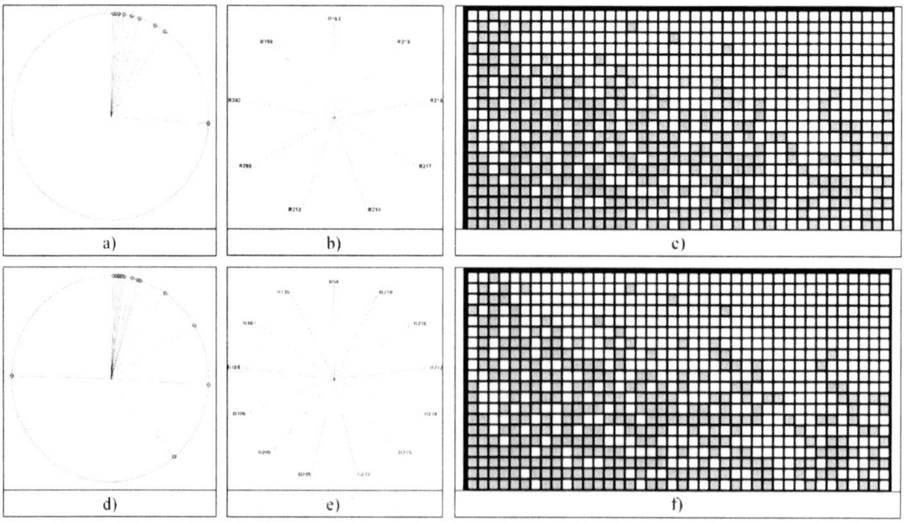

xxxxxxxxxxxxxxxxxxxxxxxxxxxxxxxxxxxxxxxxxxxxxxxxxxxxxxxxxx0xxxxxxxx
xxxxxxxxxxxxxxxxxxxxxxxxxxxxxxxxxxxxxxxxxxxxxxxxxxxxxxxxxxxxxxxxxxxxxxx
xxxxxx0xxxxxxxxxxxxxxxxxxxxxxxxxxxxxxx0xxxxxxxxxxxxxxxxxxx3xxxxxxxxxx
xxxx0xxxxxxxxx20x0xx00100

As can be seen from (Figure 3f), this self-replicator too, is present in many rules (299 out of 800: 37,375% of the total).

It is interesting to note that the two structures have many bases in common.

Evolution with I_1

The third rule in (Figure 1e) was obtained using the following parameters

- States number = 5
- Neighborhood radius = 1
- Generations = 20
- Population = 40
- Size x = 40
- Size y = 40
- Steps of simulation = 350
- Ignore initial steps = 50
- Seeds = 1
- Elite (%) = 50.0
- Mean mutation (%) = 2.0
- Generation mode: crossover + mutations

The image represents the rule with the highest fitness in the last generation. In this experiment, unlike the previous one, mean and elite fitness increase very rapidly. The fitness of the best rule in any given generation is not very distant from these values (see Figure 4a). The best rule is discovered very late (in generation

19). In the early steps of the evolutionary process, fitness is low and even individuals with relatively low fitness enter the elite. Fitness improves over the following steps. The second and third best rules are discovered in generations 14 and 13 respectively, after which fitness continues to improve (see Figure 4b). The rule that ultimately takes second place is in first place until the last generation. Mean and elite λ values converge towards a single value. The λ value for the best rule is not far away from these mean values (Figure 4c). However the convergence is an illusion. Once again, we observe fine structure in the distribution of $\bar{\lambda}$, suggesting the presence of a large number of evolutionary attractors (Figure 4d).

Let us now consider the structure corresponding to the code below, a self-replicator with a uniform state of (0).

0001000
0001000
0004000
0000000
0004000
0001000
0001000

Figure 4. The first image (a) shows the dynamics of maximum, mean and elite fitness for 20 steps. The second image (b) shows a phylogenetic tree in which we observe the discovery of the best, second best and third best rules. The third image (c) shows the dynamics of λ during the evolutionary process. The last image (d) displays the distribution of $\bar{\lambda}$ in the last generation.

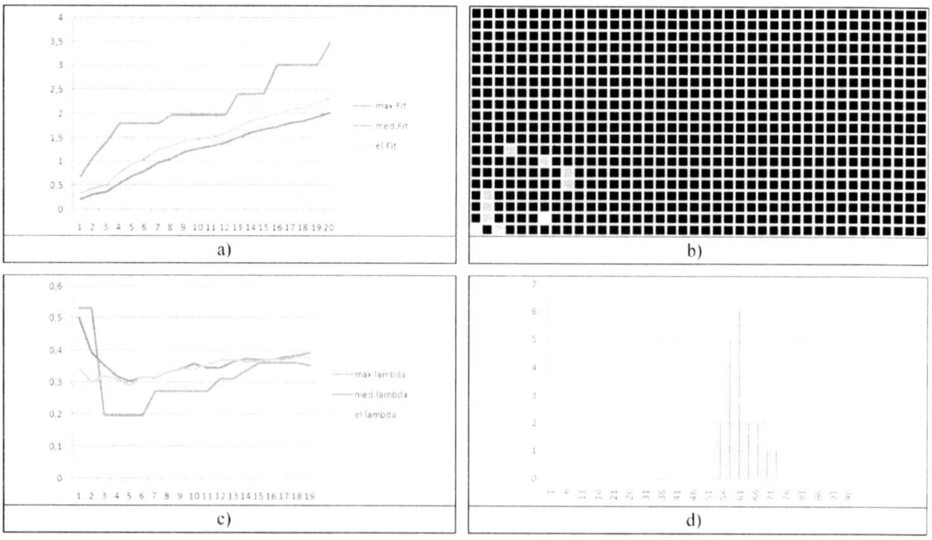

As can be seen in (Figure 5a) and (Figure 5b), the structure uses 19 elementary rules. The reference string is:

xxxxxxxxxxxxxxxxxxxxxxxxxxxxxxxxxxxxxxxxxxxxxxxxxxxxxxxxxxxxxxx
xxxxxxxxxxxxxxxxxxxxxxxxxxxxxxxxxxxxxxxxxxxxxxxxxxxxxxxxxxxxxxxxxx
xxxxxxxxxxxxxxxxxxxxxxxxxxxxxxxxxxxxxxxxxxxxxxxxxxxxxxxxxxxxxxxxxx
xxxxxxxxxxxxxxxxxxxxxxxxxxxxxxxxxxxxxxxxxxxxxxxxxxxxxxxxxxxxxxxxx
xxxxxxxxxxxxxxxxxxxxxxxxxxxxxxxxxxxxxxxxxxxxxxxxxxxxxxxxxxxxxxxxx
xxxxxxxxxxxxxxxxxxxxxxxxxxxxxxxxxxxxxxxxxxxxxxxxxxxxxxxxxxxxxxxxx
xxxxxxxxxxxxxxxxxxxxxxxxxxxxxxxxxxxxxxxxxxxxxxxxxxxxxxxxxxxxxxxxx
xxxxxxxxxxxxxxxxxxxxxxxxxxxxxxxxxxxxxxxxxxxxxxxxxxxxxxxxxxxxxxxxx
xxxxxx4xxxxx0xxxxxxxxxxxxxxxxxxxxxxxxxxxxxxxxxxxxxxxxxxxxxxxxxxxx
xxxxxxxxxxxxxxxxxxxxxxxxxxxxxxxxxxxxxxxxxxxxxxxxxxxxxxxxxxxxxxxxx
xxxxxx0xx0xxxxxxxxxxxxxxxxxxxx3xx0xxxxx1xxxxxx0xxx0x0x000xx420000

In this experiment, unlike the previous one, the two structures have only 2 rules in common: 0.25% of the total (see Figure 5c).

Evolution with I2

The second rule in (Figure 1) (code C7.2) was obtained using the following parameters:

- States number = 4

Figure 5. The first two columns show the dynamic networks for the two structures present in experiments 2 and 3. The last column identifies rules in which both structures are present.

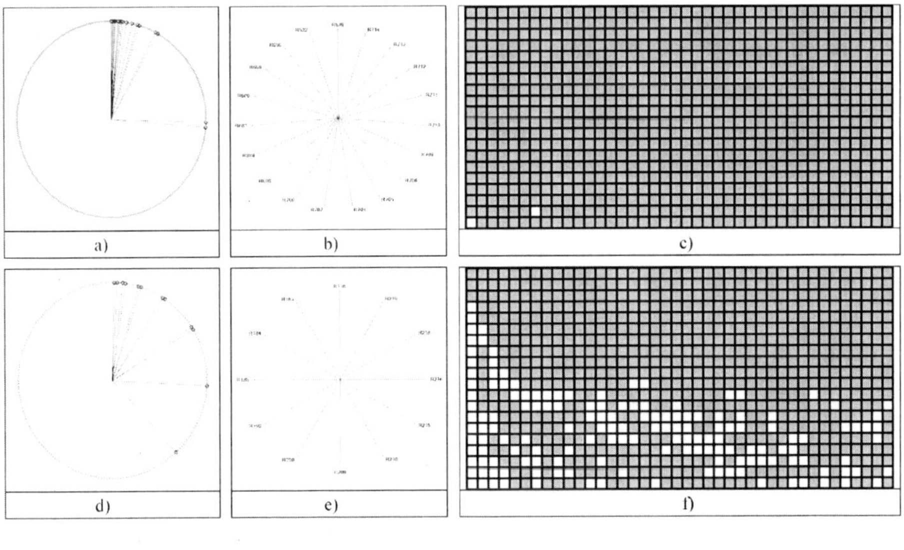

- Neighborhood radius = 1
- Generations = 20
- Population = 40
- Size x = 40
- Size y = 40
- Steps of simulation = 350
- Ignore initial steps = 50
- Seeds = 1
- Elite (%) = 50.0
- Mean mutation (%) = 2.0
- Generation mode: mutations

In this case, the fitness function is based on the I_2 index. We observe that while mean and elite IE increase only slowly (see Figure 6a), the fitness of the best rule jumps rapidly in generation 16. It is important to note that this rule is discovered early-in generation 11-after which it becomes a permanent member of the elite. In later simulations, we improve this initial value. The second and third best rules are discovered after 18 and 15 generations respectively.

In this experiment, unlike the first one, where we observe a rapid convergence between mean λ for the whole population and for the elite, the λ value for the best rule is very low and very distant from the mean value (see Figure 6c). This is further evidence that any link between CA complexity and λ is weak. Analysis of the distribution of $\overline{\lambda}$ confirms the presence of fine structure and several evolutionary attractors (see Figure 6d).

Let us now consider the structure corresponding to the code below, a self-replicator with a uniform state of (0).

00010000
01101100
01101100
00010000

Figure 6. The first image (a) shows the dynamics of maximum, mean and elite fitness for 20 steps. The second image (b) shows a phylogenetic tree in which we observe the discovery of the best, second best and third best rules. The third image (c) shows the dynamics of lambda during the evolutionary process. The last image (d) displays the distribution of $\bar{\lambda}$ in the last generation.

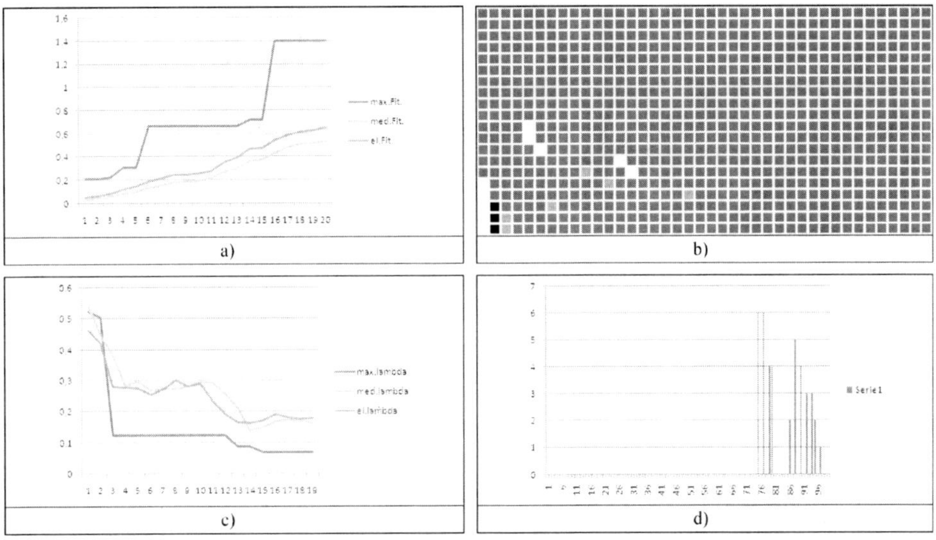

This structure uses 12 elementary rules. (Figure 5d) and (Figure 5e) shows the corresponding dynamic networks. The reference string is:

xxxxxxxxxxxxxxxxxxxxxxxxxxxxxxxxxxxxxxxxxxxxxxxxxxxxxxxxxxxxxxx
xxxxxxxxxxxxxxxxxxxxxxxxxxxxxxxxxxxxxxxxxxxxxxxxxxxxxxxxxxxxxxxxxx
xxxxxxx0xxxxxxxxxxxxxxxxxxxxxxxxxx0xxxxxxxxxxxxxxxxxxxx00xxxxxxxxx
xxxx31xxxxxxxxx01xxxx00x00

This structure is present in 154 out of 800 rules: 19.25% of the total. (Figure 5f).

Other Evolutionary Simulations

The behaviors described in the previous sections are not special cases. Rather, they exemplify the general behavior of CAs. Figures 7 through 9 show results from 9 experiments using various values of *k, r* and several different fitness functions. We observe the presence of emergent structures. During the evolutionary process, these structures appear in several different rules. Some of these rules are listed in the following chapter. The first column shows the values of *k and r*, and the fitness function used in the experiment, together with the code for the structure; the second column provides a 3-D view of the structure as it develops in time and space; the third column shows the genetic network corresponding to the structure and the number of rules used; the last column shows the CAs in which the structure is present.

Experiment 1:

In the first experiment, we identified two structures with generative code (2) and (222). The first develops as a sequence of squares; the second follow a linear growth pattern. The first structure, present in 299 CAs out of 800, uses 13 bases. It is interesting to note that the first structure uses 13 elementary rules while the second uses only 9. 8 bases are common to the two structures. As we will see in Chapter 12, this finding is a harbinger of major developments to come.

Experiment 2:

In the second experiment ($k = 4$, $r = 1$), with fitness function I_1, we identified a glider $A = \begin{pmatrix} 0220 \\ 3003 \end{pmatrix}$.

The structure is present in 368 CAs out of 800 and uses 19 elementary rules.

Experiment 3:

In this experiment ($k = 4, r = 1$), with fitness function I_2, we identified 2 self-replicators $A = \begin{pmatrix} 33 \end{pmatrix}$ and $B = \begin{pmatrix} 02120 \\ 20002 \\ 20002 \\ 02120 \end{pmatrix}$

The self-replicators used 12 and 14 elementary rules respectively. The first is present in 150 CAs out of 800, the second in just 32, a very low number. The second self-replicator appears only at a very late stage in the evolutionary process.

Experiment 4:

The self-replicator (11011) that appears in this experiment is much more complex than its predecessors. The fact that it uses 70 bases confirms this finding. It is present only in a very small number of CAs (8) and appears very late in the evolutionary process.

Experiment 5:

This experiment generates two gliders from the same initial seed. The two gliders move in opposite directions. The generation of the two gliders involves a very high

Figure 7. The figure shows the presence of several emergent structures. During the evolutionary process, the structures appear in several different rules. The first column provides a 3-D view of the structure as it develops in time and space. The second column shows the genetic network corresponding to the structure and the number of rules used. The last column shows the CAs in which the structure is present.

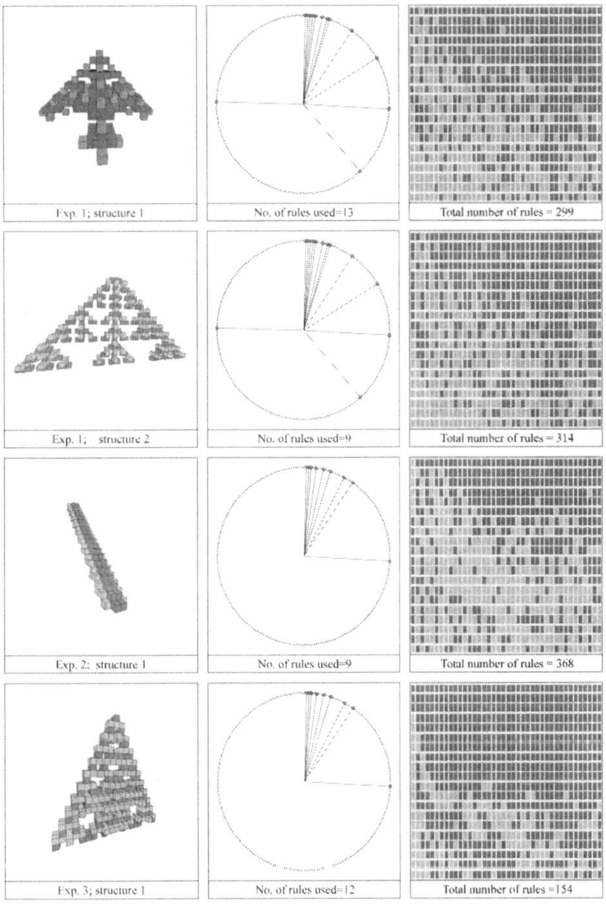

number of bases (93). The corresponding gene is present only in a limited number of CAs (6 out of 800).

Experiment 6:

In this experiment, we identified two very beautiful self-replicators. The first uses 42 bases and is present in 14 local rules, including the second best rule in the experiment. As the self-replicator develops in time and space, it displays a highly complex central structure. By contrast, the first self-replicator uses just 18 bases

Figure 8. The figure shows the presence of several emergent structures. During the evolutionary process, the structures appear in several different rules. The first column provides a 3-D view of the structure as it develops in time and space. The second column shows the genetic network corresponding to the structure and the number of rules used. The last column shows the CAs in which the structure is present.

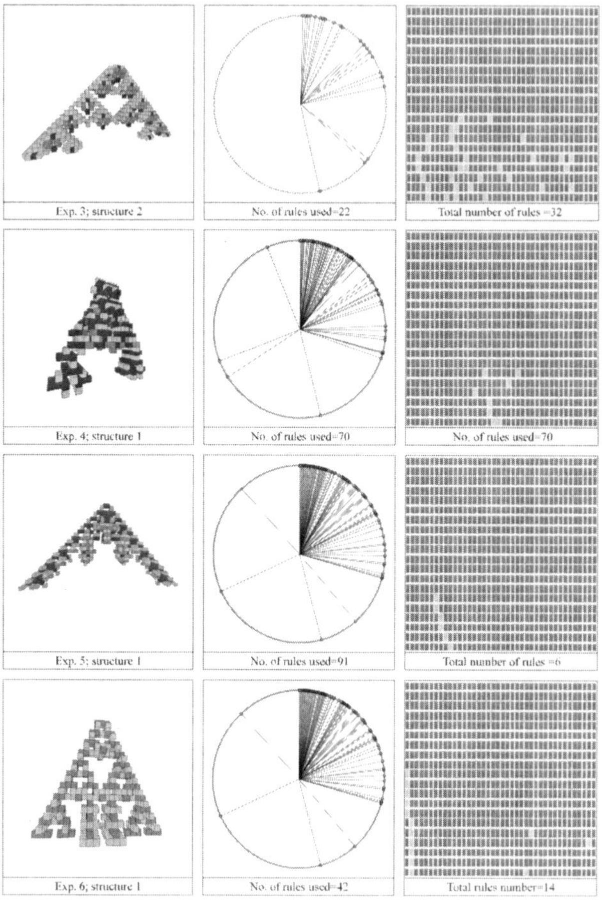

and is present in a larger number of rules, including the best rule produced by the experiment.

Experiment 7:

This experiment produced one very simple self-replicator. The structure uses 9 bases and is present in many rules (288).

Experiment 8:

This experiment generated a complex self-replicator that uses 52 bases and

Figure 9. The figure shows the presence of several emergent structures. During the evolutionary process, the structures appear in several different rules. The first column provides a 3-D view of the structure as it develops in time and space. The second column shows the genetic network corresponding to the structure and the number of rules used. The last column shows the CAs in which the structure is present.

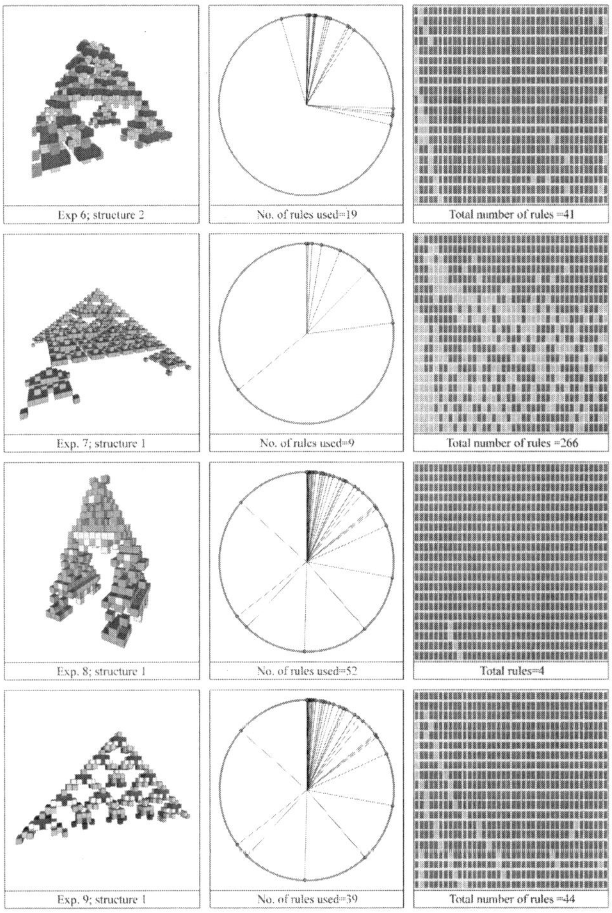

is present only in very few rules (4). The self-replicator appears at a late stage in the evolutionary process and is not capable, on its own, of making any significant contribution to the fitness of the rule.

Experiment 9:

This experiment identified a self-replicator with extremely spectacular patterning. The self-replicator uses 39 elementary rules and is present in 44 CAs, including the second best rule in the last generation. It appears very early during the evolutionary

Table 1. Bases used and number of CAs in which the structure is present

Structures	Bases	CAs
1	13	299
2	9	314
3	9	368
4	12	150
5	22	32
6	70	8
7	93	6
8	42	14
9	19	41
10	9	266
11	52	4
12	39	44

process

If we compare the number of bases used by each structure and the number of CAs in which it is present, we observe that the probability of finding a structure in a CA is much lower in structures that use a large number of bases (see Table 1).

A Few Considerations on Evolutionary Processes

In our experiments, we observe that Input Entropy produces pure self-replicators. Many are very simple and are present in nearly all rules. Each self-replicator has many variants that also appear in a large number of rules. It is as if evolution involved a struggle by self-replicators to appear in all rules in all generations. We can represent this process by observing the number of rules in which different self-replicators are present. In the experiments we have just analyzed, it is white structures, (corresponding to state 0) that prevail: especially one variant which is present in many rules in many generations. By contrast, red self-replicators seem to be rare. When they do appear, they vary very little. Self-replicators belonging to this type have a parsimonious genetic structure. Either they use just a limited number of local rules or they select their rules from a variable subset of the rules present in the genome (the CA rule). In other words, they are primitive organisms. They often appear very early in the evolutionary process, as if they were emerging directly from artificial matter. Their functional characteristics depend on a specialized sub-set of local rules. Some of these rules can be modified with no effect on the shape or behavior of the structure; others do exert an influence; in other cases any change destroys both shape and function. We also observe second order aggregations that are themselves self-replicators. In these cases, two basic self-replicators combine to form a genetic structure that remains extremely parsimonious, even if it now contains double the information present in the original structure.

It appears that the IE-based index can discover basic self-replicators that are not immediately apparent to the observer during the simulation. The index discovers specific variations in the behavior of the CA, extracting the significant information from the background noise. During the evolutionary process, the IE-based fitness function successfully identifies gene groups with high levels of prevalence and that are highly adaptive to their environment. It distinguishes these groups from pure types, eliminates sporadic variants and conserves only the winning variants: those that appear most frequently. In this way, the function splits the population of possible self-replicators into classes, creating a competition in which only the fittest survive. This fitness function does not seem to find major changes in the shape and behavior of the self-replicators.

From a functional point of view, the I_1 -based fitness function appears to detect a subset of possible variations in the phenotype. This implies it may be more selective than the IE-based function and may not detect basic self-replicators or pure types. What it does do is detect variants with a richer structure. In other words, it acts as if it were a pre-processor for this kind of structure. It is as if the fitness function considers a population of rules in which major and chaotic variants have already been eliminated. In our simulations, the patterns produced with this fitness function appear to be more orderly, even in the initial stages of the simulation. We observe for the first time the variations that identify self-replicators or gliders. It is as if the index filters out the redundancy in the primordial soup, leaving phenomena that are already semantically interesting. In other words, we find variants with specific highly structured behaviors and that use a large number of rules. In sum, the fitness function seems to capture the complication inherent even in the simplest self-replication process, extending it until it involves almost the complete gene pool for a family of self-replicators. In quantitative terms, the fitness function produces many more variants, many of which are hard to classify in terms of shape or the gene pool, since they have chaotic patterns and a very high number of rules. There is no struggle for the survival of the fittest. When we observe self-replicators and other complex multi-structured configurations, we find simultaneous growth processes at several different levels and unexpected geometrical transformations we have probably never seen before. Simple self-replicators are rare and often disappear or degenerate during the evolutionary process.

Fairly often, we find derived self-replicators (combining data from different sources). These appear to be relatively stable and relatively common. Often their key traits are inherited from their ancestors. Alternatively their function is to transfer information within the artificial environment. In this case, their functions combine self-replication with information transport. Fitness functions based on I2 generate CAs with similar characteristics to those produced by I_1.

FUTURE TRENDS

In this chapter we have seen that evolutionary processes represent an effective tool for the discovery of complex rules and that they can be used to evolve structures of differing levels of complexity. Many of these structures are self-replicators. Their presence raises a number of issues:

1. How does their presence influence the development of the CA?
2. What is the relationship between these structures and our indexes of complexity?
3. Why is it that some of these structures appear even in the earliest phases of the evolutionary process?
4. How is it that several structures can share a large number of their bases?
5. Is there any correlation between the bases underlying a structure and the probability that the structure will emerge during the evolutionary?

To attempt a reply to these questions, we need to take a closer look at the nature of these structures, and especially at self-replicators. This is what we will do in the coming chapters, in particular in Chapters 8, 10, 11 and 12.

What we will find is that there is no special correlation between the complexity of self-replication and the complexity of the rules used by self-replicators, as measured by our complexity indexes. On the contrary, self-replication seems to be an intrinsic feature of CAs. The only role of complexity is to ensure that a self-replicator can be expressed and survive inside a rule. In Chapter 11, we will see that self-replicators express special symmetries in their phenotypes; in Chapter 12, we will see that these symmetries are conserved even in the presence of mutations; in Chapter 13, we will see how self-replicators can join together to form more complex structures.

CONCLUSION

In previous attempts to design artificial self-replicating systems, we encountered many difficulties. These have now been overcome. We have invented a method that allows us to generate these systems in a process of artificial evolution. Our experiments have generated huge volumes of data. We have found self-replicators whose growth seems to be based on models very different from those we find in biology; some have the ability to change not only their shape but their underlying structure. Obviously each class of self-replicator plays a distinct role in pattern generation. In this chapter, we have presented the self-replicators we found using fitness functions based on Input Entropy and on two indexes - I_1 and I_2 we defined ourselves. Each of these indexes performs a particular kind of search of rule space; each generates a wide range of complex self-replicating behaviors, never seen in previous work. In the text, we have provided a number of indications concerning the kind of self-replicators our indexes find and the way they behave. The behaviors, shapes and functions of these emergent self-replicators differ according to which index is used in the fitness function. This suggests that different fitness functions explore different regions of rule space. Some self-replicators seem to have properties similar to those of stem cells. So long as the initial data display certain basic spatial relationships, they can reproduce any kind of pattern (see Chapter 12). These are what we have called *universal constructors.* Universal constructors could be an ideal artificial laboratory for the exploration of functional analogies with biological stem cells.

REFERENCES

Adami, C. (1998). *Introduction to Artificial Life*. New York: Springer-Verlag.

Bak, P. (1996). *How Nature Works: The Science of Self-Organized Criticality*. New York: Copernicus Press.

Bilotta, E., Lafusa, A., & Pantano, P. (2003b). Life-like self-reproducers. *Complexity*, *9*(1), 38–55. doi:10.1002/cplx.10118

Bilotta, E., & Pantano, P. (2006). Structural and Functional Growth in Self-Reproducers. *Complexity*, *11*(6), 12–29. doi:10.1002/cplx.20131

Bilotta, E., & Pantano, P. (2008). Biological traits in artificial self-reproducing systems. In Orsucci, F., & Sala, N. (Eds.), *Complexity Science in Information Communication Technologies: Interaction Between Living Systems and Information Technologies* (pp. 109–123). Hershey, PA: IGI Global.

Byl, J. (1989). Self-Reproduction in small cellular automata. *Physica D. Nonlinear Phenomena*, *34*, 295–299. doi:10.1016/0167-2789(89)90242-X

Chou, H. H., & Reggia, J. A. (1997). Emergence of self-replicating structures in a cellular automata space. *Physica D. Nonlinear Phenomena*, *110*(3-4), 252–276. doi:10.1016/S0167-2789(97)00132-2

Chou, H. H., & Reggia, J. A. (1998). Problem solving during artificial selection of self-replicating loops. *Physica D. Nonlinear Phenomena, 115*(3-4), 293–312. doi:10.1016/S0167-2789(97)00237-6

Codd, E. F. (1968). *Cellular Automata.* New York: Academic Press.

Gardner, M. (1970). Mathematical games: the fantastic combinations of John Conway's new solitaire game "Life." . *Scientific American, 223*, 120–123. doi:10.1038/scientificamerican1070-120

Gell-Mann, M. (1994). *The Quark and the Jaguar*. London: Abacus.

Langton, C. G. (1984). Self-Reproduction in Cellular Automata . *Physica, 10D*, 135–144.

Langton, C. G. (1995). *Artificial life: an overview*. Cambridge, MA: The MIT Press.

Lohn, J. D., & Reggia, J. A. (1995). *Discovery of self-replicating structures using a genetic algorithm*. Paper presented at IEEE International Conference on Evolutionary Computation (ICEC'95), Perth, Australia.

Lohn, J. D., & Reggia, J. A. (1997). Automatic discovery of self-replicating structures in cellular automata. *IEEE Transactions on Evolutionary Computation, 1*(3), 165–178. doi:10.1109/4235.661547

Mitchell, M. (1996). *An Introduction to Genetic Algorithms*. Cambridge, MA: The MIT Press.

Mitchell, M., Crutchfield, J. P., & Hraber, P. T. (1994). Dynamics, Computation, and the "Edge of Chaos": A Re-examination . In Cowan, G., Pines, D., & Melzner, D. (Eds.), *Complexity: Metaphors, Models, and Reality* (*Vol. 19*, pp. 497–513). Reading, MA: Addison-Wesley.

Morin, E. (2001). *La méthode* (*Vol. 5*). Paris: Seuil.

Morita, K., & Imai, K. (1996). Self-reproduction in a reversible cellular space. *Theoretical Computer Science, 168*, 337–366. doi:10.1016/S0304-3975(96)00083-7

Perrier, J. Y., Sipper, M., & Zahnd, J. (1996). Toward a viable, self-reproducing universal computer. *Physica D. Nonlinear Phenomena, 97*, 335–352. doi:10.1016/0167-2789(96)00091-7

Prigogine, I., & Stengers, I. (1984). *Order out of Chaos: Man's new dialogue with nature*. London: Flamingo.

Reggia, J. A., Armentrout, S. L., Chou, H. H., & Peng, Y. (1993). Simple systems that exhibit self-directed replication. *Science*, *259*, 1282–1287. doi:10.1126/science.259.5099.1282

Salzberg, C., & Sayama, H. (2004). Complex genetic evolution of artificial self-replicators in cellular automata. *Complexity*, *10*(2), 33–39. doi:10.1002/cplx.20060

Sayama, H. (1998). Introduction of Structural Dissolution into Langton's Self-reproducing Loop. In C. Adami, R. K. Belew, H. Kitano, & C.E. Taylor (Eds.), *Artificial Life VI: Proceedings of the Sixth International Conference on Artificial Life* (pp. 114-122). Cambridge, MA: MIT Press.

Sayama, H. (2000). Self-Replicating Worms that Increase Structural Complexity through Gene Transmission. In M. A. Bedau, J. S. McCaskill, N. H. Packard, & S. Rasmussen (Eds.), *Artificial Life VII: Proceedings of the Seventh International Conference* (pp. 467-476). Cambridge, MA: The MIT Press.

Sipper, M., Tempesti, G., Mange, D., & Sanchez, E. (Eds.). (1998). Artificial Life: Special Issue on Self-Replication. *Artificial Life*, *4*(3), 237–257. doi:10.1162/106454698568576

Tempesti, G. (1995). A new self-reproducing cellular automaton capable of construction and computation. In F. Morán, A. Moreno, J. J. Merelo, & P. Chacón (Eds.), *ECAL'95: Third European Conference on Artificial Life* (LNCS 929, pp. 555-563). Heidelberg, Germany: Springer-Verlag.

Vitanyi, P. M. B. (1973). Sexually reproducing cellular automata. *Mathematical Biosciences*, *18*, 23–54. doi:10.1016/0025-5564(73)90018-7

von Neumann, J. (1966). *Theory of Self-Reproducing Automata* (Burks, A. W., Ed.). Urbana, IL: University of Illinois Press.

Wolfram, S. (1983). Statistical Mechanics of Cellular Automata. *Reviews of Modern Physics*, *55*, 601–644. doi:10.1103/RevModPhys.55.601

Wolfram, S. (1984). Universality and Complexity in Cellular Automata. *Physica*, *10D*, 1–35.

Wolfram, S. (2002). *A New Kind of Science*. Champaign, IL: Wolfram Media.

Wuensche, A. (1999). Classifying Cellular Automata Automatically: Finding gliders, filtering and relating space-time patterns, attractors basins and the Z parameter. *Complexity*, *4*(3), 47–66. doi:10.1002/(SICI)1099-0526(199901/02)4:3<47::AID-CPLX9>3.0.CO;2-V.

APPENDIX

Rule Codes

C7.1: $D = 2, k = 4, r = 1$

0001210333000000123101100201312200031100000003100000100000000221
200000020000001000002010000003000101200020000000000000000300000030020
00001003100000323000300300010000000000000000221033020000300000000001
0200003313032220003000100

C7.2: $D = 2, k = 4, r = 1$

0000000000000000000000000000030000000000000000000000000000000000000
000000000100000001000000023110000020000000000000000000000000100000000
00000000000000000000000000000000000000000000000000300000000000000
002031000100000100000000

C7.3
$D = 2, k = 5, r = 1$

0002001004000200000400000400000024100000001400000000200000000030
0020100000002000200000100010040000001000000000000200200000000200401
41043040000304004200000000000000001001000000000000000400000000001001030
000001030000000002020000000000000040000001000000000200030400100000000
0020010000300000000000001000000300010000000300000000000000300000012003
00024010000040100030400140000040002442434000000001430141000030242
24312412400330300000130001010032401021042100130421222103043200033
44140210000300023300123343200403243122442120301400032442304023112
2001004413000211000240440000000021242023240002020200400010020000200
000200010423012044140010344243301101300131043043131000103000001000
00000002000000003020000330000230000003014003000000000000030420000

C7.4 $D = 2, k = 6, r = 1$

542000503420000000000510000000000000004000020400000003000000020440
000400000000000000505420000000500000500532050102000522012000000003
0010000030551000000000010300000000040000035050300350002200005003020
000200340001003143030000000425000100520100000044300100000000100001
100034025000000325000000004341100200013000053011434002400004000000
200400000032400110100000000024511250400000050300013000500020003000
000204000500000000003410000043004000000500200000000000010200300000
200000100005000400000000450002004000051340000000000500430000002400
20000003000000000502002051100000050000010000000043340000100000000000
0200000200003300000100001502000000503100000001000000300040034000030
300000300020100500300000000020000000025514400000050400142000003000
400000020405000000004000203300030000450323400004000020000000140402
530000300022000052510000000000300140040004002030000000000000500040
310000002145400302043202121542022000054420000011005530000000000100
020500003000040040002400000200004200000400005000053040000000000000
530031042000000000003303000100213400000030000000001000302000005200
000041300103232000000010022100500001000000003000000000030000330000
1153403223300000100000000000000500050000001031400120004042000100054
00013201000010000001244350300030002503000000000000000023000000104000
0100020000000000010003002003000032000000001002000501201100002020201
42153000250001030004003000001003000000000000000000400025000000520000
100502340000500000000002000415350000041200304001502000000000010000
0000000200000004015241000000003000024005030000030000320003000000040
000000004000140000010100000250040000000010200000235500100001100000
0000100130013100000000000000000045123003314415000344552401044050000
040040031132530005340000040000400000100000000300000000411000100041
505451303200000100100041545050020000000300000020000000000000400035
00000000030400040015000000500401004312005502000000003000120000000
500000434300000000005000000002040320000205000004000044253000540000
00005421300050400000000000500000000044000000005000053042501010000000
3000004150042005001000000

Chapter 8
Life-Like Self-Replicators

INTRODUCTION

The concept of a cellular automaton derives from John von Neumann's studies of the logic of life. In these studies, von Neumann focused on self-replicating structures with universal computational capabilities. Given the appropriate initial conditions, a universal computer can perform any finite computation, reproducing even the most complex biological behaviors. It is well known that the ECAs described in (Wolfram, 2002) and the Game of Life (Gardner, 1970; Evans, 2003; Adachi et al., 2008) have universal computational capabilities. It has been shown, furthermore, that certain one-dimensional CAs can generate structures that are equivalent to the components of an idealized digital computer, and that, by connecting these components in different ways, it is possible to implement any kind of algorithm. In brief, these CAs are equivalent to the better known - and simpler - Turing machine and share its ability to perform universal computation (Smith 1971).

The basic idea underlying the design of CAs as models of biological systems is that, in some cases continuous models (e.g. models based on differential equations) are unable to provide an adequate representation of the process. One example is self-replication of discrete units such as cells. In the 1940s and early 1950s, this problem led von Neumann to pose a crucial question: what kind of logical organiza-

DOI: 10.4018/978-1-61520-787-9.ch008

tion is required for an automaton to self-replicate? When he talked about automata, he was referring to artificial systems that could reproduce like living organisms. The theoretical backdrop was Turing's theory of computation and Gödel's work on the foundations of mathematics. More specifically, he was interested in Gödel's Theorem of Incompleteness, according to which "there exist arithmetical truths that, in principal, can never be proved". One of the key elements in Gödel's proof is the role of self-referential propositions such as "this sentence is false". Gödel shows that this kind of proposition gives rise to contradictions. According to Schuster & Sigmund (1983), biological self-replication is a particularly good example of self-referentiality. A genetic instruction such as "make of copy of yourself" ought to do nothing more than produce a copy of itself (self-reference). This would imply an unlimited lengthening of the genome, but not the construction of an organism. How can we escape the dichotomy between self-reference and self-replication? In von Neumann's initial *thought experiment,* in 1948, he does not imagine a fixed lattice; the components in the system are free. The model uses the genetic information available in the system in two distinct ways - as un-interpreted (syntactical) and interpreted (semantic) data. The automaton consist of two parts: a flexible structure, and a set of replication instructions that allows the replication of both parts - i.e. replication of the structure and replication of the genome that gives rise to the structure (Schuster & Sigmund, 1983). Burks, one of von Neumann's students, noted that the first process describes the dynamics of the system and referred to it as a "kinematic" model. Ulam introduced the idea of cells, suggesting that the self-replicating system could consist of discrete spatial units, distributed across a lattice. Von Neumann's first model, developed between 1952 and 1953, had 29 states. Kinematic processes were replaced by the exchange of information among neighboring cells. Von Neumann pointed out that once the universal copier was in place, the system could evolve machines of ever increasing complexity. However, what interested him was not the simulation of biological or energy processes (Aspray and Burks 87), but the logic allowing a self-replicating machine to evolve "increased complication".

Many years later, Codd and Langton drastically simplified Von Neumann's original model (Codd 1968; Langton 1984), and Vitanyi (1973) proposed a CA model of sexual reproduction. Von Neumann's work first became known to the general public when A. Burks created a collection of his works, at the end of the 1960s (von Neumann 1966) and when John Conway proposed his CA-based, Game of Life, a highly enjoyable, easily programmable metaphor for life and death, that produced beautiful simulations on the computer screen (Gardner, 1983). Meanwhile, scientific interest in self-organizing systems was growing rapidly. Many in the new disciplines of Artificial Life and Complexity Science saw CA as a paradigm for universal computation and self-organization (Varela, Maturana, and Uribe 1974). And so, over the last half century von Neumann's work has given rise to new research in a wide range of fields, from biological modeling to computer science to numerical methods (Casti 1989; Emmeche 1994; Toffoli & Norman, 1987; Wolfram 1984; 2002). Even today many scholars are picking up von Neumann's challenge (Sayama, 2004; Salzberg et al., 2006; Bilotta & Pantano, 2006; 2008). The most fundamental step forward came when Langton (1984), built a 2D, 8-state CA, and used it to prove not only that self-replicators exist but that self replication does not require a Universal Constructor. Since then, Byl (1989) and other authors (Tempesti, 1995; Chou & Reggia, 1997; 1998) have produced self-replicators that are simpler than Langton's. However, these systems are all purpose-built. In our own work, we have recently demonstrated a different route to self-replicator design. In a series of publications (Bilotta et al., 2002; 2003; Bilotta & Pantano, 2005), we show that self-replication can emerge spontaneously from GA-generated rules. We observe, furthermore, that several of these rules are not "complex" in terms of the Wolfram classification. As we will see later, these observations represent a significant step forward in understanding the complexity of CAs. In this Chapter, therefore, we will begin by provide a formal definition of self-replication. We will then proceed to survey the classical literature, focusing on the self-replicators proposed by Langton and by Byl. Finally, we will examine the large family of self-replicating structures, generated by k-totalistic rules, and offer a detailed account of self-replication in 9 new structures. In each case, we will describe the characteristic periodicity of the self-replication process, offer a description of its various phases and illustrate the huge range of behaviors to which it gives rise.

BACKGROUND

Periodicity and Phases of the Self-Replication Process

Before we consider details of self-replicators' behavior, it is useful to define this behavior in formal terms. The states **x (t)** of a $2d$ CA $2d$ with M cells, can be represented by matrix mxm, where M=mxm, and each element has a value between 0 and k-1. Consider now the initial states:

$$c(0) = \begin{pmatrix} 0 & 0 & 0 \\ 0 & A & 0 \\ 0 & 0 & 0 \end{pmatrix} \tag{8.1}$$

where A is a matrix $m'xn'$, whose elements have values between 0 and $k-1$, and all elements of c (0) that do not belong to A have a value of 0. We say that self-replication has taken place, if, after a certain time t, $c\ (t)$ contains more than one sub-matrix A and the elements separating the sub-matrices all have a value of zero. The replication period τ is the minimum time required for self-replication. During period τ, matrix A will go through several transformations. These are the phases of the self-*repli*cation process. As an example, consider the rule $k = 5, r = 1$ (code C8.1 in the Appendix of this chapter. The initial data is the same as in (8.1). $A = \begin{pmatrix} 1 & 1 & 1 \\ 1 & 1 & 1 \end{pmatrix}$ is a 2x3 sub-matrix of c(0). In the next steps of the simulation, the state of the automaton changes as shown below:

$$c(1) = \begin{pmatrix} 00.............00 \\ .... \\ 0..0010100..0 \\ 0..0134310..0 \\ 0..0134310..0 \\ 0..0010100..0 \\ .... \\ 00.............00 \end{pmatrix},\ c(2) = \begin{pmatrix} 00.............00 \\ .... \\ 0..00001000....0 \\ 0000000000000 \\ 0..101343101..0 \\ 0..101343101..0 \\ 0000000000000 \\ 0..00001000....0 \\ .... \\ 00.............00 \end{pmatrix},\ c(3) = \begin{pmatrix} 00.............00 \\ .... \\ 0..111000111..0 \\ 0..111000111..0 \\ .... \\ 00.............00 \end{pmatrix}.$$

After 3 steps, we clearly observe two new structures, similar to the initial.

$$A_2 = \begin{pmatrix} 01010 \\ 13431 \\ 13431 \\ 01010 \end{pmatrix},\ A_2 = \begin{pmatrix} 000010000 \\ 000000000 \\ 101343101 \\ 101343101 \\ 000000000 \\ 000010000 \end{pmatrix},\ A_3 = \begin{pmatrix} 111000111 \\ 111000111 \end{pmatrix}$$

As a consequence $\begin{pmatrix} 00000 \\ 0A0A0 \\ 00000 \end{pmatrix}$, The replication period τ is 3.

Langton's Self-Replicator

Let us now examine Langton's self-replicating structure. Langton's CA has 8 states (d=2 *e* k=8) and a 5 element von Neumann neighborhood consisting of the cell itself, and the cells immediately above, below, to the right and to the left of the cell. The Appendix of this Chapter provides the rules table for the system. The first column in the table shows the code for the elementary rule. The second column represents the state of the evolving automaton. The next four columns represent the state of the cells immediately above it, to its right, below it and to its left. The last column gives the state of the automaton after the next time step. Altogether there are 219 elementary rules. This is much smaller than the number (85=32768) required for an exhaustive list. This is because Langton treats rotations of the neighborhood as equivalent i.e.

$$(a_{ij},a_{i-1j},a_{ij+1},a_{i+1j},a_{ij-1}) \equiv (a_{ij},a_{ij-1},a_{i-1j},a_{ij+1},a_{i+1j}) \equiv (a_{ij},a_{i+1j},a_{ij-1},a_{i-1j},a_{ij+1}) \equiv (a_{ij},a_{ij+1},a_{i+1j},a_{ij-1},a_{i-1j}) \tag{8.2}$$

In this way, he reduces the number of rules to 857. It is important to note that Langton constructed the rules for his self-replicator by hand, starting from Codd's

Figure 1. Self-replication in Langton's Loop. (a) The basic self replicator; (b) different stages in the replication process; (c) and (d) Formation of a colony. Note that some loops have lost the ability to replicate.

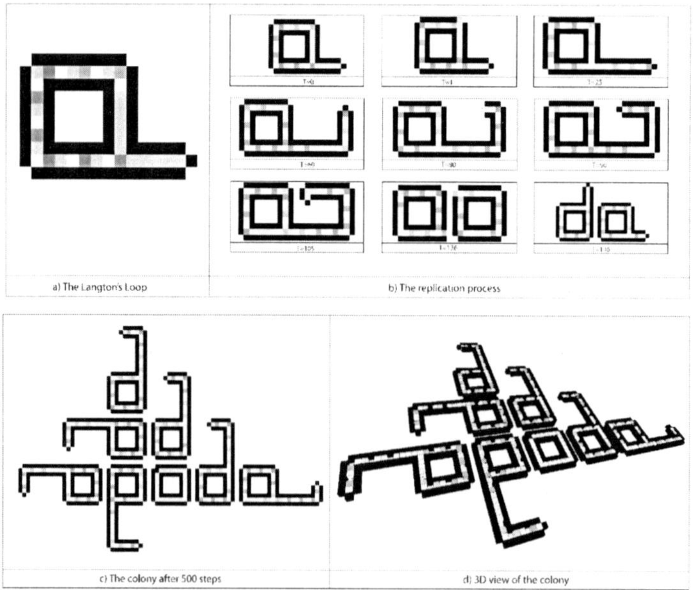

periodic emitter (Codd,1968). The codes for this self-replicator, also known as Langton's Loop, are:

022222222
2170140142
2022222202
2720000212
2120000212
2020000212
2720000212
21222222122222
207107107111112
02222222222222

The states of neighboring cells (above, below, to the right, and to the left of the cell) have values of 0. (Figure 1a) provides a graphical representation of the self-replicator; (Figure 1b) depicts the 126 step self-replication process. If we examine the process carefully, we observe that it begins when the right hand arm of the structure starts to grow. After 35 steps, it begins to fold; after 60 and 80 steps, it folds again. After 90 steps, all four phases are complete. After 146 steps, the arms of the two self-replicators reform and the process begins again: the "parent" self-replicator replicates upwards and the "child" replicates towards the right. (Figure 1c) and (Figure 1d) depict the continuation of the process and the creation of a colony. Observe that the self-replicators at the center of the colony have no space in which they can replicate. As a result they are sterile, in a situation of permanent equilibrium.

Langton's Loop initially consists of 86 automata, each with a non-zero state. The number of cells with non-zero states increases gradually with time. After 146 steps, the number doubles and continues to increase as the system develops. At the beginning of the self-replication process, the system uses 31 elementary rules. This number increases until it reaches 49 (after 146 steps). Between 150 and 500 steps, the number of rules used ceases to grow, oscillating between 40 and 50 (see Table 1). At step 500, the system uses 47 rules, at step 600 it still uses 47. After 700 steps

Table 1. Number of rules used after different numbers of steps. In the initial phases of the process the number of rules used grows. It then stabilizes, oscillating between 40 and 50.

Time	**0**	**50**	**100**	**146**	**150**	**200**	**250**	**300**	**350**	**400**
n. rules used	31	33	38	**49**	46	36	44	49	42	50

Figure 2. Row 1 (a and b) all the rules used during the first 2 and the first 10 steps of the simulation. Row 2 (c) frequency of rule use over the whole 146 steps of the self-replication process; (d) activation (in white) of specific rules over time. Row 3 (e) frequency of use of rules during the first step (31 rules used) and (f) the second (36 rules used). Row 4 (g and h) dynamic networks created by the first two steps.

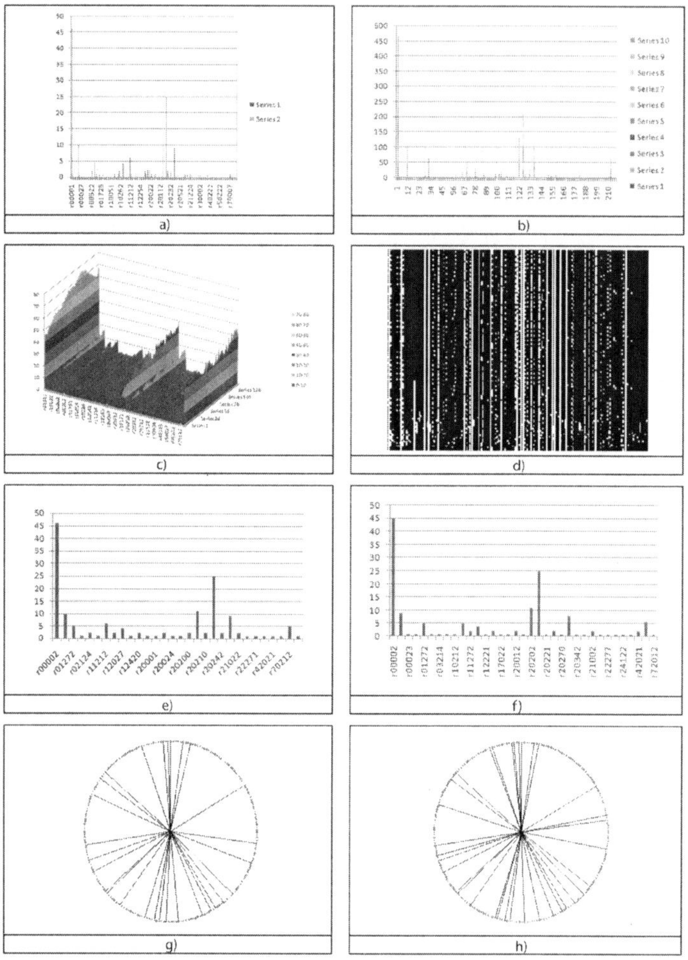

it is using 56.

In reality, Langton's rules are never used all together. Some (39) are never used; others only rarely. (Figure 2) shows the frequency with which individual rules are used, at step 2 and step 10. From the data given in the top row, it is clear that the two rules used most often are R00002 followed by R20272. Other rules are used more rarely. The second row shows the frequency with which rules are used over the whole self-replication process. We observe that as the process proceeds, R70212 is used ever more frequently, until it reaches a high level where it stabilizes. The right panel in the second row provides an overview of the occasions on which the various rules are used. Time is represented from the bottom up; the rules are listed from left to right. We observe that some rules are consulted all the time, others at irregular intervals, some very rarely. The third row shows the frequencies at which rules are consulted at step 2 and step 3. As was already clear in the previous figure, these frequencies change from one step to the next. The last row provides an alternative view of the rules. Here the graph represents the rules used at steps 1 and 2. In this way, we realize that the rules constitute a *dynamic network*, in which only certain nodes are activated at certain times.

(Figure 2) shows that the self-replication process has a characteristic period (the time taken to complete the process), a rhythm (determined by the dynamic networks that regulate the frequency with which specific rules are used at specific times) and a geometry (created during the formation of colonies). The geometry brings to mind a spiral symmetry, with a radius that increases over time, and a tendency of the colony to occupy all available empty space. As available living space falls, the self-replicators lose their ability to reproduce and become sterile. Langton's self-replicator consists of a nucleus of code, contained within an outer shell. The code contains on the one hand, the information necessary for the arm to grow and for the system to self-replicate (translation), on the other the code necessary for its

Figure 3. a) Byl's Loop ; b) The Loop after replication; c) and d) rules used after 100 and after 200 steps

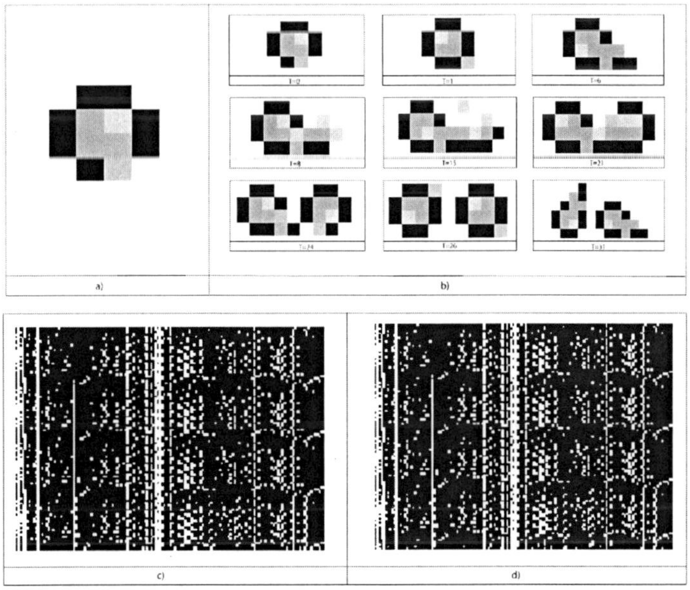

own self-replication. Internal data paths have the ability to transmit data as signal packets in which each packet contains the state of two associated units. The signals contained in the Loop include a compressed version of the instructions required for its self-replication. In this way, the Langton genome emphasizes the way the information is used: either in interpreted or non-interpreted form - equivalent to the biological processes of transcription and translation. In Langton's Loop, translation takes place at the end of the arm and depends on collisions between the signal and other signals. Transcription depends on the duplication of the signal at the crossroads between the two branches.

Designing CAs for Self-Replication

Langton's work gave rise to one line of research that focused on the search for new self-replicating structures and another that led to the discovery of simpler ones. For example, Byl (1989) discovered a self-replicating loop in a 2D CA with $k = 5$ and a Moore neighborhood. This loop had 12 rather than 86 cells, and a period of 24 steps. For the code of the loop, see (Figure 3). The rules are given in the Appendix of this Chapter.

$$A = \begin{pmatrix} 0220 \\ 2312 \\ 2342 \\ 0250 \end{pmatrix}$$

Like the Langton Loop, Byl's loop changes its direction of growth after the completion of a replication cycle. In (Figure 3), it is easy to observe both the replication period and the characteristic periods for the use of rules. Other loops were discovered by Chou and Reggia (1997; 1998) using CA with $k = 6$ and $k = 8$ and neighborhoods of 5 and 9 cells respectively. Unlike Langton's Loops, these self-replicators lack the distinction between a shell and internal code. The only function of Langton, Byl and Reggia's loops is self-replication. Tempesti (1995), by contrast, has developed a loop similar to Langton's, but with the ability to attach an executable program. Every time the loop replicates, the program is duplicated and attached to the new copies. The program is conserved inside the Loop, where it is connected to the self-replication code. Tempesti's Loop uses a 10 state CA with a Moore neighborhood. Up until 1995, the majority of Loop models were hand-designed - a laborious and time-consuming process. In that year, Lohn and Reggia (1995; 1997) introduced Genetic Algorithms as a way of discovering CA rules with emergent self-replicating capabilities. Sayama (1998, 2000) implemented a loop similar to Langton's but with the ability to self-destruct. Many of these systems replicate a previously defined structure. By contrast, Chou e Reggia (1997; 1998) explored the possibility of a universe of CAs in which self-replicating structures emerge spontaneously from a "primordial soup". The self-replicators they obtained were highly satisfactory and demonstrated the kind of results that Genetic Algorithms could produce. In our own work (Bilotta et al., 2002; Bilotta et al., 2003; Bilotta & Pantano, 2005; Bilotta & Pantano, 2006), we have demonstrated the possibility of

Figure 4. A single seed self-replicator generated by rule C8.1 The images in the first 3 rows show various phases in the replication process. The images in the last row show (left) IE values in different phases of the replication process: (center), The IE trajectory in phase space, and (right) use of rules.

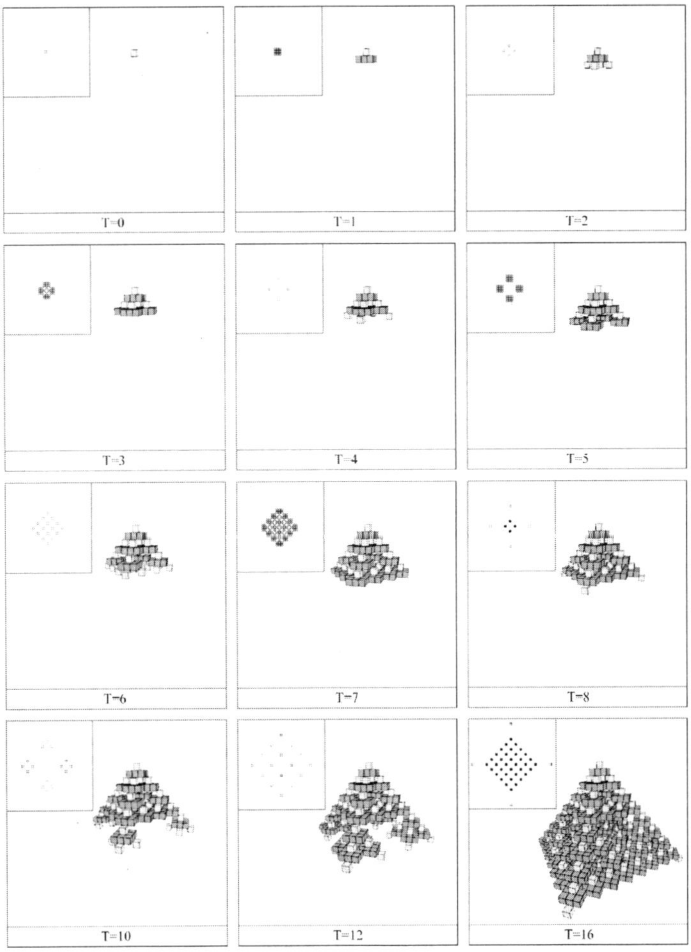

directly evolving self-replicators with Genetic Algorithms. It is this theme we will discuss in this chapter.

MAIN FOCUS

As we have seen in the previous section, the literature describes how, after half a century of scientific progress, we have finally succeeded in producing artificial self-replicators. In the previous chapter, we introduced GA systems with a fitness function based on IE, described how they spontaneously produce self replicators and showed that self-replication is a capability innate to artificial life (Bilotta et al., 2002). In our experiments, we detected several different categories of self-replicator (distinguished by their shape or behavior) and showed that by modifying their genetic code, we can obtain families of self-replicators (Bilotta & Pantano, 2006). In nearly all our experiments, we observed that:

1. Evolution produces systems that are highly differentiated in terms of behavior, complexity and the number of rules used.
2. Self-replicators appear not just at the end of the evolutionary process but also in rules that have not had time to evolve.

These are important observations. What they show is that self-replication is an inherent, pervasive feature of CAs, which does not strictly depend on rule complexity,

Figure 5. A single seed self-replicator generated by rule C8.1 The images in the first 3 rows show various phases in the replication process. The images in the last row show (left) IE values in different phases of the replication process: (center), The IE trajectory in phase space, and (right) use of rules.

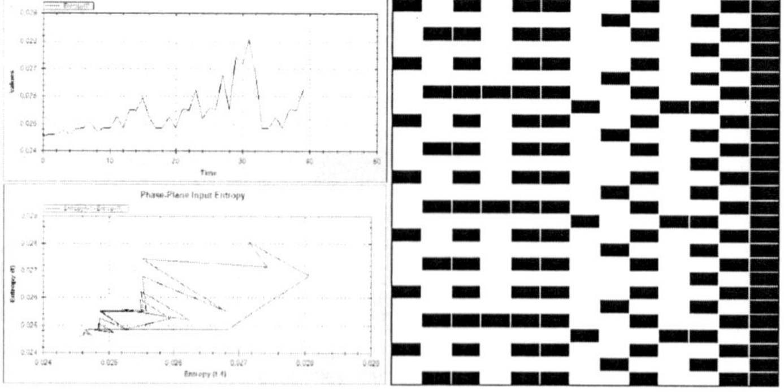

as measured by complexity indexes. Furthermore, there are many different kinds of self-replicator. In this Chapter therefore, we will present a gallery of self-replicating systems evolved using GAs. These self-replicators are nothing more than the tip of an iceberg, most of which has yet to be explored. In Chapter 13, we will attempt an initial classification, based on their behavioral characteristics.

Figure 6. A self-replicator generated by rule C8.1. The self-replicator consists of 3 seeds on a single row. The images in the first 3 rows show various phases in the replication process. The images in the last row show (left) changes in IE values during the replication process, (center) the IE trajectory in phase space, and (right) use of rules. The last image on the right shows the fundamental replication period and the different periods at which system uses specific rules.

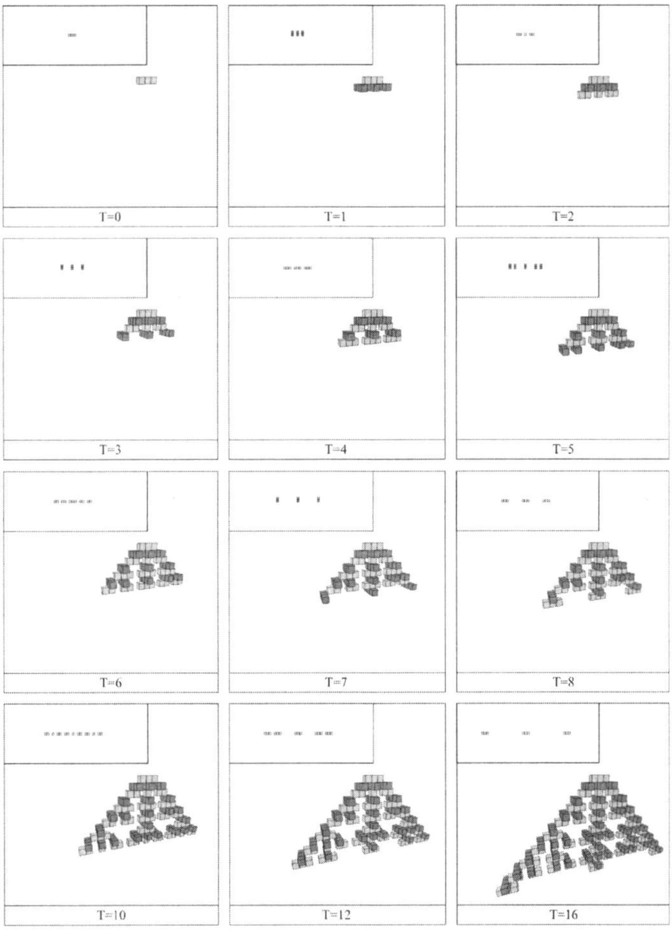

The Behavior of Self-Replicating Systems

In what follows, we will analyze the behavior of the self-replicators generated by the experiments described in the previous chapter.

Self-replicator 1:

Self-replicator 1 was detected during the first experiment using rule C8.1. In order to allow replication of the experiments, we have listed the rules for the self-replicator in the Appendix to this Chapter. The matrix (8.1) A corresponding to the self-replicator is:

$$A = (2)$$

In other words, the self-replicator consists of a single cell, in state 2. The images in the first three rows of (Figures 4 and 5) show the state of the system at steps 1, 2, 3, 4, 5, 6, 7, 8, 10, 12 and 16. In the first step, the self-replicator evolves into 9 cells in state 1. The second step completes the first stage in the self-replication process, generating 4 self-replicators separated by cells in state 0. The emergent system is rhombic in form. After an intermediate step (step 3), the 4 self-replicators move apart. After further intermediate steps, step 8 sees the appearance of active cells in the center of the system, all in state 3. At step 10, the replication process repeats. The system now includes 16 self-replicators, organized in groups of 4. At

Figure 7. A self-replicator generated by rule C8.1. The self-replicator consists of 3 seeds on a single row. The images in the first 3 rows show various phases in the replication process. The images in the last row show (left) changes in IE values during the replication process, (center) the IE trajectory in phase space, and (right) use of rules. The last image on the right shows the fundamental replication period and the different periods at which system uses specific rules.

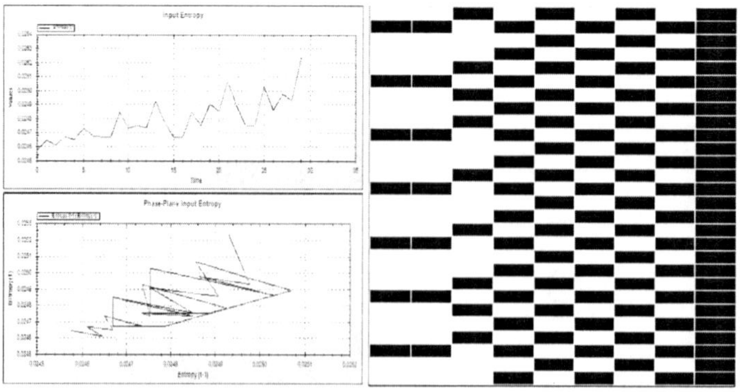

Figure 8. A self-replicator generated by rule C8.2. The self-replicator consists of 2 seeds on a single row. The images in the first 3 rows show various phases in the replication process. The images in the last row show (left) changes in IE values during the self-replication process, (center) the IE trajectory in phase space, and (right) use of rules. The last image on the right shows the characteristic period of the replication process and the characteristic periods at which the system uses specific rules.

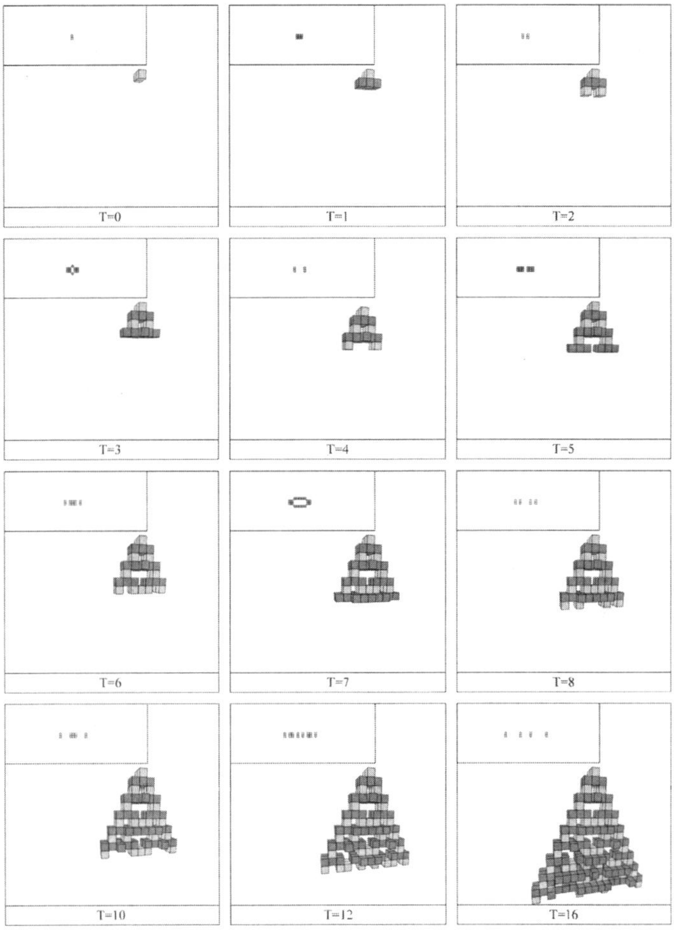

step 16, the number of self-replicators returns to 4 and we observe the presence of many cells in state 3.

This self-replicator takes two steps to increase its numbers from 1 to 4 and 8 steps to pass from 4 to 16. As can be seen from the last image in (Figures 4 and 5), this appears to be the fundamental replication period for the structure. The images

Figure 9. A self-replicator generated by rule C8.2. The self-replicator consists of 2 seeds on a single row. The images in the first 3 rows show various phases in the replication process. The images in the last row show (left) changes in IE values during the self-replication process, (center) the IE trajectory in phase space, and (right) use of rules. The last image on the right shows the characteristic period of the replication process and the characteristic periods at which the system uses specific rules.

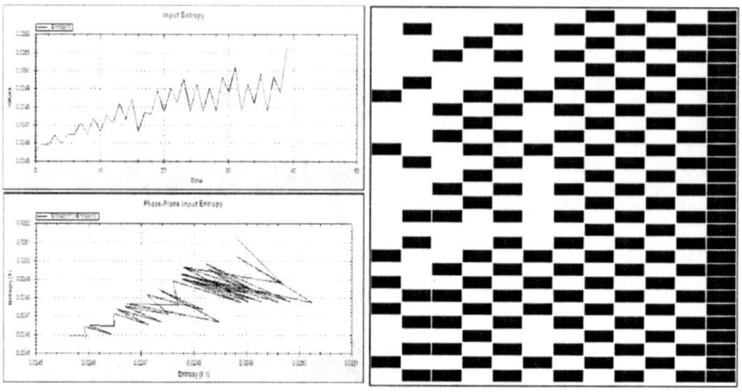

to the left on the last row show changes in IE over time and the trajectory of IE in phase space.

Self-replicator 2:

Self-replicator 2 was also found during the first experiment with rule C8.1. The matrix (8.1) A corresponding to the self-replicator is:

$A = (222)$

The self-replicator consists of a strip of 3 automata in state 2. After 4 transitional steps, the structure generates 2 new structures identical to the original (Figures 6 and 7). At step 8, the structures move apart. At step 12, the number of structures increases to 5. At step 16, it falls to 3. The process whereby new structures appear and disappear follows a precise rhythm which we will examine in the next chapter. The right hand image in the last row shows the presence of other periodical phenomena, with periods that are multiples of 2. Periods of 16 are especially common.

Self-replicator 3:

Self-replicator 3 was detected during experiment 3, using rule C8.2 (see Appendix I). The matrix (8.1) A corresponding to the self-replicator is given by:

Figure 10. A self-replicator generated by rule C8.3. The self-replicator consists of 20 automata arranged on four rows. The images in the first 3 rows show various phases in the replication process. The images in the last row show (left) changes in IE values during the replication process, (center) the IE trajectory in phase space, and (right) use of rules. The last image on the right shows the replication period and the characteristic periods at which the system uses specific rules.

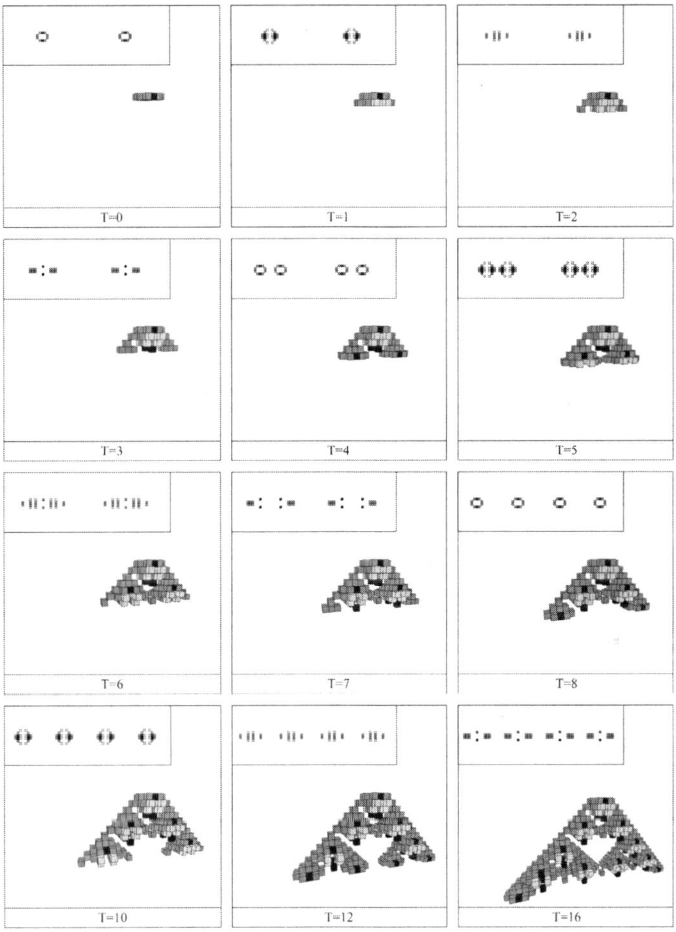

$A = (33)$

The matrix consists of a strip of two connected cells, both in state 3. After two steps, the structure doubles. After a series of intermediate steps, marked by the emergence of new structures, it doubles again in step 8. As the replication process proceeds, it gives rise to structures of unsuspected complexity (Figures 8 and 9). As in

Figure 11. A self-replicator generated by rule C8.3. The self-replicator consists of 20 automata arranged on four rows. The images in the first 3 rows show various phases in the replication process. The images in the last row show (left) changes in IE values during the replication process, (center) the IE trajectory in phase space, and (right) use of rules. The last image on the right shows the replication period and the characteristic periods at which the system uses specific rules.

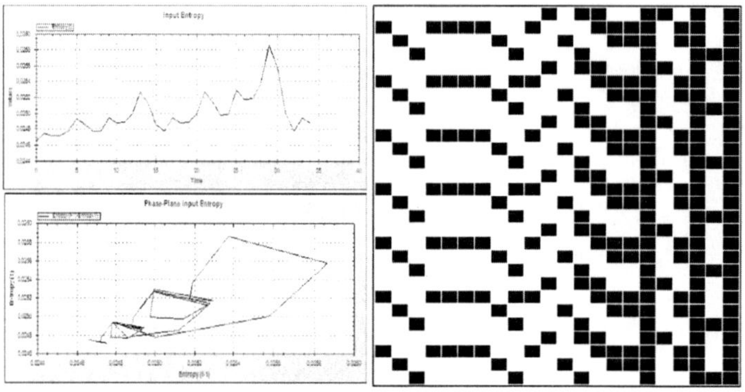

the previous examples, different aspects of the structure have different characteristic periods. This tendency is particularly evident in the data for use of specific rules.

Self-replicator 4:

Self-replicator 4 was detected during experiment 3, using rule C8.3 (see Appendix I). The matrix (8.1) A corresponding to the self-replicator is:

$$A = \begin{pmatrix} 02120 \\ 20002 \\ 20002 \\ 02120 \end{pmatrix}$$

The matrix describes a structure of 20 cells, on four rows. After a sequence of intermediate phases, the structure doubles in step 4, and quadruples in step 12. In step 16, the number of structures drops back to 2. The dynamics of this self-replicator, while more complex than those of the previous example, are fundamentally similar. The characteristic period for growth in the number of structures is a multiple of the replication period. The initial growth in the number of structures is followed by a fall (Figures 10 and 11). The characteristic periods at which the system consults the rules are multiples of 4.

Figure 12. A self-replicator generated by rule C8.4. The self-replicator consists of 5 cells, arranged along a single row. The images in the first 3 rows show various phases in the replication process. The images in the last row show (left) changes in IE values during the -replication process, (center) the IE trajectory in phase space, and (right) use of rules. The last image on the right shows the sequence in which the system uses specific rules. As can be seen, the characteristic period for the use of the elementary rules is 16.

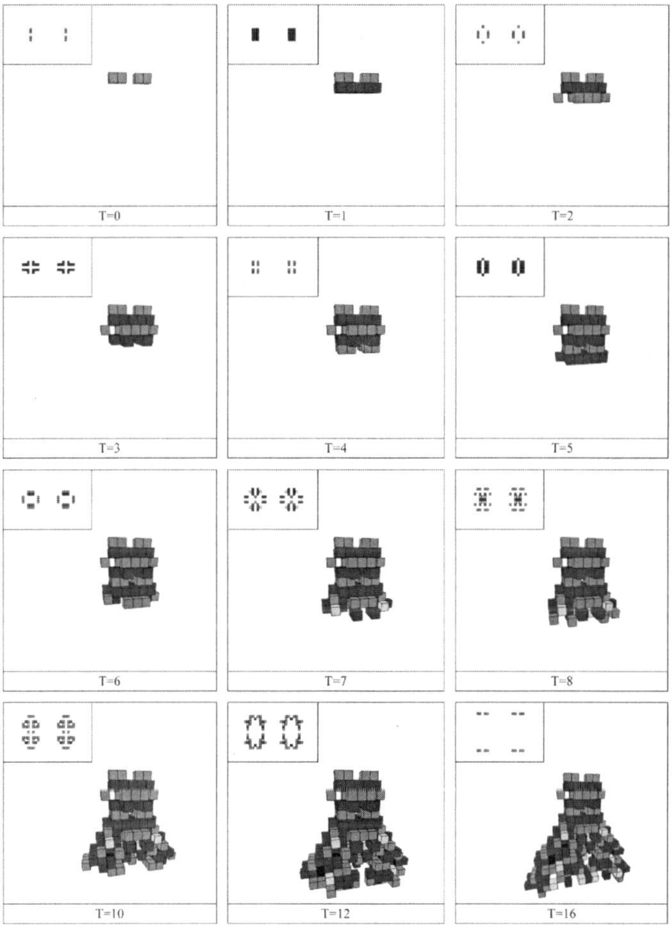

Self-replicator 5:

Self-Replicator 5 was detected during experiment 4, using rule C8.4 (see Appendix I). The matrix (8.1) A corresponding to the self-replicator is:

$A = (11011)$

Figure 13. A self-replicator generated by rule C8.4. The self-replicator consists of 5 cells, arranged along a single row. The images in the first 3 rows show various phases in the replication process. The images in the last row show (left) changes in IE values during the -replication process, (center) the IE trajectory in phase space, and (right) use of rules. The last image on the right shows the sequence in which the system uses specific rules. As can be seen, the characteristic period for the use of the elementary rules is 16.

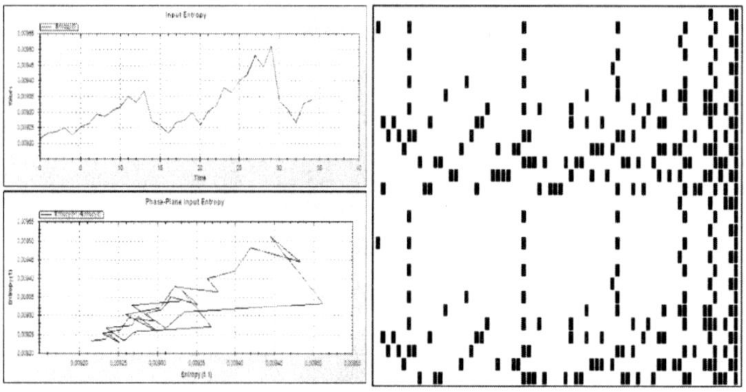

The self-replicator consists of 5 contiguous cells. After 4 steps, the system generates 4 structures arranged around the vertices of a square (see Figures 12 and 13). The self-replication process is more complex than the processes described in the previous examples. At step 16, we discover the same four structures, this time rotated by 90° and arranged around the corners of a rectangle. As can be seen in the last image in (Figures 12 and 13), there is a degree of periodicity in the use of specific rules ($\tau = 16$).

Self-replicator 6:

Self-Replicator 6 was detected during experiment 4, using rule C8.5 (see Appendix I). The matrix (8.1) A corresponding to the self-replicator is:

$A = (40204)$

The self-replicator consists of 5 cells arranged along a single row. After just 2 steps, the structure doubles and rotates by 90° (see Figures 14 and 15). In the following steps, the two structures disappear, though some of their components are conserved. In step 8, the original 2 structures reappear, quadrupled, with the same orientation as in step 1. In step 10 we observe 8 structures, with an orientation orthogonal to the original structure. In step 16, the number of structures drops back to 4, this time with the original orientation. In steps with odd numbers, we observe the presence of two new structures:

Figure 14. A self-replicator generated by rule C8.5. The self-replicator consists of 5 cells, arranged along a single row. The images in the first 3 rows show various phases in the replication process. We observe the emergence of two new structures, one in opposite phase to the other. As the replication process proceeds, the two structures replicate, rotate and move apart. The images in the last row show (left) changes in IE values during the replication process, (center) the IE trajectory in phase space and (right) use of rules. The last image on the right shows the sequence in which the system uses specific rules. As can be seen, the characteristic period for the use of the elementary rules is 8.

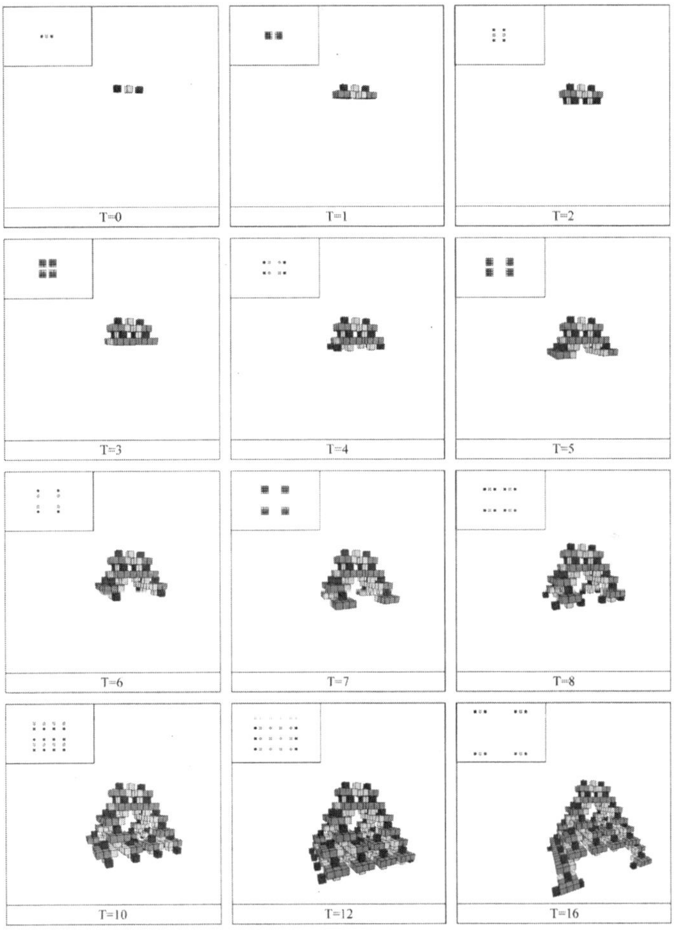

Figure 15. A self-replicator generated by rule C8.5. The self-replicator consists of 5 cells, arranged along a single row. The images in the first 3 rows show various phases in the replication process. We observe the emergence of two new structures, one in opposite phase to the other. As the replication process proceeds, the two structures replicate, rotate and move apart. The images in the last row show (left) changes in IE values during the replication process, (center) the IE trajectory in phase space and (right) use of rules. The last image on the right shows the sequence in which the system uses specific rules. As can be seen, the characteristic period for the use of the elementary rules is 8.

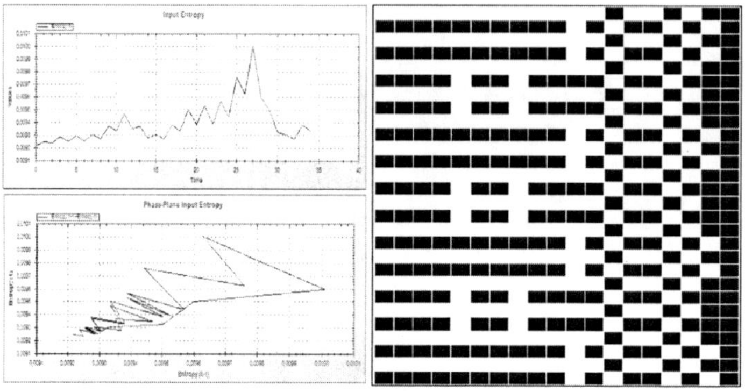

$$A' = \begin{pmatrix} 3 & 3 & 2 \\ 3 & 3 & 2 \\ 3 & 3 & 2 \end{pmatrix} \text{ and } A' = \begin{pmatrix} 2 & 3 & 3 \\ 2 & 3 & 3 \\ 2 & 3 & 3 \end{pmatrix}$$

The two structures always appear together, with the first in opposite phase to the second. As the self-replication process proceeds, they rotate, move apart and replicate. As we will see in Chapter 13, they are far more stable than the original structure. From the last image in the diagram, it is evident that the characteristic period for the use of rules is 8.

Self-replicator 7:

Self-replicator 7 was detected during experiment 8, using rule C8.6. The matrix (8.1) A corresponding to the self-replicator is:

$$A = (202)$$

The self-replicator consists of 2 cells with state 2 separated by 1 cell with state 0 (see Figures 16 and 17). In this example, duplication takes places after 4 steps. We

Figure 16. A self-replicator generated by rule C8.6. The self-replicator consists of 2 cells, separated by a cell in state 0. The images in the first 3 rows show various phases in the replication process. We observe the emergence of two new structures, one in opposite phase to the other. As the self-replication process proceeds, the two structures replicate, rotate and move apart. The images in the last row show (left) changes in IE values during the replication process, (center) the IE trajectory in phase space, and (right) use of rules. The last image on the right shows the sequence in which the system uses specific rules. This image confirms the presence of a 9 step characteristic period that appears to regulate the replication process.

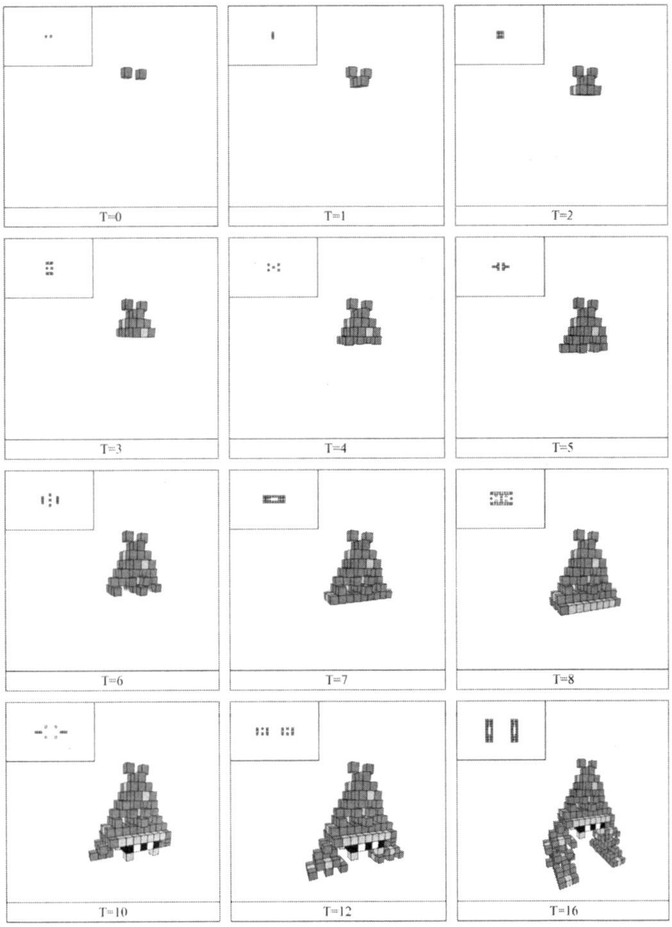

observe that the new structure is contaminated by the addition of new elements. In step 3, we see the emergence of a new structure with better defined behavior. This structure doubles in step 9. The 9 step replication period matches the characteristic period for the use of specific rules, as shown in the last image in (Figures 16 and 17).

Figure 17. A self-replicator generated by rule C8.6. The self-replicator consists of 2 cells, separated by a cell in state 0. The images in the first 3 rows show various phases in the replication process. We observe the emergence of two new structures, one in opposite phase to the other. As the self-replication process proceeds, the two structures replicate, rotate and move apart. The images in the last row show (left) changes in IE values during the replication process, (center) the IE trajectory in phase space, and (right) use of rules. The last image on the right shows the sequence in which the system uses specific rules. This image confirms the presence of a 9 step characteristic period that appears to regulate the replication process.

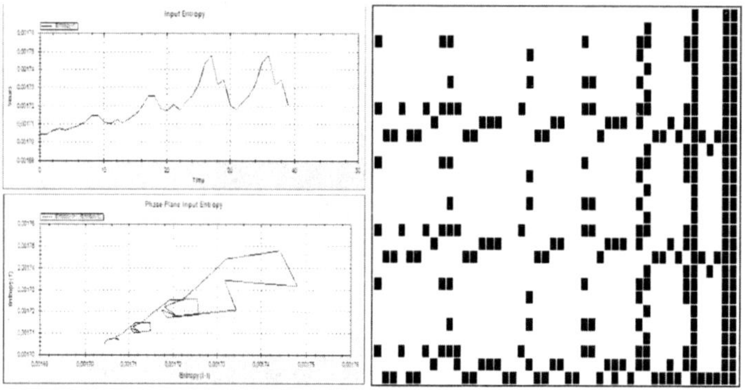

Self-replicator 8:

Self-replicator 8 was detected during experiment 9, using rule C8.7. The matrix (8.1) A corresponding to the self-replicator is:

$A = (12)$

The self-replicator consists of two cells, in states 1 and 2 respectively. The first replication takes place after 4 steps when the structure doubles; it then quadruples at step 12 (see Figures 18 and 19). The replication period is 8. The importance of this period is confirmed by the way in which the system uses specific rules, as show in the last image in (Figures 18 and 19).

FUTURE TRENDS

In this chapter, we have seen that self-replication is a pervasive feature of CAs. And yet, the structures we observe are all the result of specific logic regulating their replication and orientation and the way they give birth to new structures. Furthermore

Figure 18. A self-replicator generated by rule C.7. The self-replicator consists of 2 contiguous cells, in states 1 and 2 respectively. The images in the first 3 rows show various phases in the replication process. We observe the emergence of two new structures, one in the opposite phase to the other. As the replication process proceeds, the two structures replicate, rotate and move apart. The images in the last row show (left) changes in IE values during the replication process, (center) the IE trajectory in phase space, and (right) use of rules. The last image on the right shows the sequence in which the system uses specific rules. This image confirms the presence of an 8 step period that appears to regulate the replication process.

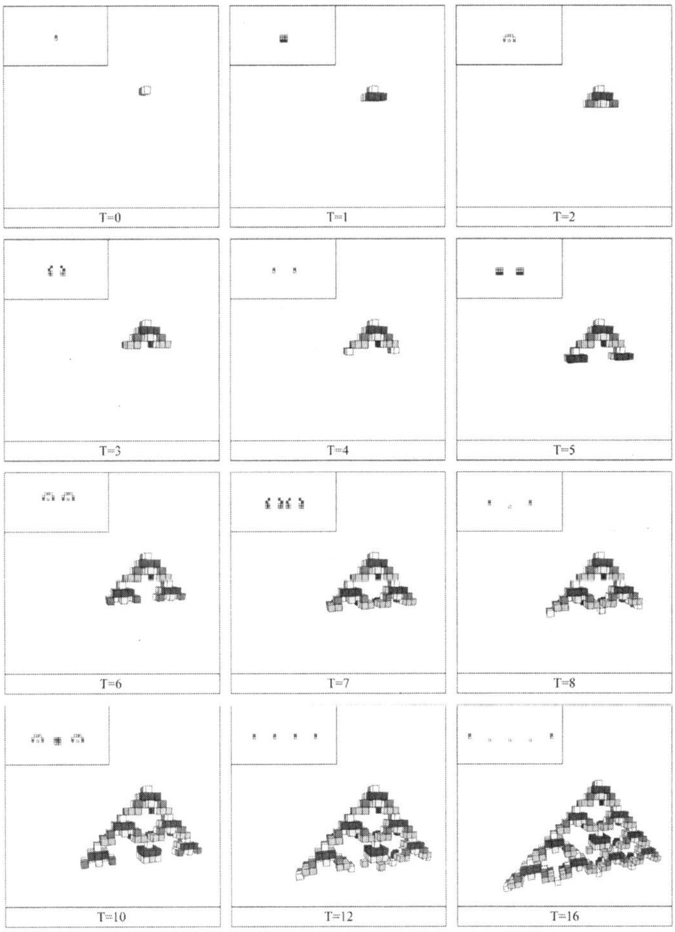

replication and growth rates differ enormously from structure to structure. These observations pose several questions:

Figure 19. A self-replicator generated by rule C.7. The self-replicator consists of 2 contiguous cells, in states 1 and 2 respectively. The images in the first 3 rows show various phases in the replication process. We observe the emergence of two new structures, one in the opposite phase to the other. As the replication process proceeds, the two structures replicate, rotate and move apart. The images in the last row show (left) changes in IE values during the replication process, (center) the IE trajectory in phase space, and (right) use of rules. The last image on the right shows the sequence in which the system uses specific rules. This image confirms the presence of an 8 step period that appears to regulate the replication process.

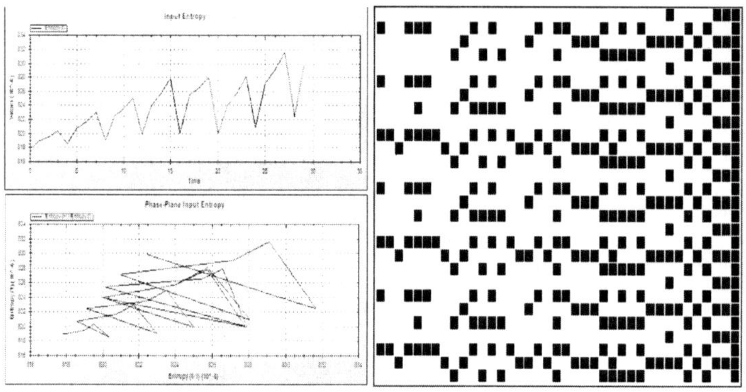

1. Does self-replication follow a rhythm?
2. Can we group the rhythms of different processes into broader categories?
3. Can we describe the growth of colonies in formal terms?
4. Can we categorize different kinds of colony growth?
5. Can we somehow measure the complexity of a self-replicating structure?
6. Are the shapes in phase space, characterized by specific symmetries?

As far as concerns the first question, we believe that at least some self-replication processes do have specific rhythms that we can group into categories. We will discuss this issue further in Chapter 10. However, we also believe that as the process becomes more and more complex, it will become ever more difficult to identify the underlying mechanisms. This difficulty is already apparent in the examples we have described in this chapter. As colonies begin to form, we observe various periodic behaviors, linked to the period of the original replication process - as if the fundamental replication period were necessarily associated with additional, shorter periods. This is an issue we will investigate further in Chapters 14 and 15, when we discuss musical rhythms, and the representation of growth dynamics in music and sound. As far as concerns colonies, we will see in Chapter 10 that they can indeed be grouped into general categories and that these can help us to answer, at least in part, our first two questions about the rhythm of self-replication processes. We will also show that general categories are much easier to identify for colonies than for individual self-replication processes. One way of measuring the complexity of a replicator is to analyze IE values for the replication process. However, it may be more meaningful to examine the code for the use of elementary rules, and the number of rules used. It is these elements which provide us with the best insights into the replication process. This is another issue we will discuss in greater detail in Chapter 10. As for the last question, some self-replicators have very simple shapes. In these cases symmetries in phase space are easy to observe. However, in other cases, trajectories in phase space are highly complex, and the resulting symmetries are complex themselves and difficult to identify. This is a problem that deserves further work.

CONCLUSION

In this chapter, we began with a brief analysis of the history of self-replicating systems and a discussion of the main systems discovered in the last 50 years of research. We went on to present some examples of self-replicators generated using GAs and a fitness function based on input entropy. Our experiments demonstrate that far from being an isolated phenomenon, self-replication is a pervasive feature of the majority of 2-dimensional CAs and an innate characteristic of artificial life. When we analyzed self-replicators, we discovered that each replicator grows at its own characteristic rate and that, as multi-state systems, self-replicators display a range of the emergent synchrony. The presence of different kinds of synchrony means that different self-replicators are regulated by different models of growth, each reflecting specific temporal dynamics in the replication process. This phenomenon is particularly evident during the formation of complex organisms such as colonies, a process that involves an increase in the number of active cells and the emergence of fractal structures, and that appears in all the self-replicators we have considered in this chapter. We will return to these issues in Chapter 11. The gallery we have shown in this chapter provides nothing more than an idea of the practically unlimited variety and complexity of emergent structures that CAs can produce. Testing the internal consistency of our data, analyzing the detailed behavior of specific systems, and grouping their behavior into macro-categories (beyond those we have already developed) will require further experimentation (Bilotta & Pantano, 2006; 2008). We have shown that some structures use a small number of rules which they consult very frequently. Others use a much larger rule set in which some rules are used very often and others much more rarely. This observation offers a way of measuring the complexity of the replication process. As von Neumann predicted in his famous thought experiment in 1948, the self-replicators we have discovered use their genetic information twice, as a non-interpreted genetic code, and as interpreted code, generating a phenotype - the organism itself.

REFERENCES

Adachi, S., Lee, J., Peper, F., & Umeo, H. (2008). Kaleidoscope of life: A 24-neighbourhood outer-totalistic cellular automaton. *Physica D. Nonlinear Phenomena*, *237*(6), 800–817. doi:10.1016/j.physd.2007.10.015

Aspray, W., & Burks, A. (Eds.). (1987). *Papers of John von Neumann on Computing and Computer Theory*. Cambridge, MA: The MIT Press.

Bilotta, E., Lafusa, A., & Pantano, P. (2002). Is self replication an embedded characteristic of Artificial/Living matter? In Standish, R. K., Bedau, M. A., & Abbass, H. A. (Eds.), *Artificial Life VIII* (pp. 38–48). Cambridge, MA: The MIT Press.

Bilotta, E., Lafusa, A., & Pantano, P. (2003a). Life-like self-reproducers. *Complexity*, *9*(1), 38–55. doi:10.1002/cplx.10118

Bilotta, E., & Pantano, P. (2005). Emergent Patterning phenomena in 2D Cellular automata. *Artificial Life Journal*, *11*(3), 339–362. doi:10.1162/1064546054407167

Bilotta, E., & Pantano, P. (2006). Structural and Functional Growth in Self-Reproducers. *Complexity*, *11*(6), 12–29. doi:10.1002/cplx.20131

Bilotta, E., & Pantano, P. (2008). Biological traits in artificial self-reproducing systems . In Orsucci, F., & Sala, N. (Eds.), *Complexity Science in Information Communication Technologies: Interaction Between Living Systems and Information Technologies* (pp. 109–123). Hershey, PA: IGI Global.

Byl, J. (1989). Self-Reproduction in small cellular automata. *Physica D. Nonlinear Phenomena*, *34*, 295–299. doi:10.1016/0167-2789(89)90242-X

Casti, J. (1989). *Alternate Realities: Mathematical Models of Nature and Man*. New York: Wiley.

Chou, H. H., & Reggia, J. A. (1997). Emergence of self-replicating structures in a cellular automata space. *Physica D. Nonlinear Phenomena*, *110*(3-4), 252–276. doi:10.1016/S0167-2789(97)00132-2

Chou, H. H., & Reggia, J. A. (1998). Problem solving during artificial selection of self-replicating loops. *Physica D. Nonlinear Phenomena*, *115*(3-4), 293–312. doi:10.1016/S0167-2789(97)00237-6

Codd, E. F. (1968). *Cellular Automata*. New York: Academic Press.

Emmeche, C. (1994). *The garden in the machine. The emerging science of artificial life*. Princeton, NJ: Princeton University Press.

Evans, K. M. (2003). Larger than Life: threshold-range scaling of Life's coherent structures. *Physica D. Nonlinear Phenomena*, *183*(1-2), 45–67. doi:10.1016/S0167-2789(03)00155-6

Gardner, M. (1970). Mathematical games: the fantastic combinations of John Conway's new solitaire game "Life." . *Scientific American*, *223*, 120–123. doi:10.1038/scientificamerican1070-120

Gardner, M. (1983). *Wheels, Life and other Mathematical Amusements*. San Francisco, CA: W. H. Freeman.

Langton, C. G. (1984). Self-Reproduction in Cellular Automata . *Physica*, *10D*, 135–144.

Lohn, J. D., & Reggia, J. A. (1995). *Discovery of self-replicating structures using a genetic algorithm*. Paper presented at IEEE International Conference on Evolutionary Computation (ICEC'95), Perth, Australia.

Lohn, J. D., & Reggia, J. A. (1997). Automatic discovery of self-replicating structures in cellular automata. *IEEE Transactions on Evolutionary Computation*, *1*(3), 165–178. doi:10.1109/4235.661547

Nowak, M. A., & Sigmund, K. (1993). *Chaos and the evolution of cooperation*. Paper presented at the National Academy of Sciences.

Salzberg, C., Antony, A., & Sayama, H. (2006). Visualizing evolutionary dynamics of self-replicators: A graph-based approach. *Artificial Life*, *12*, 275–287. doi:10.1162/artl.2006.12.2.275

Salzberg, C., & Sayama, H. (2004). Complex genetic evolution of artificial self-replicators in cellular automata. *Complexity*, *10*(2), 33–39. doi:10.1002/cplx.20060

Sayama, H. (1998). Introduction of Structural Dissolution into Langton's Self-reproducing Loop. In C. Adami, R. K. Belew, H. Kitano, & C.E. Taylor (Eds.), *Artificial Life VI: Proceedings of the Sixth International Conference on Artificial Life* (pp. 114-122). Cambridge. MA: The MIT Press.

Sayama, H. (2000). Self-Replicating Worms that Increase Structural Complexity through Gene Transmission. In M. A. Bedau, J. S. McCaskill, N. H. Packard, & S. Rasmussen (Eds.), *Artificial Life VII: Proceedings of the Seventh International Conference* (pp. 467-476). Cambridge, MA: The MIT Press.

Schuster, P., & Sigmund, K. (1983). Replicator dynamics. *Journal of Theoretical Biology, 100*, 533–538. doi:10.1016/0022-5193(83)90445-9

Smith, A. R. III. (1971). Simple Computation-Universal Cellular Spaces. [JACM]. *Journal of the Association for Computing Machinery, 18*(3), 339–353.

Tempesti, G. (1995). A new self-reproducing cellular automaton capable of construction and computation. In F. Morán, A. Moreno, J. J. Merelo, & P. Chacón (Eds.), *ECAL'95: Third European Conference on Artificial Life* (LNCS 929, pp. 555-563). Heidelberg, Germany: Springer-Verlag.

Toffoli, T., & Norman, M. (1987). *Cellular Automata Machines: A New Environment for Modelling*. Cambridge, MA: The MIT Press.

Varela, F. J., Maturana, H. R., & Uribe, R. (1974). Autopoiesis: the organization of living systems, its characterization and a model. *Bio Systems, 5*, 187–196. doi:10.1016/0303-2647(74)90031-8

Vitanyi, P. M. B. (1973). Sexually reproducing cellular automata. *Mathematical Biosciences, 18*, 23–54. doi:10.1016/0025-5564(73)90018-7

von Neumann, J. (1966). *Theory of Self-Reproducing Automata* (Burks, A. W., Ed.). Urbana, IL: University of Illinois Press.

von Neumann, J., & Burks, A. W. (1970). von Neumann's self-reproducing automata. In A. W. Burks (Ed.), Essays on Cellular Automata (pp. 3-64). Urbana, IL: University of Illinois Press.

Watson, J. D., & Crick, F. H. C. (1953). A Structure for Deoxyribose Nucleic Acid. *Nature, 171*, 737–738. doi:10.1038/171737a0

Wolfram, S. (1984). Universality and Complexity in Cellular Automata. *Physica, 10D*, 1–35.

Wolfram, S. (2002). *A New Kind of Science*. Champaign, IL: Wolfram Media.

APPENDIX

Langton's Rule Table

Table 2.

1. 00000-->0	39. 01625-->1	77. 11126-->1	115. 20052-->2	153. 20742-->2	191. 50023-->2
2. 00001-->2	40. 01722-->1	78. 11127-->7	116. 20057-->5	154. 20772-->2	192. 50027-->2
3. 00002-->0	41. 01725-->5	79. 11152-->2	117. 20072-->2	155. 21122-->2	193. 50052-->0
4. 00003-->0	42. 01752-->1	80. 11212-->1	118. 20102-->2	156. 21126-->1	194. 50202-->2
5. 00005-->0	43. 01762-->1	81. 11222-->1	119. 20112-->2	157. 21222-->2	195. 50212-->2
6. 00006-->3	44. 01772-->1	82. 11224-->4	120. 20122-->2	158. 21224-->2	196. 50215-->2
7. 00007-->1	45. 02527-->1	83. 11225-->1	121. 20142-->2	159. 21226-->2	197. 50222-->0
8. 00011-->2	46. 10001-->1	84. 11227-->7	122. 20172-->2	160. 21227-->2	198. 50224-->4
9. 00012-->2	47. 10006-->1	85. 11232-->1	123. 20202-->2	161. 21422-->2	199. 50272-->2
10. 00013-->2	48. 10007-->7	86. 11242-->4	124. 20203-->2	162. 21522-->2	200. 51212-->2
11. 00021-->2	49. 10011-->1	87. 11262-->1	125. 20205-->2	163. 21622-->2	201. 51222-->0
12. 00022-->0	50. 10012-->1	88. 11272-->7	126. 20207-->3	164. 21722-->2	202. 51242-->2
13. 00023-->0	51. 10021-->1	89. 11322-->1	127. 20212-->2	165. 22227-->2	203. 51272-->2
14. 00026-->2	52. 10024-->4	90. 12224-->4	128. 20215-->2	166. 22244-->2	204. 60001-->1
15. 00027-->2	53. 10027-->7	91. 12227-->7	129. 20221-->2	167. 22246-->2	205. 60002-->1
16. 00032-->0	54. 10051-->1	92. 12243-->4	130. 20222-->2	168. 22276-->2	206. 60212-->0
17. 00052-->5	55. 10101-->1	93. 12254-->7	131. 20227-->2	169. 22277-->2	207. 61212-->5
18. 00062-->2	56. 10111-->1	94. 12324-->4	132. 20232-->1	170. 30001-->3	208. 61213-->1
19. 00072-->2	57. 10124-->4	95. 12327-->7	133. 20242-->2	171. 30002-->2	209. 61222-->5
20. 00102-->2	58. 10127-->7	96. 12425-->5	134. 20245-->2	172. 30004-->1	210. 70007-->7
21. 00112-->0	59. 10202-->6	97. 12426-->7	135. 20252-->0	173. 30007-->6	211. 70112-->0
22. 00202-->0	60. 10212-->1	98. 12527-->5	136. 20255-->2	174. 30012-->3	212. 70122-->0
23. 00203-->0	61. 10221-->1	99. 20001-->2	137. 20262-->2	175. 30042-->1	213. 70125-->0
24. 00205-->0	62. 10224-->4	100. 20002-->2	138. 20272-->2	176. 30062-->2	214. 70212-->0
25. 00212-->5	63. 10226-->3	101. 20004-->2	139. 20312-->2	177. 30102-->1	215. 70222-->1
26. 00222-->0	64. 10227-->7	102. 20007-->1	140. 20321-->6	178. 30122-->0	216. 70225-->1
27. 00232-->2	65. 10232-->7	103. 20012-->2	141. 20322-->6	179. 30251-->1	217. 70232-->1
28. 00522-->2	66. 10242-->4	104. 20015-->2	142. 20342-->2	180. 40112-->0	218. 70252-->5
29. 01232-->1	67. 10262-->6	105. 20021-->2	143. 20422-->2	181. 40122-->0	219. 70272-->0
30. 01242-->1	68. 10264-->4	106. 20022-->2	144. 20512-->2	182. 40125-->0	
31. 01252-->5	69. 10267-->7	107. 20023-->2	145. 20521-->2	183. 40212-->0	
32. 01262-->1	70. 10271-->0	108. 20024-->2	146. 20522-->2	184. 40222-->1	
33. 01272-->1	71. 10272-->7	109. 20025-->0	147. 20552-->1	185. 40232-->6	
34. 01275-->1	72. 10542-->7	110. 20026-->2	148. 20572-->5	186. 40252-->0	
35. 01422-->1	73. 11112-->1	111. 20027-->2	149. 20622-->2	187. 40322-->1	
36. 01432-->1	74. 11122-->1	112. 20032-->6	150. 20672-->2	188. 50002-->2	
37. 01442-->1	75. 11124-->4	113. 20042-->3	151. 20712-->2	189. 50021-->5	
38. 01472-->1	76. 11125-->1	114. 20051-->7	152. 20722-->2	190. 50022-->5	

Byl's Rule Table

Table 3.

```
000000 000010 000031 000420 000110 000120 000311 000330 000320 000200 000242 "+"000233 004330
005120 001400 001120 001305 001310 001320 003020 003400 003420 "+"003120 002040 002050 002010
002020 002420 002500 002520 002200 002250 002220 "+"040000 050000 050500 010052 010022 010400
010100 020055 400233 405235 401233 "+"442024 452020 415233 411233 412024 412533 432024 421433
422313 501302 502230 "+"540022 542002 512024 530025 520025 520442 523242 522020 100000
100010 100033 "+"100330 101233 103401 103244 111244 113244 112404 133244 121351 123444 123543
"+"122414 122434 300100 300300 300211 300233 304233 301100 301211 303211 303233 "+"302233
344233 343233 351202 353215 314233 311233 313251 313211 312211 335223 "+"333211 320533 324433
325415 321433 321511 321321 323411 200000 200042 200032 "+"200022 200442 200515 200112 200122
200342 200332 200242 200212 201502 203202 "+"202302 244022 245022 242042 242022 254202 255042
252025 210042 214022 215022 "+"212052 212055 212022 230052 234022 235002 235022 232042 232022
220042 220020 "+"220533 221552;
```

Rule Codes

C8.1: D=2,k=4,r=1

030120033000000012310110020130200003110000000010000000000001221
000000000002001000002010002003000000200000000000000000300000030000
000010030000003230003003000100000000000000002000000200003000000001
00000033100322200000000100

C8.2: D=2,k=4,r=1

000000000000000000000000000030000000000000000000000000000000000
000000000100000001000000023110000020000000000000000000000100000000
000000000000000000000000000000000000000000000000003000000000000000
0020310001000001000000000

C8.3: D=2,k=4,r=1

100100000000000000003000100000000000000020000032322103200012000
000000000100003000000000010003020202003003000300000200020020000021 1
030200102000000000000000001020000301123001000002000000000010003330
0000000010000000022300000

C8.4: D=2,k=5,r=1

002144411000002402000000000020003233100342040021000000000002044
422102000210124121420000000000000210032114434134410410000001444222
002103420012022430000002400000200000340004301000001000000120200000
300010022200000200020413200211400000000301000400000000000000000040

1200400000000000012302400034241100010000020230200000000200102003 00
344333144000132400002020410200200000000030000041110230000001230004
130003323210112312031131300000000000000000040002233240000400000000
0100400000000000021024322131100000000000000000010401000400403030003 0
000100010002030243433221333100231433242413221000000002000030421200
300004210000000002040110040320000023430302104000000004000000300031
00000000010040044103342300032001100110010220000000000000000243400000

C8.5: D=2,k=5,r=1

04000000000334002000200120003100022010300000000000000001000001003
300040000000000000003012000300030100004200000100000000200003000000
002000000040000000004000030000030000200000000400000100012120000 03200
030040030000001000000000000000000300003400000000000002000000000040000
10000000031040200040001300000400410000000000000000003000110004002 30
000000000000000010000000200010100010000000000000000000400004000000 0
2000400000404000000000000000100000000000000000000430003340000000 00200
00000010000000200300004000000000430000321210000110404000300430000 0
000001010000000000003000020000000000000000000104000100000000000001000
00002042000000000000000000000000040004001000000000403000000000310000 0
0000000240030000000000100000200041000400000000000000200002000004300 00

C8.6: D=2,k=7,r=1

020100400006360000000000000000000000000000000000000050600000000 0
00000000000050050000000000000000000024000000106000000005400000000 4003
0001000005001500002502000000040000000050000000000000602600000000000006
0010000000000002400525000000000000600000020050000000000000000000000 10
000050004000000000000000000000000000050000000000000216600000013300100 1
0341000050000000002000000004460536203040011050000600002000000000000
00635200010123000050000000000000000000000000000000300000002022000 6
00000000000002610031600030000000000004220000000000000000000000000004
23004001000401525000000010000030540000000000000000003500000000500000000
000636520000000000000000003000000000000000200202040000000002000000000
00000000000000000000000000006000000000065400004000000000504004000000
0000000000000000000000000000000000000000005000010000000000000000
0000000000004000000000000000200010000000000005000000000000000000000002
00300000006060040050000000000000000000000000000000000050000000000 3
200000000050000000404006000002100004200000000000000050000000000000 0
00000000000000050000000040500450000002006600000000000000000000000000 0
50015400040000302523542000000000600003164621130300000000000000000421
0000000001000306020010000100000060005244340000000000600020000 5030

1002200020060000000000020000030000020000000011200000000020000000004
0000030134300000030000500003106000100000061300040656164200401362 6
30500000000000006000000003000000002526000000050000000000000000000
31050000220000365000000000400630060060042050005400000000030101250 0
00600000001400012600000500000002000000000000000000000300000000000 0
00000000032000000000000000000060420600030000200003003211200000650 0
00050000000000053000000000000010000000600040000000000000000006500000 1
05005000002000220001000040010400200000005006400000000000463503320 0
56120000000000000000520000000000000650233000000000000000000000040453
10000503600400000000203200032000000460000060060320000100006000000
03000060012000000006500002001000235302005000000000000140000003062 00
03000006200000000010200500000000000000000000000000000006016500220 00
00444006000000000000000200000000000000000300000000000000000000000
20000001005000003000000000330006025251001242030000000000000004000
60000000002000500060000006001060056430000020000351000031425451621
00000000003200000006000040066602001004000000000005200450000020 40
00003000000000000000000002000000100000000000300400000100005060004
10000033000000000000000000000000062414440400000000000064000100000
30000000000000000000005000005000000000000000000000051115000042252 6
00000030100000000000400030006000000003000000003000600000602431510 00
00000000000006000000000000000000000006143000002000000000000000003 00
00001003000123040000040000054503160003000004032043210000001640000
30000000000000050000100400000002000004000000000014040000400005000 00
00000460006200000100003020002400000000402000000000000000000000060
02300500000002000000000000000000000000030050000000000000000000061
03564252000004000050000003000000060000000000010000020043040000000
00000000250000011050300000000000000000451005002150400010503000030 0
00000000000000120000101000020000000200000000046530100066400060400
03000000000000000000000004000000000000000000000050000000000000000
30040000000000000000100000000000000000600000040302540000003000005 4
00000023600000000000500001002100500060564045101000000050000000000
00000000000000000050130004000300000020014010600043055400000000000
20000003604000000000004000000000000000000003000000000500000601230 00
00000000000160040000050400405000000000510405000000010000500100002 6
00002100000000000000000000004000000000001650300040000050351000000 5
00000030020000000000300010056500301005040630323040000000300002035 00
15230000030500002040100000200000000050003000000000000400420000000
00646200204240000000620000000000000360020000000000040000060000000 0
00000000000000000000020000000000400000000320000006000000000500400 0
00311033000000206400430452455000610000000613000214000000652500000 0

0003000050400000000000006500030000044000000000000302000102000000000
4000061100302213000063660460000000006302000000000200000000062000000
0000000000640003000010000000000000000000000000500000000000000000000
0030000000001000000000000000000000404602226400000602000000001001003
0333000000000000000000000000000000030000000000000000000064500454050
0000000100000000600000365436000000301466100000000620101256300010 0
3505100000562434600000000000000000600500300000000000004400012430000000
0130002000000000000000614151000230000300000000000500000020000020600
0045005002004000000000000000000000100004030600041200000005000000000
0000000503000003305022310030600000000000000000000000103040000000000
6252300000000030000000000000000003000000000500364202000000054000000006
0003003203000030000000000010200000302020000000651000012000020011 00
0002000000023500000000114461000000300000000000000000630000000000000003
0060004600000000300000000000003000010000020063412300000000500000 6500
0010300501430000003040000000600012000300000000000000000400000000040
000500630060010000500030066005130000044150000000000030304000253622 6
050010010000406011003000000062300060000060004000000000000400002000 02
000004131650000000046040026004100034000004063000054355000000000003000
000000000000040600000000600020416450000300000000000000206500000000 00

C8.7: D=2,k=8,r=1

5047253030040541001177500000040100070000006504660600000100450 05
30154101306034055301052542500000610572000014073301003000000000000
00100000000000050376050063117670303030000000001450572100000004014 00
3400037006074701344160200250034000000700450042205101075202070043 3
0730600200000704074505400100000000000000061335044362771000106000003
3371241140055377000000000000002006020000000000000470000000001200000437
50720000044467700000000000500670010005003000000000000200000000042 26
01000750000066416000451141340460207000026000000100004200000400000 0
0000000400000315626243072360204620010000303100000060300000000000000
0000005000000400000070020000557000000000000500070300000004025000002 0
01350203200000600120050000056000000000000000000025770000000500000030 64
6000505626000020140710630770030000060000050000005000002000500500 24
65022520120006046767475363604004000603040070000000207376764500007 3
0302060165230000003010655626200000000000000070211000000000000050000000
0000001342507000034002000000000000000523000464152530005007000360404
71034511030066010500005560035742175550000605622000010015405000016 0
0000044002300000000550212037220055325631730652220010050000000011030 0
0020300072000774301300400057000340500026100063500305107000472466 2
500002000062364044005002300302000000640000000005340003000000000500

06645140050030005777670061004232707447660000010720235004000000054
01420006004700000300050000000322221773300137470001030000035460001
705076105517730624513006220067225744474372113137001675400002073000
102336054071170010230200000061600000007050200000500030031000003003
000004677100000700400720060021740520000000000000000601000656725020
233415300001210000000050071003640036604001000003600000005600540000
04000000000736176450000001000070470302270340000030301674000004000
000000325000004510260000200743103420000000006007001123505020621000
02010001057275312400000000233056125214060063160000072600000000000
000330076630323075001413707271127311457715151745727700056500524750
60061020000575045000371010300000000000000000040100000570011204400060
000106300060010000000300000000060700000004331046003032100000043000
003000660004240667376000001147600555207272402300130070507300300000
6410407101712000205000000000000000000020000000000070006542170000005
25023260350010206000000372000000630000006055550067000001401075000
00000000001013050570003000510037564000155345470060000000040070040
2020000000500064206540010000000000000000030145003451000700021160002
07400000000000331000000045000361000077000000000503401114141315156260
00000007000240000000000000600000000000307615415440505000000004620000
070657517477500004000605070226074050027700000000000001003240001607
710000000000000007413620570270000057070670200000404754550020244002
077555000061770150025606700000020030000003714000477657351307700000
00050160502006200005000077100017410450000200001300076054402324100
04300030030400040000075270005000577426725013320706130020400210007
01375170060250000031000000612756040002102061000040000076724000101
44562370000003000714072000000300020356557145003157000474452700605
26315007300003000000000000000001750000306475100000000000000020576253
013510004005030000000002577000000000023507000035437725402160700633
6710624000000000000070300300020000602000045177533644570100525347000
50400043471626062000004060432242004700225650265000000001405043460
06000231003352620006501000064042070000002303002003000051700103013
3002000402064351677600350000000000000000005504000642203460066646150
00000540061400353127760077357000630551560070600600736400024005070
000116000267202613000065002000000600022203006200000005000000066000
33734736375250004620200473000020427512570345613310540044050000023
740160010000430015351750100762000000334100003002022611100006650200
0100240400000546000340340000000000000072300001133305244270007473410
117104121100055601130461110006065622000245000003000103321736000720
055010653003007445071127250424040003000004005150602000367003003046
000000000620000721651570003070000500140000001100206000006000505000

5020527564310030060000000552304010000604004700402216100000000015037
7060600020450040201777000560766040502006150023001070000000000550000
0000076421667500002000000000000000001053153200444427000301054607300
0055000000007021200560720000000000005700000040160600000062000322100
7470000760000022212160144000645120000004074010010060700000006700500
0000000000000000000070000000000003734000160000033006303000000020000
0000000000000014145646000055030400240042000000000720000300307627 1
0000070675000007750000000000020100000073051051200000037041473001 34
3000600000150522230060000260330000500470004040070626320000700313 2
7677416600000077000000462501227120340275752200000000000000001650000
7064006200000000461612770000306010001000070606050600064000000000
0034550000000705016050002164605100020000000003000000004000000606356
5006300002467373160063002741401040546254765060003000000000060243
0000047100000005012433400000615040033600000000070000400000014 2266
0060000000100201270300600174050000000405777000000006500000106 00700
0200000000004600000000600000000020000020500605200300570000002 1655
2070005570204000660003070043015515400726361064000076150215 0000030
000400000731554267000000200761500173750001003000000206010000000020
0304310000000020670005254671262450700362311450206400072000770 40050
0000001100040071773561300000000001246400025710000600003005707 13420
0000000527100700753160012004104406000004006600000000000000000000 1
0400205505000132077024500000002000005000706004435711771454410 00600
0100101304700034200002772614050070040577300304512500000052 1000000
3006603063132317110000641006740343740675005070001000000444000060 40
00400006200040000000000000000000000000214002060077037450453620 5
4130206000075300000555472430010411062000000000700000000000300232400
0000002030001600000000716651003000500000220271665000000267002 54000
0043422004025000004500020040051140210222007000210604000000000651 14
1661103755730401457000402674001400303000000500300720000103263700 00
0004073504033100000000400373650006000700051000100046000000037440 00
0010000375740000001200320200050000500057616005300011000000030000 03
02006720003014000000006001303500000001200010010006200370000000060
2000003000504000011000000470005575057000040045056331650037260100 00
3420010003100020030070000300000000300007050005220010000622726000 02
0000000007637220000000007020700706600000102300000000077111600075 47
0000000015311536150250000300000010000000000000000601001030030030 03
0000000061612114712457303744743007500600000400027706000007050 0070
0000000000000000000000000000006100040627070000007624221005700000
0010503000024341402200000000000000100100006000000774170400602 0000
3371050700000076103160024675150050000402720304005013000000000041 17

0000303714030003600000000030000060000127010000000000000000000001500
0016612600714300000304554740630000001000000073730000000004600355555
0000063010323200000035746000034744000006000000000000205563000000000
0000000000003600730000310675623074200000006063322161657347000202
3021170063010670007500072020000050706032160560005256130001464215466
6043267120005605100050100335235300030053006633077500000200271010000
1005000000000507050137750000000000000400003000000000500000061070000
0000300010243552000420001066570001072100020000000632000000000000014
0004407520502005550000000000641100037032552000300005140002533000000
0343040056303402150034500024000005430000350215360346000074600000011
6000000000005151554634141041700577001642374105030703200030026703000
0341422544004042740002007060000041230056655000100744500000771335000
0005000000000000200000626030440000310001050000010003006060600300000
0154600100000057105147116340760741322752617600020000400001064000000
0040405370030000000001000034100400000557571532004033400000000000000
0000000000000000036000072004320326400040000000000001000067642424000
0720000000060000000000000000200030200736022326003700704030030600011
2410023460025700325024424000023051003000426001375000000100000070000
7400004720006031103074000001007006002030005030404000005055320670000
5400000510007205040000301600002662020005100750334600604120121603000
0060000750173000005070000000000000000000542123333357010005047710000
2731530641325670100500001000053000100000233000020124000667070700000
4716230006000250000066454055000600160000032000300654700200017065757
3000077030042700000000050400377061060074300035341726404400000000000
1441625400060000050301000522502060000040007040000520000000000070000
3000000064004221302250020470000004004040000000005074742502005000575
1725006030000000746017334327047000005220000204003552735050000000000
0000003200710452740000103000510600070700000050150000024000075700000
5322445375306674640040006005564612644466556006000000000001440470055
0010000500003650602614061060640500000060000020000102000006200000000
0000000000000000000006000664325000300005412200662034220002710020
0000040420000100100024446414560100000000010000450660000000000700044
0064061020744307501300704000000000000000000000000000700040050000000
0000000051200005040054000000000000000000000102003760536000007705000
7151000000000006464600000276351644465236025655635400047430000000600
2261453237430020700000017061700015200000000100000013777123000000233
2732000002015106006141751557700001000030600443066775400300760000000
0000000363700000000000000000000000000000363034000050000340206246333
2000000042072372244215126020000000000751551375401060170000200000222
5000000000000021003430066000200064733256003605260000000000030000000

00501000161003070000604026706555000035700001000205506361000450036
02004400100004070300054006641301000000020435410600640020306530020
00040604002203500010047646160000070004560000070000000000000103006
40203000001570020700072005000500002000527642500020000000000000000
06006020060000000060066002730000000405073770025073000220000520200
30630617767542503005000000700400051770100465244500010700010100060
37562641070100015100007000323000000030700020024410570424001066006
02000027000400002600000034222003104700000300052530000040030240000
05714020060005657500070016006750005450060073675152100010000476000
06655563016140640000040000000000432000022100110162020000230031100000
70500000160036252045000003026302200320000005000340000000007000000
00000524400050700000060000300000000000000000130220500000070700404 07
03600500020004000070060200303131063754035400701000066050150200000
20700000000200000050600500000400706070706764405003100010000533530
07057775472060432436000000474063000000000000042100000002045000070
07160710000001536500060000000000000000007776600561001060001054001 00
623771264577052000010300600500000000000700300000125000110407000000 0
30600063000000000004500005000141004240357036005060000600010700000 0
00075240000113017001205510056006000037600072007600005007000005420 3
00051427126153005060000003004000061677302006200205005100000004130
02072360006000700747744465274460000034354500000660005645000004055
67610000000000060121110000056273544126575000037000000100000000002 00
00006107000335740200060076003260177700310320003506100567343503000
13474617731300000000000000760000674760000000300336537020000305326 7
31732000010027060051017300005000053000000000000000000003500264050 0
00060000001023030000000023741007000000000000720000070002070005000 0
30050005004002317277060004600020465500317305070002101000004021204
00000000002420001005476654173506000700670000000006000077012000070 0
70100000300000500050020006000000000100030004300746000004360000010 6
00000301100000020000007601301030000002230000543636620117370053675 00
00601003000600000460000000060000000310300000000000000000000000000 20
01767007615000040005000410050040005000000500300000000055340000000 5
53730615140477064155400037005000000432400000006034325343070000020
04531456700000000000000000000300675467123641424125354403722706 23
33317200300050040647350211624500000000000440640010003500005030000 20
00000400000703500004010600003006364654000002001002350650400120006
000010000000006000000040

Chapter 9
Language Structures in Cellular Automata

INTRODUCTION

The ingenuity of nature and the power of DNA have generated an infinite range of languages - including human language. The existence of these languages inspires us to design artificial cognitive systems whose dynamic interaction with the environment is grounded, at least to some extent, on the same basic laws. Modern scientific knowledge provides us with new opportunities to investigate and understand the logic underlying biological life. We can then use this logic to derive design principles and computational models for artificial systems. The technologies we apply in these studies provide us with new insights into the complexity of the processes underlying the evolutionary success of modern species. We have yet to fully penetrate the mysteries of these natural languages. Nonetheless, the literature suggests (Chomsky, 1957; Aronof & Rees-Miller, 2003; Bilotta & Pantano, 2006) that while the superficial features of different languages depend on different physical supports and different mechanisms, their deep structures share common rules. These constitute linguistic universals, organized at different levels of complexity, where each level has its own rules of composition. At all levels, we can consider these rules as "production rules" or even as rules of reproduction.

The use of formal languages to investigate CAs is not itself a novelty. (Minsky, 1967, Hopcroft, et al., 2007). In nearly all cases, the formal language provides an

DOI: 10.4018/978-1-61520-787-9.ch009

abstract representation of the system, based on a similar set of elements. Usually it is defined by a finite set of strings, constructed from a finite alphabet of symbols. Like a natural language (Chomsky, 2000), a formal language consists of the following elements:

1. Symbols are abstract entities with no intrinsic meaning; they are often said to be "non-interpreted".
2. An alphabet is a finite set of symbols. Alphabets are often represented with the Greek character Σ. Other letters can also be used. The formula B = (0, 1) states that B is an alphabet consisting of two symbols, "0" and "1". The formula C = (a, b, c) states that C is another alphabet made up of 3 symbols "a", "b", "c". Punctuations marks and spaces are considered as symbols belonging to the alphabet. At times, descriptions of the alphabet use them as metasymbols.
3. Words are finite sequences of symbols belonging to the alphabet. Thus, "001010" and "110" are words constructed from alphabet B; "caaabbb" e "ab" are words built from alphabet C. The beginning and end of a string are often shown with vertical bars. The length of the string is the number of characters between the bars. For example "|00111101|= 7"; "|aaabbb|= 6". Null strings contain no characters and have a length of 0. These strings are usually shown by the Greek character "E".
4. A grammar is an abstract structure providing a precise representation of the formal language. This consists of a system of rules providing a mathematical representation of a (usually infinite) set of finite sequences of symbols (strings) constructed from an alphabet which is also finite. There are two kinds of formal grammar. A generative grammar consists of a system of rules with the ability to generate all possible strings belonging to a given language. It is a computational system in which an algorithm generates language strings. The rules make it possible to iteratively rewrite strings, beginning with a pre-defined initial symbol. Generative grammars are modeled on the way a language is written. By contrast, an analytical grammar is a system of rules that starts with an arbitrary input string, and applies an algorithm to the string, determining a series of outputs. When these outputs are combined using Boolean operators they produce a Yes/No response, determining whether the string belongs to the language described by the grammar. An analytical grammar is a language parser. The first author to formalize generative grammars was Noam Chomsky (Chomsky, 1957; 1980), who proposed a taxonomy of grammars, known as the "Chomsky hierarchy". The grammars identified by Chomsky include context-free grammars and regular grammars. These respectively determine context-free languages and regular languages. The difference is that context-

free grammars have a broader range of production rules and can generate more formal languages than regular grammars.

The central argument of this chapter is that Cellular Automata have their own internal grammar. This recalls the theory of Universal Grammar. The theory does not describe any specific language and makes no claim that all languages have the same grammar. Rather, the goal is to identify a series of innate rules allowing children to acquire languages (emergence) and to learn to produce valid sentences (productivity). By analogy, the Universal Grammar for Cellular Automata does not provide a description of specific systems. Rather it describes the general mechanisms underlying the emergence of complex structures. The productivity of CAs is expressed in rules of production expressed in patterns. As in other forms of communication, we find these patterns at all levels of CA *semiotic languages*, (Sebeok, 1972; Bateson, 1968; Eco, 1984; Deely, 1980; Emmeche & Hoffmeyer, 1991). For instance, if we analyze the number of regular domains in a system, the periodicity of these domains, or the number of gliders and other basic structures it produces, we discover that the semiotic emergence of patterns in domains and more generally in space-time diagrams, is the result of intrinsic (innate) rules. Once we identify these rules and treat them as components in a genetic code, we can build semiotic structures with a syntactic and semiotic organization comparable to that of natural language. The existence of such structures demonstrates CAs' extreme productive creativity. Thus, the main aim of this chapter will be to demonstrate that 1-D CAs have an internal semiotic structure, similar to that of a natural language and that this is not the result of external formalization but of internal organizational structures "innate" to these systems. In other words, a CA and its elementary particles can be seen as a structured set of distinctive elements, comparable to the elements in a natural or formal language. These include:

1. Phonemes: individual cells, with their own specific state and characteristic functions, obeying principles of organization similar to those governing phonemes in natural language;
2. The rules allowing phonemes to combine in structured particles similar to morphemes (structures that interact to produce emergent complex behavior);
3. A lexicon (domains, gliders, emergent particles);
4. Syntactic rules making it possible to construct higher level structures from elements in the lexicon.
5. Semantics (the local and global behaviors of particles, the overall behavior of the system). Semantics could help us to classify different levels of CA behavior. One key notion is that of a *semantic field* (Trier, 1931; Eco, 1975). This is a set of concepts whose mutual relations define a whole field of related

meanings. At the lexical level, these concepts map a set of terms linked by paradigmatic relations in a "lexical field". One example of a semantic field is the naming of colors. The names of colors have been studied both by psychologists (Rosh, 1975; Lakoff, 1987) and by researchers in linguistics. Different languages and cultures use different sets of color names. From a superficial viewpoint, it would thus appear that different languages structure the domain in different ways. All, however, refer to the same chromatic range and share the same underlying organization. In terms of CAs, we can use the concept of semantic fields to describe the global dynamics of the system and its basins of attraction. In other words, we can take a specific attractor as a starting point, and study all the attractors that fall into the same basins.

In what follows, we will begin with a brief description of semantic fields in CAs. To use a linguistic metaphor, we will describe the equivalent of a set of words belonging to the same conceptual domain, for example, river, stream, brook, feeder, rivulet, delta, estuary, outlet, river mouth, tributaries, etc. We will then focus on CAs' genetic code. We will show that if we map a $k = 4, r = 1$ CA onto DNA bases, there is no need for the external rules required by the transformation and production rules of generative and transformational grammars. If we start from a simple seed - for example a seed that gives rise to a glider - and watch the evolution of the system, we observe that the ability to create words and quasi-words is an emergent property of the CA's internal rules. The same rules also give rise to replication, recursivity and palindromes, as well as to left- and rightwards shifts in groups of morphemes.

We will describe examples of the generation of regular domains and examples of experiments in which we couple domains of different types to create what we have called "sum domains". This is similar to the production rules we observe in language when we join "ash" + "tray" to produce "ashtray" or when we join "work" to "-ed" to produce "worked". We will go on to analyze the relations between basins of attraction and regular domains and the production rules for regular domains and those for gliders. The latter share the same compositional properties as the productive models underlying sum domains. Here too we find recursive features.

BACKGROUND

A 1-D CA, with k states and M cells, has k^M possible states. Given that the system is autonomous and deterministic, the system will always return to one of its previous states within a maximum of $t \leq k^M$ cycles; within a finite time, the system will always enter a limit cycle or settle into a fixed point. These fixed points and limit cycles are *attractors* for the dynamics of the CA. Given an attractor a, we can collect all states of the CA whose trajectories end in the attractor. These constitute the attractor's *basin of attraction*. Obviously any given state of the CA can only belong to one attractor. The set of all attractors and their respective basins of attraction describes the global dynamics of the CA.

Figure 1 shows the global dynamics of ECA 112, with *M*=5. In this case, the CA has $2^5 = 32$ possible states, organized into 5 basins of attraction, made up of 15, 5, 5, 5 and 2 states respectively. The first 5 elements are limit cycles each with a periodicity of 5. The last is a fixed point with code 00000.

Given a state $x(t)$ of the CA, the state $x(t-1)$, that leads to $x(t)$ in the next step is known as $x(t)$'s pre-image.

Obviously, a given state may have more than one pre-image or none. In this latter case, it is known as a "Garden of Eden". The only way in which a "Garden of Eden" can appear in the CA is if it is defined as an initial condition. The CA shown in Figure 1 has 11 "Gardens of Eden" 16 states with a single pre-image and 5 states with 3 pre-images each.

Figure 1. The global dynamics of ECA 112, with $M = 5$

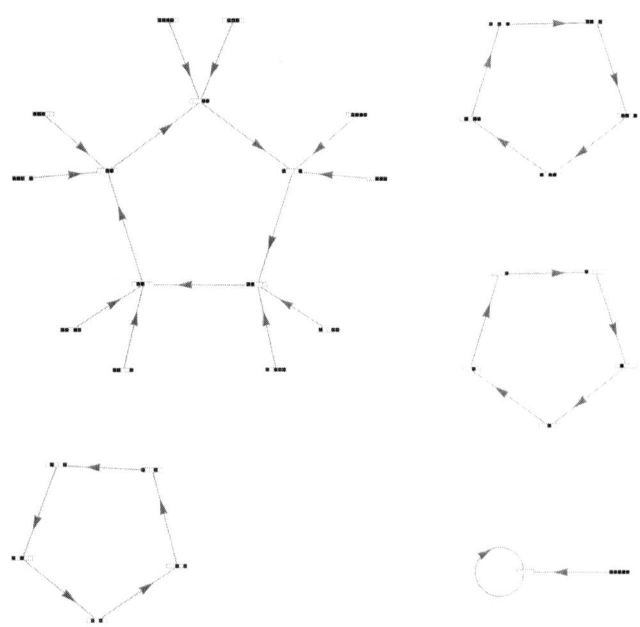

Consider a CA with $k = 4$ and $n = 3$ (i.e. a radius of 1). The code for the CA is C9.1.

We have already observed in Chapter 4, that with $M = 4$, the CA has 14 basins of attraction (see Figure 2); 13 lead to limit cycles; only one leads to a fixed point. One limit cycle has a period of 20 (domain1), 4 have a period of 4 (domains 2, 4, 7 e 8); 8 have a period of 2 (domains 3, 5, 6, 9, 10, 11, 12, 13). Appendix at the end of this Chapter describes the basic characteristics of each basin.

As we saw in Chapter 4, the number of basins of attraction changes rapidly with the number of cells in the CA. In general, when the number of cells increases from an odd to an even number, we observe a large increase. When it increases from an even to an odd number, the number of basins remains constant or falls. The size of the basins increases very significantly, as we can see if we observe the effect of changes on the largest basins of attraction in the previous example.

Figure 3 shows images of basins of attraction for rule C9.1, with $M = 5$, $M = 6$ $M = 7$, $M = 8$, $M = 9$, $M = 10$.

There are many possible ways of representing these basins and the images produced in this way are very spectacular (Wuensche, 2000).

Figure 2. Basins of attraction for rule C9.1 with M=4

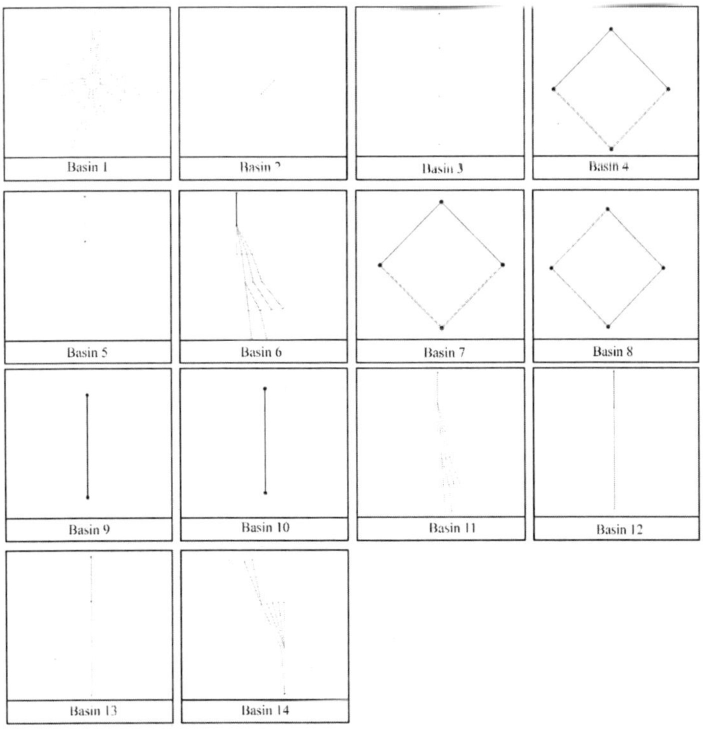

Figure 3. Large basins of attraction for C9.1 with $M = 5, 6, 7, 8, 9, 10$

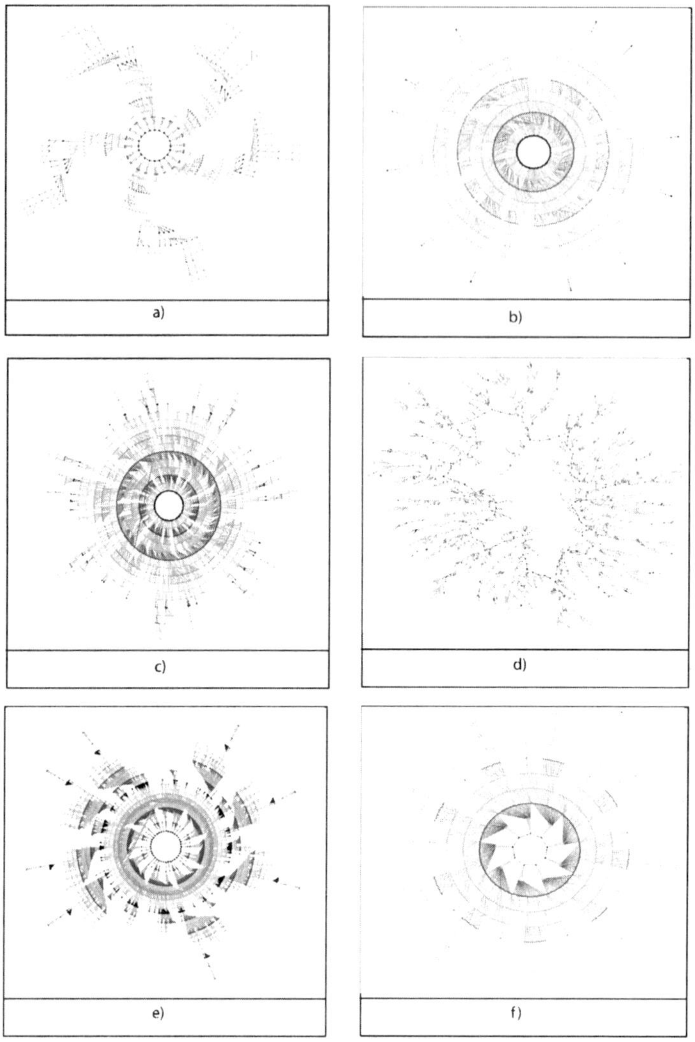

MAIN FOCUS

Let us now focus on issues relating to the formal representation of regular domains and gliders and the rules governing their composition. The catalog of a CA represents the set of all its regular domains and gliders and all their interactions. It thus supplies a high level description of the CA that is independent of the local rules that regulate the development of individual cells (Crutchfield & Young, 1989; Crutchfield, 1994). With this approach, the construction of a catalog is computationally very expensive and is only possible for CAs with a limited number of cells. Later on, we will provide rules, showing how it is possible to generate elements of a catalog for large lattices and describe a number of laws, regulating the relations between regular domains and between regular domains and gliders, by using our own method.

Genetic Approach and the Emergence of Language Structures in CAs

To gain new insights into the relationships between these elements, we examine them as through they were elements in a language, identifying the equivalent of characters, words, quasi-words, palindromes, and left and right repetitions. In this analysis, gliders whose data generate regular domains are considered as words; quasi-words are generated from two words; words are governed by rules of composition and come together to create sentences. For the purposes of genetic analysis, (see later chapters), we will represent the characters in our code as if they were bases in a sequence of DNA. In this code:

$$0 \rightarrow A \text{ (Adenine)}$$
$$1 \rightarrow T \text{ (Thymine)}$$
$$2 \rightarrow C \text{ (Cytosine)}$$
$$3 \rightarrow G \text{ (Guanine)}$$

We can thus represent the CA with the code: AGTATTATCTGTTTCTAGAAGCCATTTACCAAGAATAATCAAGTTACGT-TAGCCTCACTGTCCG.

Let us analyze some of the CA's basic structures. In what follows, we use the terms word and quasi word to indicate elements of a cellular automaton which give rise to distinctive behavior in the pattern (such as gliders and regular domains)

The first glider (A), undergoes the following transformations:

- **A** word: a union of more than 1 A would become a periodic domain (see the first couple of images in Figure 4.
- **AA** word: this sequence generates a regular domain of white and yellow horizontal stripes (see the second couple of images in Figure 4.
- **CA** word: this sequence creates a regular domain.

- **CAA** word: this sequence creates a regular domain.
- **GA** quasi word: this sequence does not generate a regular domain; it is the equivalent of a sequence of characters that only acquires a meaning when it is attached to the end or the beginning of another word. First, the sequence generates a transient, then a permanent state (see Figure 4) fifth couple of images).
- **T** quasi word: this sequence is not a glider but generates a glider.
- **TC** quasi word: like T, this sequence is not a glider but generates a glider.
- **TG** quasi word (as above).
- **TTA** word: this sequence generates a regular domain.
- **TTT** quasi-word: this sequence generates a transient which is then transformed into a regular domain.
- **ACC** word: this sequence generates a glider identical to the glider produced by word A. It then becomes a regular domain of diagonal stripes.
- **ACCA** word: this becomes a regular domain. The word A is repeated.
- **GATC** variant of quasi word GA - a sequence that is not a glider but generates glider TC.
- **TTTG** quasi word: this generates first a yellow transient and then a regular domain (it seems to be repetitive).
- **ACCAA** quasi word: this sequence is produced by combining word ACCA + word A, or words ACC + A+ A. It does not produce a new word. The sequence is a palindrome of the following sequence. In the following sequence (a palindrome of the first) we see the same glider on the left.
- **AACCA** quasi word: this structure includes a repetition of AA on the left
- **ACCCA** word: the presence of an additional C generates a broader strip in the regular domain. Despite this modification, the basic structure ACCA is conserved. If we add large number of Cs, we eliminate A's ability to replicate.

The end result is the single glider AA. If we add additional Cs to glider ACCCA the two elements rotate around each until they join in a single structure.

- **ACCGA** quasi-word (variant) (or a combination between word ACC and quasi-word GA). If we add additional Cs, the two gliders ACC and GA stand out clearly. By adding 51 Cs, we obtain glider GA .

We can continue this process for the other gliders. Figure 5 provides a schematic representation of the process. Its two key features are the way gliders interact to

Figure 4. Words and the domains they generate

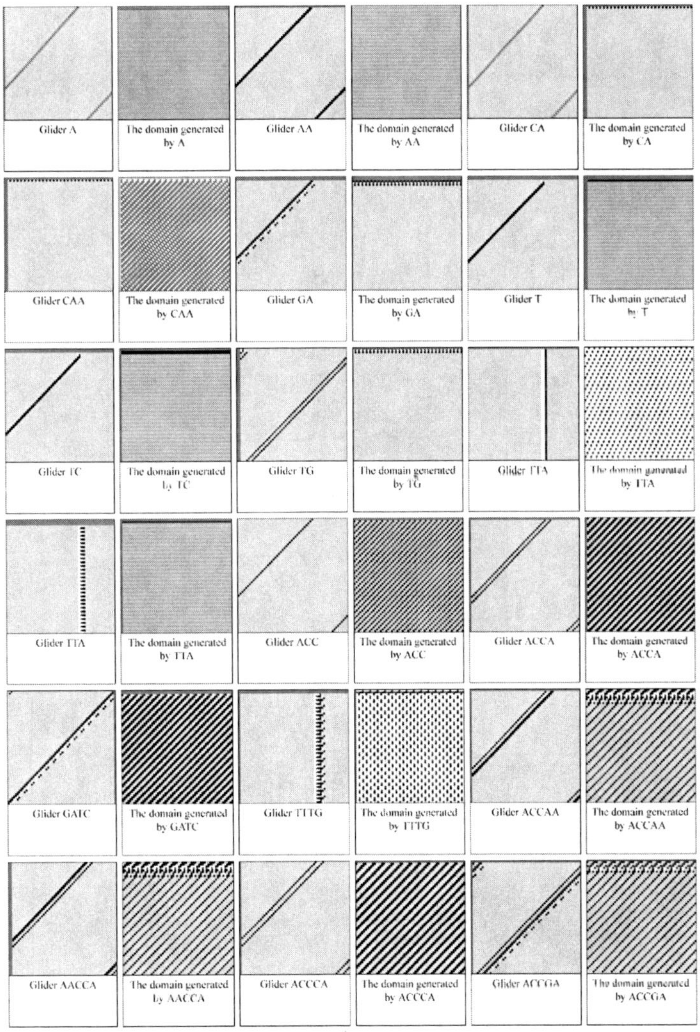

form emergent structures more complex than themselves and the tight correlation between gliders and regular domains, which are themselves carriers of information.

This means that the emergence of patterns in domains and, more generally, in space-time diagrams is the result of intrinsic (innate) rules of composition as they were linguistic elements, at different levels of complexity. We can use this knowledge to build semiotic structures with a syntactic and semiotic organization comparable to that of natural language. Just as spoken language gives people the

Figure 5. Generation and self-reproduction of gliders of gradually increasing complexity

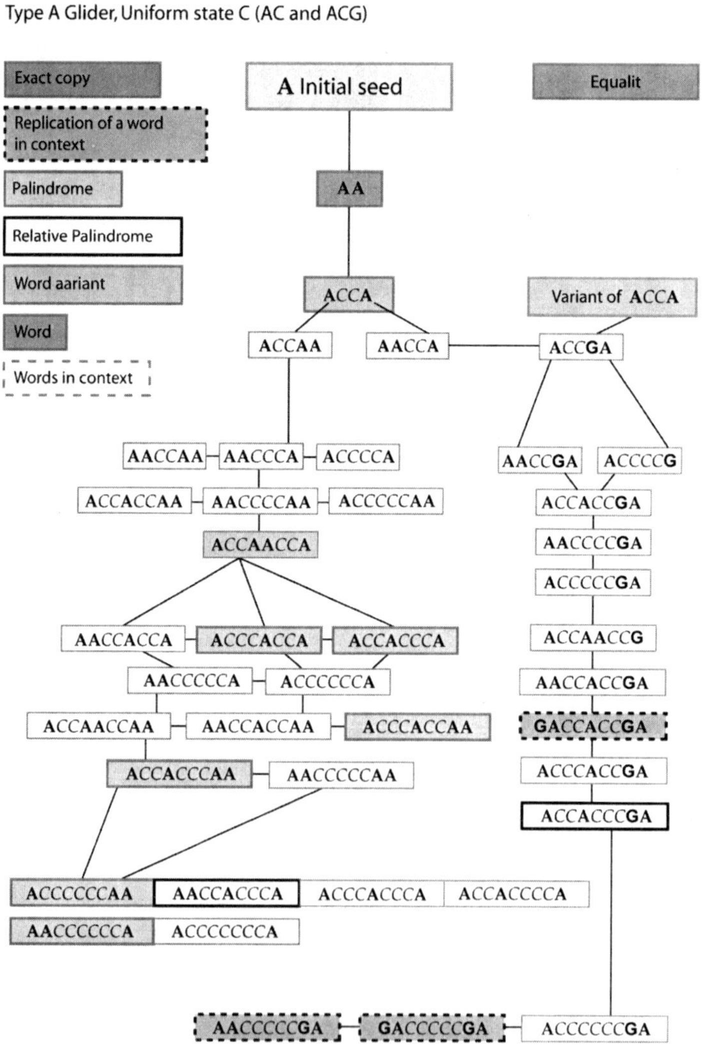

ability to create an infinite variety of sentences, CA languages give us the power to create an infinite variety of configurations. An analysis of a CA that identifies its main components produces common vocabulary that allows us to organize CA behavior in a productive way.

Regular Domains

As an example, let us apply our linguistic metaphor to the case of regular domains.

Let $S = \{0, 1, \ldots, k-1\}$, be an alphabet representing the possible states of a cell in the CA and let λ_m be a string of length m composed of symbols in S. Let Ω be the set of all possible strings (or words) composed of symbols in S. Then:

$$\Omega = \{\lambda_m / \lambda_m = s_1 s_2 \ldots . s_m, s_i \in S\} \tag{9.1}$$

Now consider a function:

$$f_{\mathrm{mod}\,h} : \Omega \to \Omega \tag{9.2}$$

The function shifts the characters of λ_m .by positions. The boundary conditions mean that a character originally in position $i < h$ will shift to position $m + i$ In this way, the length of the string is still m. In brief, function f defines a translation of *length h* or a *translation module h*. If we consider two strings, λ_m, ρ_m of the same length, and an application (9.2) such that:

$$\rho_m = f_{\mathrm{mod}\,h}(\lambda_m) \tag{9.3}$$

The characters belonging to string ρ_m will be shifted by positions with respect to the characters in string λ_m. For example, with h=3, the characters of $\rho_6 = 032120$ shift by 3 positions with respect to $\lambda_6 = 120032$.

Now consider a string λ_m composed of the concatenation of n substrings λ_h, such that:

$$\lambda_m = \lambda_h \lambda_h \ldots \lambda_h \tag{9.4}$$

Obviously, $m = hn$. Now consider a1-D CA with *k* possible states for each of its *m* cells. The overall state can thus be described by a string λ_m, as in (9.1). After time *p*, we observe (9.3):

$$\rho_m = f_{\mathrm{mod}\, s}(\lambda_m)$$

where ρ_m is the state of the CA after time *p*. The characters of the string have shifted by positions. In this case, we say that λ_m is a *regular domain modules*, with *p phases* π_i, where $i = 0,1,..p-1$, and p is the period of the regular domain. Obviously, each phase can be represented by a string of length *m* over alphabet *S*. Let C_S be the set of all phases of the regular domain.

The **Shift Velocity** for the domain is given by:

$$v_s = \frac{s}{p} \tag{9.5}$$

Suppose that for each phase there exist a set of strings c_h^i in the form given in (9.4). Thus:

$$\pi_i = c_h^i c_h^i \ldots c_h^i \tag{9.6}$$

In this case, the c_{ih} are the generating strings for the different phases and we can restrict our analysis to these strings alone. If each of the generating strings is made up of a single character, we say it is *periodic.* When all the phases for a regular domain are the same, we say it is *homogeneous.* Obviously, a rest state generates a homogeneous domain.

To clarify, consider C9.2 (listed in the Appendix of this Chapter), which has $k = 4$ and $r = 1$. Figure 6 shows the spatio-temporal pattern produced by a CA with 300 cells ($m = 300$) after 300 time steps, beginning with random initial conditions. The patterns contain several regular domains.

The generating strings for these domains have different shapes and sizes. Figure 7 shows images of the domains contained in this rule.

The lengths of the generating strings vary between 4 and 9, the period between 1 and 3 and the module between 1 and 2. The shift velocities are 1 (maximum), 2/3

Figure 6. Pattern produced by a 4-state rule (C9.2)

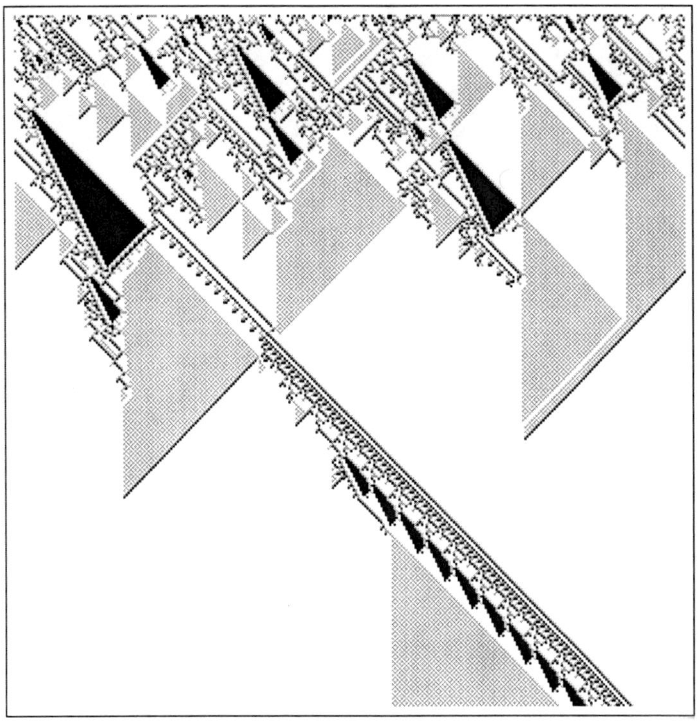

and ½ (minimum).

The characteristics of the domains shown in Figure 7 are reported in the Appendix.

If we carefully analyze the structure of the strings, we observe that they contain repeating elements. It is these elements that allow them to generate new domains. For instance, we can obtain new domains from domain 312020200 by inserting string (20) or by using the string (33), obtained from domain 33120200. In Figures 8a and 8b, we see two new domains (31202*0202020202*00 and *333333*1202*0202020202*00) obtained in this way. (The italics show the new elements in the strings).

Domains 1121010 and 111200 have a shift velocity of ½. By introducing string (111), we can use them as the basis for new domains. For example we can obtain the domains:

1121010*111*, 111200*111*, 1121010*111*111200.

The last domain is obtained by concatenating the strings for the first two domains, using sting (111) as glue (see Figure 8d).

Obviously, this kind of behavior is not restricted to this particular CA. Rather it

Figure 7. Regular domains produced by the C9.2 rule

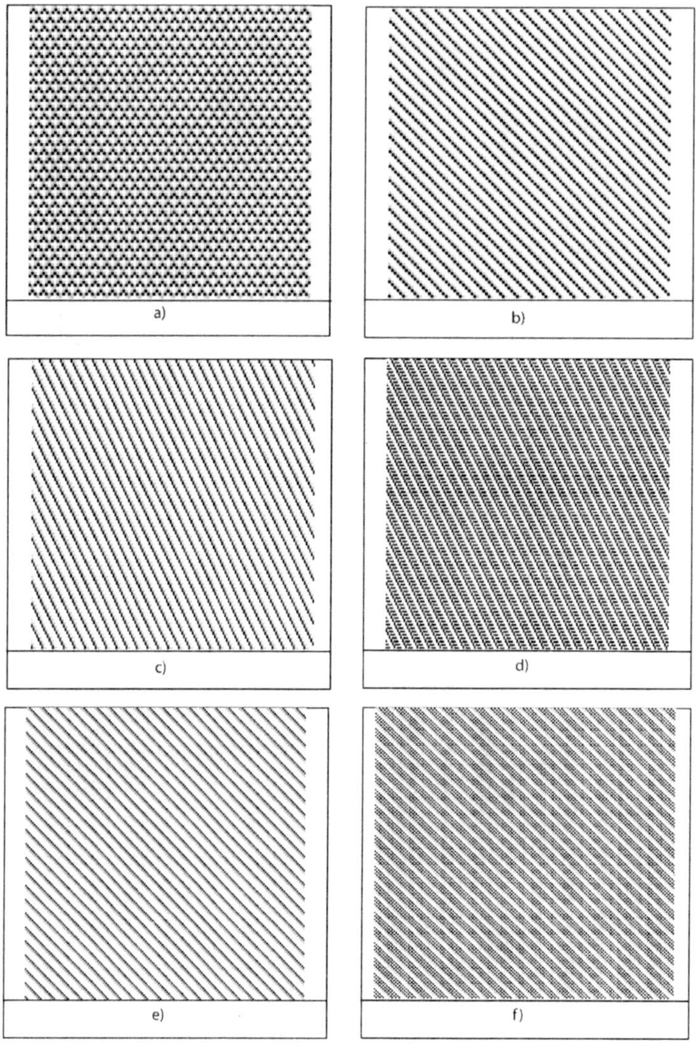

is a general characteristic of complex rules.

For all these rules, we observe that:

1. There exists a large range of regular domains, with different sized generating strings;
2. The shift velocities, module and period of the domains vary;
3. The strings representing the domains contain substrings that can be used as lower level elements in the composition of new domains or as "glue" tying together domains with the same shift velocity

Figure 8. Row 1: the left panel shows the domain generated by string 31202020202020200 (a); on the right we see the domain generated by string 333333312020202020200 (b). Row 2: the left panel shows the domain generated by string 1121010111(c), the image on the right displays the domain generated by string 1121010111111200(d).

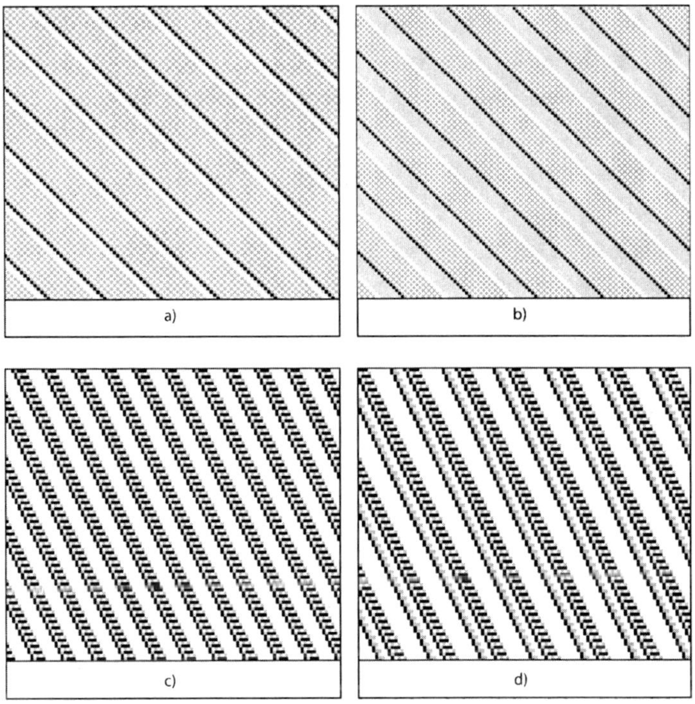

These observations suggest that we can identify basic components underlying complex behaviors and rules for the composition of these components into larger units.

Basins of Attraction and Regular Domains

The global dynamics of CAs are tightly associated with the presence of regular domains. For example, a homogenous domain represents a fixed point in CA dynamics. Limit cycles can also be used to obtain regular domains. Consider a CA composed of m cells, with a limit cycle of period p. Let l_c be the state of the limit cycle for a given basin of attraction. In this case, we can use (9.4) $\lambda_m = \lambda_h \lambda_h \ldots \lambda_h$ to generate a regular domain for the CA, using:

$$\lambda_m = l_c l_c \ldots l_c \tag{9.7}$$

In the Appendix of the Chapter, we highlight the domains for the basins of attraction, described in the background to this chapter.

It is interesting to note that basin 9 generates the same domain as basin 10, that basin 11 generates the same domain as basin 6 and that basins 12 and13 generate the same domains as basin 5. Basin 14 (a fixed point) gives rise to a homogeneous domain. All in all, the basins generate 10 independent domains.

Compositions of Regular Domains

As we have seen in Chapter 4, regular domains depend on basins of attraction whose number depends on the number of cells in the CA. As a result, the number of possible regular domains is enormous, making it impossible to create a complete catalog. The identification of basins of attraction is a time-consuming task. As we saw in the previous section, we can compose regular domains to create new domains, using strings ("enzymes") to glue them together.

We have already shown that in many CAs it is possible to compose regular domains by introducing an application and a so-called composition operation:

$$\begin{aligned} &\& : \Omega^2 \to \Omega \\ &(a,b) \in \Omega^2 \Rightarrow c = a \,\&\, b \in \Omega \end{aligned} \tag{9.8}$$

This makes it possible to join two words in Ω to make a new compound word. For example, $a = 1003421$ and $b = 0034233$; $a \,\&\, b = 10034210034233$.

Ω represents the set of all possible words constructed from alphabet . This includes words of any length, and not just *m*. As a result (10.1) becomes:

$$\Omega = \left\{ \lambda_m \,/\, \lambda_m = s_1 s_2 \ldots . s_m, s_i \in S, m \in N \right\} \tag{9.9}$$

Given a CA, we consider the subset $\Sigma \subset \Omega$ containing all generating strings for the CA's regular domains. Σ is thus composed of all $c_s^{\,i}$ in Ω. What is the nature of Σ? Are there laws making it possible to compose elements in Σ into new elements?

It is obvious that such laws exist (Bilotta & Pantano, 2001). For example, it is possible to join a phase in a regular domain to itself, obtaining a phase in another regular domain, i.e.

$$c_s^{\ i} \,\&\, c_s^{\ i} \in \Sigma, \forall i : 0 \le i \le p-1 \quad \text{and } \forall c_s^{\ i} \in \Sigma \tag{9.10}$$

However, if the phases are different, or if they belong to different domains, the composition operation will not always produce a phase in Σ In other words, Σ is not closed with respect to the composition operation $\&$.

Given a regular domain, consider set $c_s = \{c_s^{\ i}, i = 0,1....,i-1\}$, containing all phases of the domain. Obviously, $c_s \subset \Sigma$. If $c_s^{\ i} \,\&\, c_s^{\ j} \in \Sigma, \forall i : 0 \le i \le p-1$ and $\forall i : 0 \le j \le p-1$ We say that this kind of domain is *complete.*

Consider a uniform domain entirely composed of symbols representing the rest state. This domain is complete.

If $\exists c_s^{\ i}, c_s^{\ j} \in \Sigma, i \ne j$ such that $c_s^{\ i} \,\&\, c_s^{\ j} \in \Sigma,$ we say the regular domain is *partially complete.* Phases $c_s^{\ i}$ and $c_s^{\ j}$ are said to be *compatible.*

It is important to note that the relationship is not symmetric. Even if $c_s^{\ i} \,\&\, c_s^{\ j} \in \Sigma$ there is no guarantee that $c_s^{\ j} \,\&\, c_s^{\ i} \in \Sigma$.

The ratio between the number of compatible phases and the total number of phases represents the *completeness* of a compatible domain.

Consider two distinct domains. If $\exists i, j : c_{s_1}^{\ i} \,\&\, c_{s_2}^{\ i} \in \Sigma$, we say that domains $c_{s_1}^{\ i}$ and $c_{s_2}^{\ i}$ are *compatible.* The domain generated by $c_{s_1}^{\ i} \,\&\, c_{s_2}^{\ i}$ is called the *sum domain.* We can compose a domain many times and obtain sum domains of different lengths, containing different phases.

If a regular domain, c_s is obtained from the composition of two regular domains (i. e. if $\exists c_{s_1}^{\ i}, c_{s_2}^{\ i} \in \Sigma : c_s^{i} = c_{s_1}^{\ i} \,\&\, c_{s_2}^{\ i}$), c_s is called a *compound domain.*

s_1 and s_2 are called the left and right difference domains: $s_1 = s - s_2$ and $s_2 = s - s_1$.

Gliders

As we have already seen in the previous chapters, one of the most startling manifestations of emergent complexity in CAs is the generation of gliders. As can be seen in Figure 7, gliders come in different shapes and sizes. In the figure, we see

4 gliders which either have very long periods (Figures 9a and 9b) or are very large (Figures 9c and 9d. The rule codes are C9.3 – C9.6 and the generating codes are:

1202011 (see Figure 9a)

200201 (see Figure 9b)

1303032331322301303032033313033113230030300133330322000133123032303133003312300133001323323300022303222300130333313111323032223303001 in the uniform state (3) (see Figure 9c)

1010010100023232221032320232323222100102023232322222222220232322221010102023232322222220100130322221010102101010 between regular domain (10100) (left) and (1010) (right) (see Figure 9d).

From one point of view, we can see gliders as a perturbation within a regular domain or between two regular domains. We can express the perturbation as follows.

Let λ^1_m and λ^2_m be two strings representing two regular domains in the CA. Let λ_1 and λ'_1 be two substrings of λ^1_m and let λ_2 and λ'_2 be two substrings of λ^2_m. Let us now suppose that the overall state of the CA is represented by a string

Figure 9. The top two figures (a and b) represent gliders with a long period; the lower ones (c and d) show very large gliders

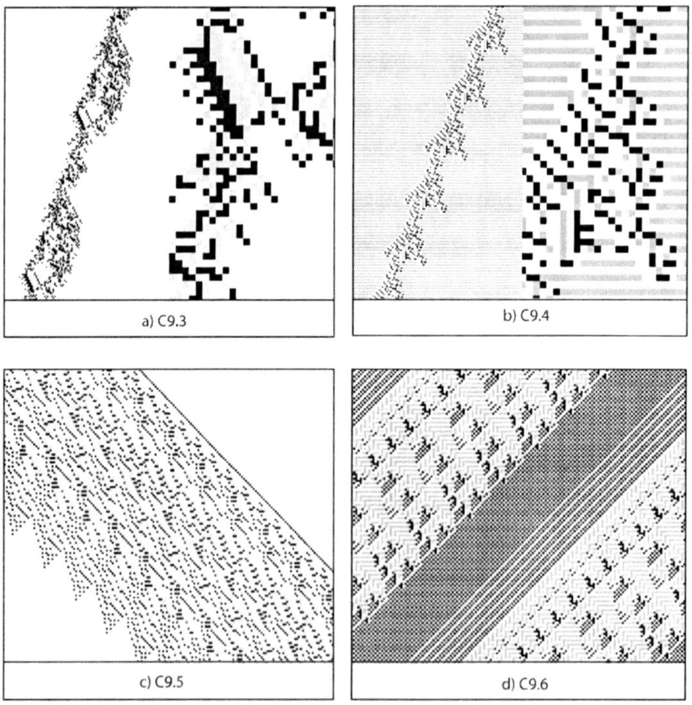

λ_m such that $\lambda_m = \lambda_1 \& \lambda_G \& \lambda_2$ and that the state of the CA after time *p*, is represented by string $\lambda_m{}' = \lambda_1{}' \& \lambda_G \& \lambda_2{}'$. In this case, λ_G is a glider belonging to the CA. The glider moves with $v_G = \frac{h}{p}$, where represents the shift of the glider during time *p*.

Over time p λ_G will go through p phases represented by λ'^{i}_{G}. This is the glider's time dimension (see Figure 9). If λ_G has a length of zero we say that it is *degenerate*. In general, gliders can move either towards the left or the right. We can thus assign them either a positive or a negative velocity. In other cases, the velocity will be zero. In these cases, we will say that the glider is s*tationary*. Collisions between two gliders, traveling at different velocities, can produce one or more new gliders, or the annihilation of both gliders. Alternatively it can create a situation of chaos.

Like regular domains, gliders can join together to form new gliders or more complex structures. Formally we can define the composition rule as follows:

$$f : AxA \rightarrow A \tag{9.11}$$

where *A* indicates the set of gliders belonging to a given CA. If we represent the composition operator with the symbol $\oplus$, **we obtain**

$$z = f(x, y) = x \oplus y \text{, with } x, y \in A \tag{9.12}$$

Consider the CA with code C9.7. Figure 10 shows the spatio-temporal pattern produced by the CA.

Now, set *A* comprises a series of gliders:

$$A = \{\alpha, \beta, \gamma, \delta, \ldots\} \tag{9.13}$$

in which $\alpha = (10001)$, $\beta = (20001)$, $\gamma = (310320020001)$, $\delta = (2202000000011220001)$,. The composition operation over the set, can be defined in the same way as in (9.12):

Figure 10. The spatio-temporal pattern produced by C9.7

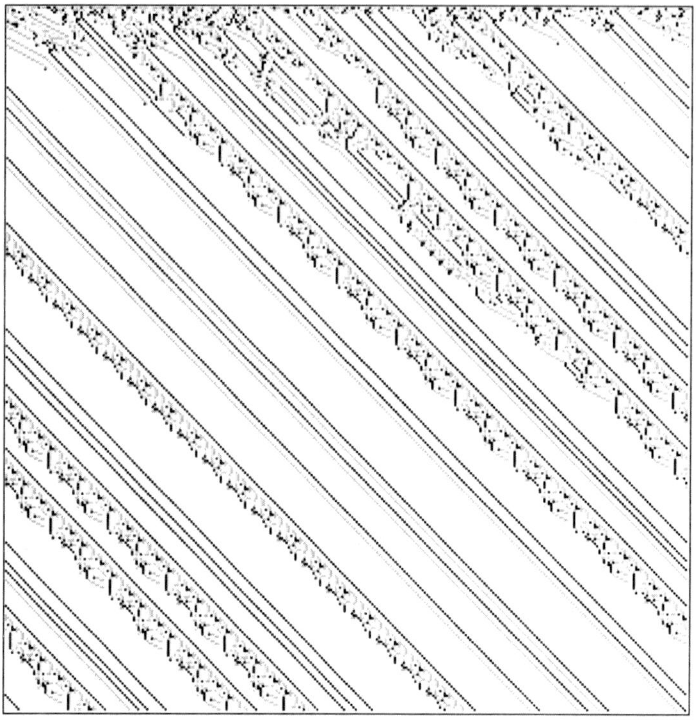

$$z = x \oplus y = (x000y)\,,\ \forall x, y \in A \qquad (9.14)$$

By iterative application of 9.14, we obtain the following gliders:

$$\alpha \oplus \alpha = (1000100010001)$$

$$\alpha \oplus \beta \oplus \gamma = (100010002000100031032002 0001)$$

$$\alpha \oplus \beta \oplus \gamma \oplus \alpha \oplus \delta = (10001000200010003103200200010001000100022020000001122 0001).$$

(9.14) tells us that the gliders cannot be joined together directly. To join them, we need the *connector* $a = (000)$.

(9.14) is not the only possible composition rule for the CA. Consider, for example another set of gliders, *B*, defined as follows:

$$B=\left\{\rho,\sigma,\tau,\nu,\omega,\zeta,,\ldots\right\} \text{ where } \rho=(22),\ \sigma=(20022),\ \tau=\left(3020022\right),$$

$$\nu=\left(31020022\right),\ \omega=\left(3102000022000001020000020300022\right)$$

$$\zeta=\left(2200003102000022000001020000020300022\right),$$

$$=\left(2200002200002200002200003102000022000001020000020300022\right) \quad (9.15)$$

(9.12) thus becomes:

$$z=x\oplus y=x0000y,\ \forall x\in B \text{ and } y=\rho,\sigma \quad (9.16)$$

For example:

$$\rho\oplus\rho=22000022,$$

$$\rho\oplus\sigma=22000020022,$$

$$\tau\oplus\rho\oplus\sigma=30200220000220000200222,$$

Now the connector is $b=(0000)$.

There are also other, more complex laws of composition. New, more complex gliders belonging to *B* can be created by adding gliders of types ρ or σ to the left hand side of the initial glider, followed by four or five zeros.

It is important to note that the ability to join gliders together is not just a property of this specific rule. On the contrary, it is shared by a very large number of complex rules. Gliders tend to join together whenever they travel at the same speed.

If we repeat the coupling many times, we obtain regular domains, as seen in Figure 11.

This latter phenomenon confirms that gliders and regular domains share a common nature and that certain forms of interaction can transform one into the other and vice versa.

Figure 11. Examples of regular domains created by repeated coupling of gliders

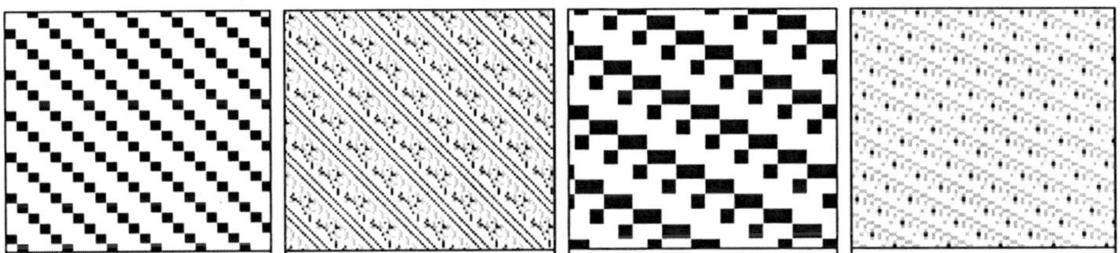

a) 100010001000 | b) 1000001230201320000 1000 | c) 200220000200220000 | d) 20310200220000

The studies described in these sections leads us a number of conclusions concerning regular domains and their composition properties:

1. The number of regular domains increases with *M*, the number of cells in the CA;
2. The period of domains is almost always a multiple or a submultiple of *M*;
3. The number of sum domains increases with *M*;
4. There are a limited number of possible periods for sum domains; these are always multiples or submultiples of *M*;
5. The level of a partially complete period decreases with *M*.

FUTURE TRENDS

The development of computational tools exploiting the proprieties of CAs is a key-point in our research program. The modularity of domains means that they share one of the key properties of natural language: they can produce a practically unlimited number of new words (neologisms) and sentences. If we consider domains as examples of emerging structure, we can state that the number of these structures increases with the number of cells in the CA. The ratio between the size of the structure and its period is usually a simple one (1:2, 1:2, 2:3, etc.). This confirms the close relationship between complexity in the spatial and temporal domains. We observe that these are the same ratios we find in Pythagoras' theory of music (Frova, 2006). Using composition rules to join base domains together, we can create sum domains which can then be used to produce new sum domains in a potentially infinite process (while in other cases, we can apply this procedure just once).We know that no catalog or high level description of CAs can be exhaustive. However, the second level descriptions, we have explored in this chapter, suggest it may still be possible to derive a number of embryonic laws. We can then use this process to derive abstract languages or to implement cognitive computational models.

CONCLUSION

In this Chapter we have described a new approach to the construction of artificial semiotic languages for CAs. The criterion we have used in constructing these languages is similar to the principles underlying natural languages. Rather than being formally derived from a computational system (e.g. a formal language constructed as a Turing machine or a generative grammar in the Chomsky hierarchy) it is based on emergent behavior and production rules implicit within the system itself.

After modeling CAs as a system with structures corresponding to those of a natural language, we have demonstrated the presence of some of the fundamental components of semiotic language. More specially the spatio-temporal patterns generated by CAs display aggregations of cells that provide the CA with specific meanings.

These rules of composition point to the presence of an *emergent language,* offering a higher level description of the CA than it is possible to obtain from the local rules. This means that we can use the CA's attractor dynamics to generate a practically unlimited number of new structures (both domains and gliders).

By identifying the components making up a CA, creating inventories for the different levels of the emergent language, analyzing the semiotic structure of the patterns and describing the morphological and syntactic rules that emerge from the analysis, we can sketch out a new approach to bio-semiotics that goes beyond the catalogs, introduced by computational mechanics.

The range of empirical methodologies used in this chapter clearly point to the emergent character of these structures.

1. An analysis of global dynamics shows that CAs are organized into structures similar to semantic fields - groups of attractors sharing the same basins of attraction. We have discovered that the greater the number of cells described by the input strings for these attractors, the greater is the size of their respective semantic fields.
2. We have introduced a code, mapping CAs to DNA bases. Using this scheme, we have shown the emergence of words, quasi-words, replication, recursivity, and palindromes, as well as left- and rightwards shifts in morpho-syntactic groups. This implies that pattern evolution involves the use of specific compositional and creative processes to manage information - as if CAs were transmitting visual messages, comparable to those produced by natural systems.
3. We have used genetic engineering to create regular domains. We have also conducted experiments in which we joined distinct domains, creating patterns that were previously unthinkable and bringing to light the hidden potential of the system. Some of the patterns created in this way would otherwise have remained invisible, even if we had performed thousands of simulations or allowed the system to evolve for incredibly long periods. The patterns we discovered reproduce the typical morphological and syntactic processes of language in which new words are created via the joining of pre-existing words, or the addition of grammatical prefixes and suffixes.
4. By comparing regular domains and gliders, we have demonstrated the tight relationship between the two phenomena, showing how gliders can give rise to regular domains and vice versa.

5. Finally, we have discovered and implemented a method showing that far from being random, the size and period of patterns are linked by simple numerical ratios - similar to the harmonic ratios we find in Pythagoras' theory of music.

We have considered the patterns generated by CAs as signs, representing real processes, similar to those we find in biology. These patterns are governed by the natural rules we have discovered and tried to explain. Much work remains to be done. We have no way of directly accessing the rules that nature shows us or the marvelous underlying design. Our studies are limited to the complex methods nature uses to represent this design - usually as they appear in patterns. In the artificial setting provided by CAs, we have found rules analogous to language structures. The discovery of these rules allows us to implement novel bio-semiotic processes of pattern creation, showing how meaning can emerge from self-organizing systems.

REFERENCES

Aronoff, M., & Rees-Miller, J. (Eds.). (2003). *The Handbook of Linguistics*. Oxford, UK: Blackwell Publishers. doi:10.1002/9780470756409

Bateson, G. (1968). Redundancy and Coding . In Sebeok, T. A. (Ed.), *Animal Communication: Techniques of Study and Results of Research* (pp. 614–626). Bloomington, IN: Indiana University Press.

Bilotta, E., & Pantano, P. (2001). Observations on complex multi-state CAs. Paper presented at Advances in Artificial Life: 6th European Conference, ECAL 2001. In J. Kelemen, & P. Sosik (Eds.), *Lecture Notes in Computer Science* (pp. 226–235). Heidelberg, Germany: Springer-Verlag.

Bilotta, E., & Pantano, P. (2006). The Language of Chaos. *International Journal of Bifurcation and Chaos in Applied Sciences and Engineering, 16*(3), 523–557. doi:10.1142/S0218127406014988

Chomsky, N. (1957). *Syntactic structures*. The Hague, Netherlands: Mouton.

Chomsky, N. (1965). *Aspects of the theory of the syntax*. Cambridge, MA: The MIT Press.

Chomsky, N. (1980). *Rules and Representations*. New York: Columbia University Press.

Chomsky, N. (2000). *New Horizons in the Study of Language and Mind*. Cambridge, UK: Cambridge University Press.

Crutchfield, J. P. (1994). The Calculi of Emergence: Computation, Dynamics, and Induction. *Physica D. Nonlinear Phenomena*, *75*, 11–54. doi:10.1016/0167-2789(94)90273-9

Crutchfield, J. P., & Young, K. (1989). Inferring Statistical Complexity. *Physical Review Letters*, *63*, 105–108. doi:10.1103/PhysRevLett.63.105

Deely, J. N. (1980). The Nonverbal Inlay in Linguistic Communication . In Rauch, I., & Carr, G. F. (Eds.), *The Signifying Animal: The Grammar of Language and Experience* (pp. 201–217). Bloomington, IN: Indiana University Press.

Eco, U. (1975). Trattato di semiotica generale. Milano, Italia: Bompiani.

Eco, U. (1984). *Semiotics and the Philosophy of Language*. Bloomington, IN: Indiana University Press.

Emmeche, C., & Hoffmeyer, J. (1991). From language to nature - the semiotic metaphor in biology. *Semiotica*, *84*(1/2), 1–42. doi:10.1515/semi.1991.84.1-2.1

Frova, A. (2006). Armonia celeste e dodecafonia. Musica e scienza attraverso i secoli. Milano, Italia: BUR Biblioteca Univ. Rizzoli.

Hanson, J. E., & Crutchfield, J. P. (1997). Computational Mechanics of Cellular Automata: An Example. *Physica D. Nonlinear Phenomena*, *103*(1-4), 169–189. doi:10.1016/S0167-2789(96)00259-X

Hopcroft, J. E., Motwani, R., & Ullman, J. D. (2007). *Introduction to Automata Theory, Languages, and Computation* (2nd ed.). Boston, MA: Addison-Wesley.

Hordijk, W., Shalizi, C. R., & Crutchfield, J. P. (2001). Upper Bound on the Products of Particle Interactions in Cellular Automata. *Physica D. Nonlinear Phenomena*, *154*(3-4), 240–258. doi:10.1016/S0167-2789(01)00252-4

Lakoff, G. (1987). *Women, fire, and dangerous things: what categories reveal about the mind*. Chicago: University of Chicago Press.

Merkle, D., & Worsch, T. (2002). Formal language recognition by stochastic cellular automata. *Fundamenta Informaticae*, *52*(1-3), 183–201.

Minsky, M. (1967). *Computation: Finite and Infinite Machines*. Englewood Cliffs, NJ: Prentice-Hall, Inc.

Rosch, E. (1975). Cognitive Representations of Semantic Categories. *Journal of Experimental Psychology. General*, *104*, 192–233. doi:10.1037/0096-3445.104.3.192

Sebeok, T. A. (1972). *Perspectives in Zoosemiotics*. The Hague, Netherlands: Mouton.

Trier, J. (1931). Der deutsche Wortschatz im Sinnbezirk des Verstandes. Die Geschichte eines sprachlichen Feldes - Von den Anfängen bis zum Beginn des 13 Jahrhunderts (Band I). Heidelberg, Germany: Winter.

Wolfram, S. (2002). *A New Kind of Science*. Champaign, IL: Wolfram Media.

Wuensche, A. (2000). Basins of Attraction in Cellular Automata; Order-Complexity-Chaos in Small Universes. *Complexity*, *5*(6), 19–25. doi:10.1002/1099-0526(200007/08)5:6<19::AID-CPLX5>3.0.CO;2-J

APPENDIX

Table 1. Basins of attraction for the rule C9.1, with M=4

Basin 1: Limit cycle Number of states in the basin: 152 States belonging to the limit cycle: 0001, 1303, 3230, 3203, 0133, 0010, 3031, 2303, 2033, 1330, 0100, 0313, 3032, 0332, 3301, 1000, 3130, 0323, 3320, 3013 Period 20 Max length of transient 7	**Basin 2**: Limit cycle Number of states in the basin: 32 States belonging to the limit cycle: 0033, 0330, 3300, 3003 Period 4 Max length of transient 4	**Basin 3**: Limit cycle Number of states in the basin: 4 States belonging to the limit cycle: 0203, 3231 Period 2 Max length of transient 2	**Basin 4**: Limit cycle Number of states in the basin: 4 States belonging to the limit cycle: 0232, 1121, 3202, 2111 Period 4 Max length of transient 0
Basin 5: Limit cycle Number of states in the basin: 4 States belonging to the limit cycle: 0302, 3132; Period 2 Max length of transient 2	**Basin 6**: Limit cycle Number of states in the basin: 15 States belonging to the limit cycle: 0303, 3131; Period 2 Max length of transient 4	**Basin 7**: Limit cycle Number of states in the basin: 4 States belonging to the limit cycle: 0333, 3330, 3303, 3033; Period 4 Max length of transient 0	**Basin 8**: Limit cycle Number of states in the basin: 4 States belonging to the limit cycle: 1112, 2320, 1211, 2023; Period 4 Max length of transient 0
Basin 9: Limit cycle Number of states in the basin: 2 States belonging to the limit cycle: 1122, 2211; Period 2 Max length of transient 0	**Basin 10**: Limit cycle Number of states in the basin: 2 States belonging to the limit cycle: 1221, 2112; Period 2 Max length of transient 0	**Basin 11**: Limit cycle Number of states in the basin: 15 States belonging to the limit cycle: 1313, 3030; Period 2 Max length of transient 4	**Basin 12**: Limit cycle Number of states in the basin: 4 States belonging to the limit cycle: 1323, 3020 Period 2 Max length of transient 2
Basin 13: Limit cycle Number of states in the basin: 4 States belonging to the limit cycle: 2030, 2313; Period 2 Max length of transient 2	**Basin 14**: Fixed point Number of states in the basin: 10 Fixed point: 3333 Max length of transient 3		

Table 2. Basins of attraction for the rule C9.1 from $M = 5$ *to* $M = 10$

Basin of attraction for the rule C9.1, with $M = 5$ Basin: Limit cycle Number of states in the basin: 525 States belonging to the limit cycle: 00313, 03032, 31331, 03103, 30031, 20303, 13133, 30310, 13003, 32030, 31313, 03031, 31300, 03203, 33131, 10303, 03130, 30320, 13313, 31030 Period 20 Max length of transient 18	Basin of attraction for the rule C9.1, with $M = 6$ Basin: Limit cycle Number of states in the basin: 1356 States belonging to the limit cycle: 000022, 333223, 330132, 200002, 333322, 233013, 220000, 233332, 323301, 022000, 223333, 132330, 002200, 322333, 013233, 000220, 332233, 301323; 000032, 333331, 133303, 313030, 032313, 332032, 201332, 300101, 200300, 233102, 320002, 213321, 031003, 300131, 200003, 133333, 313330, 031303, 303231, 233203, 220133, 130010, 020030, 223310, 232000, 121332, 303100, 130013, 320000, 313333, 031333, 303130, 130323, 323320, 322013, 013001, 002003, 022331, 023200, 212133, 030310, 313001, 032000, 331333, 303133, 030313, 313032, 032332, 332201, 101300, 300200, 102233, 002320, 321213, 003031, 131300, 003200, 333133, 330313, 303031, 231303, 203233, 133220, 010130, 030020, 310223, 000232, 332121, 100303, 013130, 000320, 333313, 333031, 130303, 323130, 320323, 013322, 001013, 003002, 331022, 200023, 133212, 310030, 001313; Period 84 Max length of transient 20	Basin of attraction for the rule C9.1, with $M = 7$ Basin: Limit cycle Number of states in the basin: 3794 States belonging to the limit cycle: 0000233, 0332120, 3300323, 3003322, 0033013, 0330002, 3300322, 2003313, 1333030, 0130310, 0023001, 1210003, 3001330, 1000103, 0130330, 0023300, 3212003, 0032333, 0332230, 3301300, 3000203, 0032233, 0331320, 3303013, 3031001, 2300100, 1000312, 0133030, 0010310, 3033001, 2330000, 1200332, 3233300, 3223003, 0130033, 0020330, 3223300, 3132003, 0301333, 3100130, 0010023, 0031210, 3303001, 1031000, 3300130, 3000023, 0033212, 3330032, 2300332, 3003301, 2033000, 2330032, 3200331, 0133303, 0013031, 1002300, 3121000, 0300133, 3100010, 0013033, 0002330, 3321200, 3003233, 0033223, 0330130, 3300020, 3003223, 0033132, 3330301, 1303100, 0230010, 2100031, 0013303, 0001031, 1303300, 0233000, 2120033, 0323330, 3322300, 3013003, 0002033, 0322330, 3313200, 3030133, 0310013, 3001002, 0003121, 1330300, 0103100, 0330013, 3300002, 2003321, 2333003, 2230033, 1300330, 0203300, 2233003, 1320033, 3013330, 1001303, 0100230, 0312100, 3030013, 0310001, 3001303; Period 105 Max length of transient 13	Basin of attraction for the rule C9.1, with $M = 8$ Basin: Limit cycle Number of states in the basin: 8848 States belonging to the limit cycle: 00001333, 03300130, 33000020, 30033223, 00330132, 33300001, 13003300, 02033000, 22330033, 13200330, 00133300, 30013003, 00002033, 03322330, 33013200, 30000133, 00330013, 03300002, 33003322, 20033013, 13330000, 01300330, 00203300, 32233003, 01320033, 00013330, 33001300, 30000203, 00332233, 03301320, 33000013, 30033001, 20330000, 23300332, 32003301, 01333000, 00130033, 00020330, 33223300, 30132003; Period 40 Max length of transient 26

continues on following page

Table 2. continued

Basin of attraction for the rule C9.1, with $M = 9$ Basin: Limit cycle Number of states in the basin: 65961 Period 18 Max length of transient 47	Basin of attraction for the rule C9.1, with $M = 10$ Basin: Limit cycle Number of states in the basin: 113550 Period 50 Max length of transient 42

Regular Domains Produced By the C9.1 Rule

Domain1 (see Figure 7a)

$$\lambda_{40}(0) = \lambda_4\lambda_4\lambda_4\lambda_4\lambda_4\lambda_4\lambda_4\lambda_4\lambda_4\lambda_4$$

Generating string $\lambda_4 = 3200$

$m = 4x10$

Period: $p = 3$

$s = 2, \lambda_{40}(t+3) - f_{\bmod 2}(\lambda_{40}(t))$

Phases:

$c_4^{\ 1} = 3200$
$c_4^{\ 2} = 2120$
$c_4^{\ 3} = 1312$
$c_2 = \{3200,\ 2120,\ 1312\}$

Domain2 (see Figure 7b)

$$\lambda_{40}(0) = \lambda_5\lambda_5\lambda_5\lambda_5\lambda_5\lambda_5\lambda_5\lambda_5$$

Generating string $\lambda_5 = 31200$

$m = 5x8$

Period: $p = 1$

$$s = 1, \lambda_{40}(t+1) = f_{\text{mod}\,1}(\lambda_{40}(t))$$

Phases:

$$c_5^{\,1} = 31200$$
$$c_1 = \{31200\}$$

Domain3 (see Figure 7c)

$$\lambda_{42}(0) = \lambda_6\lambda_6\lambda_6\lambda_6\lambda_6\lambda_6\lambda_6$$

Generating string $\lambda_6 = 111200$

$$m = 6x7$$

Period: $p = 2$

$$s = 1,\ \lambda_{40}(t+2) = f_{\text{mod}\,1}(\lambda_{40}(t))$$

Phases:

$$c_6^{\,1} = 111200$$
$$c_6^{\,2} = 111123$$
$$c_1 = \{111200,\ 111123\}$$

Domain4 (see Figure 7d)

$$\lambda_{42}(0) = \lambda_7\lambda_7\lambda_7\lambda_7\lambda_7\lambda_7$$

Generating string $\lambda_7 = 1121010$

$$m = 7x6$$

Period: $p = 2$

$$s = 1,\ \lambda_{40}(t+2) = f_{\mathrm{mod}\,1}(\lambda_{40}(t))$$

Phases:

$$c_7^{\ 1} = 1121010$$
$$c_7^{\ 2} = 1120222$$
$$c_1 = \{1121010,\ 1120222\}$$

Domain5 (see Figure 7e)

$$\lambda_{40}(0) = \lambda_8\lambda_8\lambda_8\lambda_8\lambda_8$$

Generating string $\lambda_8 = 33120200$

$$m = 8x5$$

Period: $p = 1$

$$s = 1,\ \lambda_{40}(t+1) = f_{\mathrm{mod}\,1}(\lambda_{40}(t))$$

Phases:

$$c_8^{\ 1} = 33120200$$
$$c_1 = \{33120200\}$$

Domain6 (see Figure 7f)

$$\lambda_{45}(0) = \lambda_9\lambda_9\lambda_9\lambda_9\lambda_9$$

Generating string $\lambda_9 = 312020200$

$$m = 9x5$$

Period: $p = 1$

$$s = 1,\ \lambda_{45}(t+1) = f_{\mathrm{mod}\,1}(\lambda_{45}(t))$$

Phases:

$$c_9^{\,1} = 312020200$$
$$c_1 = \{312020200\}$$

Rule Codes

C9.1:$D = 1, k = 4, r = 1$
3000210133003300220031133320310310022102023030013301123202233203

C9.2:$D = 1, k = 4, r = 1$
3232210133003220310230100300320213022321011310200200211022120330

C9.3:$D = 1, k = 3, r = 2$
120221211120200100200101022202202220221110110102122121200100122
111220121212210100112020202121110110120220000102212110111100210201
1110022011202102011221202121100001011222201211012101022020011112200
122100100100100211000211202200111102000121220000

C9.4:$D = 1, k = 3, r = 2$
001011020100020010011002010002002010000200021000100201000012000
020100002022020002000120000100000000022000010002000010000000010020
200012202000002001000100010020020000010000000000020112000002021002
2001000000010100000020000000020000122102110000001

C9.5:$D = 1, k = 4, r = 1$
3303210133003300210031133320310212022102023030013301123202233203

C9.6:$D = 1, k = 4, r = 1$
00000200200033310000000003000330303000301120100100100201211002012

C9.7:$D = 1, k = 4, r = 2$

0000020000333102000301010220120100200210030100000000001300130200
1003002100213310000002003000000001000002000102020000000010301000202
2030003110202202333030100000100100020330300230000332310011330120002
0013012100002300220000000000210020203221002000211033003120022200031
2001013013100000003300100210000100000110021002002003100332303003 00
0000320102000001111030000100001202200000000011000002002020023010 02
2020102320100010000031001000320323010010002211011202202200210320 02
0300320002100000000003000320120002203000001220032023020230010001301
0203022210010120031300022002001001000103010000010002200010003010 13
0000203010023000001000021210000000010000022020020003223030000200 022
1333102030103002033112223122000002330010120022303020133002100031 02
0003020000002000030031203112030002333000010000300002020102020200 32
2000031100200001001303202021000000313000200030000002001032320003 10
2010101000200003103000031000300000101000230000010011200031031000 00
1010000000020130203000233202001200002210010000301212032000100100 00
000100001303033111110000200200000000

Chapter 10
Models of Self-Replicators

INTRODUCTION

Structural models and patterns are vitally important for human beings. From birth, we base our emotional and cognitive representations of the external world on species-specific signals (the human face) and exploit these signals to structure our instinctive behavior. The creation of cognitive patterns to represent the world lies at the very heart of human cognition. It is this process that underlies our efficient use of signs, our ability to communicate with natural languages and to build cognitive artifacts, the way we organize the external world, and the way we organize external events in our memories and our flow of consciousness. Patterns are sometimes called *schemas,* or *models,* and discussed in terms of a *gestalt* (Piaget, 1960; 1970; Koelher, 1974). In the middle ages a pattern meant "*the original proposed to imitation; the archetype; that which is to be copied; an exemplar*" (from the On Line Etymology Dictionary). Modern use dates back to the XVIII century. In 1977 Christopher Alexander introduced a new way of using the term in architecture. For Alexander, a pattern was a model used to encode and organize existing knowledge, avoiding the need to reinvent the knowledge every time it was needed. For Alexander a pattern was "a three part rule, which expresses a relation between a certain context, a problem, and a solution" (Alexander et al., 1977).

DOI: 10.4018/978-1-61520-787-9.ch010

Patterns can be enormously diverse and some may be very complex. Others, however, can be very simple. Some can be described in terms of processes determining the replication of their internal components. They can be recognized by their structural characteristics, defined in terms of color, shape, and the relationships between their components. Sometimes they are organized symmetrically. Symmetry is a formal, structural characteristic of a pattern. As Pliny pointed out in his *Natural History,* (Book 34, Ch. 65; Loeb edition, Vol. 9, 174/76) the word "symmetry" is of Greek origin and has no equivalent in Latin. In the golden age of ancient Greece, "symmetry" was one of a group of terms denoting harmony, rhythm, balance, stability, fine proportions and well-organized structures. However, the sense of the word has changed over the centuries. Today, the term belongs to geometry where it is tied to the concept of bilateral or mirror symmetry. The existence of symmetry is proved - or can be proved - by applying well defined rules, defined in a formal mathematical or physical system. Yet the original uses of the word - tied to the emergent esthetic qualities of symmetry - remain as valid as ever.

There are many way of measuring symmetry and of defining it in operational terms (Stewart, 2007). We can study symmetry:

1. With respect to the passage of time;
2. As a spatial relationship;
3. As a geometric transformation such as reflection, rotation and changes in scale;
4. Through other kinds of functional transformation;
5. Through abstract elements in theoretical models of language, music and thinking strategies.

As human beings, our experiences allow us to consciously perceive the passage of time. The difference between past and future, "before" and "after" is entirely clear to us. Science represents time geometrically, along an axis. On this axis, the present is a point and the future and past are bilaterally symmetric - tightly coupled symmetric opposites. Mathematics gives us no clue to distinguish one from the other. Visually speaking, we can recognize objects even when they change their position in space or when they are transformed by reflections, rotations or changes in scale. We can even imagine how an object may rotate in space. This is because our internal mental model of the objects allows us to represent all its possible transformations in space. We can even recognize symmetric structures in abstract objects: in general theoretical concepts, in the syllogisms of logical and deductive reasoning or in the music of Bach and Haendel (Hofstadter, 1999). Yet in recent years, we have given up the harmonious vision of the world, typical of Euclidean geometry and this has forced us to change our vision of symmetry. Today, we talk in terms of fractal geometry. The term "fractal" (from *fractus* - meaning broken) was first introduced in Benoit Mandelbrot's book *The Fractal Geometry of Nature* (Mandelbrot, 1975). In this book, Mandelbrot demonstrates that the natural world is permeated by what he calls "fractal geometry". Fractal geometry describes a vast range of phenomena, from colonies of bacteria to clusters of galaxies, from the human nervous system to social networks. The theory of fractals provides methods for describing the natural irregularity of objects. In this setting, Euclid's concept of length is seen as the result of a measurement process, characterized by a constant D, known as its fractal dimension (Mandelbrot, 1982; Frontier, 1987; Schroeder, 1991; Turcotte; 1992). Fractal dimensions provide a statistical measure - the mean variation in the data structure - (Normant & Tricot, 1993) for the general complexity of a system. They tell us a lot about the fractal geometry of a system. However, some of the information is lost. The fundamental characteristic of fractal geometry is scalar symmetry. Later sections of this chapter will describe this symmetry in formal terms. Human art reflects the presence of fractal geometry in human beings and in their nervous system. So we find fractals everywhere: in music, architecture, literature, painting and other forms of art. If we enlarge a fractal object, we perceive no change in its structure. This is what has been called scale invariance or self-similarity. Fractal objects are composed of smaller copies of themselves. In his discussion, Mandelbrot uses the example of the coast of Great Britain or the line of a range of mountains against the sky. It is well known that coupled recursive processes and feedback mechanisms can produce extremely beautiful fractal structures (Flake, 1998). For example, so called L-systems give rise to images resembling trees or ferns etc. The study of fractals has produced many new mathematical structures. One example is the Mandelbrot set - perhaps the most complex mathematical object currently known. The framework provided by fractal geometry also provides us with insights into other more familiar mathematical objects such as the Cantor set, Kock's snow flakes, the Julia sets and Sierpinki's triangle. The theory of dynamic systems describes a wide range of fractal structures, including the bifurcation diagram for the logistic map (Hilborn 2000; Strogatz, 2000), the Lorenz Attractor (Hilborn 2000; Strogatz, 2000; Stuart, 2000), and the magnificent, richly diverse shapes produced by the Chua oscillator (Bilotta & Pantano, 2008a). The presence of fractal structures in spatio-temporal patterns from CAs was observed from an early stage (Wolfram, 1984; 2002) and remains a rich field of research (Chua, 2007; 2009). Symmetry - and fractal symmetry in particular - depend on morphogenetic processes. The use of mathematical models to study pattern formation or morphogenesis (a topic we will discuss in detail in Chapter 11) began with the work of D'Arcy Thompson and later Turing. Today it is one of the most important areas in mathematical biology (Crutchfield & Schuster, 2003; Murray, 1989). In this chapter, we will analyze symmetries in the patterns generated by 2-D CAs, and the fractal nature of the processes giving rise to these symmetries. Self-replicators, with their varying shapes, functions and behavior display life-like logic during the replication process. More specifically, we observe variation, increasing complexity and universal constructor capabilities, in von Neumann's sense of the term. Although they grow at differing rates, they create structures similar to the Sierpinksi triangle and its minor variants. 2-D CAs also produce other structures such as Sierpinski's arrowhead. These can serve as meta-models which we can use to describe the formation of colonies of self-replicators in k-totalistic CAs. Further on, we will provide a detailed description of the growth dynamics of these systems. Sometimes self-replicators develop in unison; sometimes each develops in its own way. In our description, we will see how self-replicators can synchronize these processes, by repeating different modules at different stages in the growth process. We will also see the similarities between 2-D self-replicators and the better known rules for Elementary CAs (ECA). This explains why certain phenomena may appear on different scales, even when they are not immediately available for observation. (Bilotta & Pantano 2006).

BACKGROUND

Sierpinsky's Triangle and Fractal Structures

The spatio temporal patterns generated by ECA rules 18 - 90 - 146 - 218 starting with a single seed display fractal structures similar to those of the module 2 version of Pascal's Triangle shown in equation 10.1. To obtain the images in the diagram, we begin with a single seed; we then calculate each number on the next row as the sum of the number directly above and to the left of the number and the number directly above and to the right (see Figure 10.1a); finally we color cells containing odd numbers in black and cells with even numbers in white (Figure 1b), obtaining a vertical pattern. This is Pascal's Triangle.

If we continue the process, we create a pattern that is scale-invariant (Figure 1c) and displays self-similarity. In other words, it always looks the same, regardless of the scale at which it is viewed. Similar patterns can be obtained from the Sierpinsky Triangle (Flake, 1998; Wolfram, 1984), one of the most interesting fractal objects. To produce the triangle, we begin with a filled triangle and divide it into four smaller triangles. Subtracting the triangle in the middle, we obtain a new triangle with a hole in the middle. If we apply the same procedure to the remaining three triangles, we obtain a new pattern containing 9 triangles. Repeating this operation several times, we obtain a pattern similar to the last image in (Figure 1). This is the Sierpiński Triangle. The Sierpiński Triangle has a fractal dimension between 1 and 2.

One way of representing the local rule for ECA 90 is to exploit the mod sum used in Pascal's Triangle. Thus:

Figure 1. a) Generation of Pascal's Triangle. The pattern is similar to those generated by ECAs with rules 18, 90, etc., with a single active cells as an initial seed; b and c) continuing the process we obtain a scale invariant pattern. The same phenomenon can be observed in Sierpiński's Triangle, which produces a similar pattern.

$$a_j\left(t+1\right) = a_{j-1}\left(t\right) \dotplus a_{j+1}\left(t\right) \tag{10.1}$$

where represents the sum of the terms mod 2. If the sum is even, the result is 0; if the sum is odd it is 1.

We can use the same mechanism with other rules. For example, ECA rule 150 can be represented as the sum of 3 terms mod 2 i.e.:

$$a_j\left(t+1\right) = a_{j-1}\left(t\right) \dotplus a_j\left(t\right) \dotplus a_{j+1}\left(t\right) \tag{10.2}$$

In this case too, a single seed produces a nested pattern with a fractal structure.

2-D CAs and Fractal Structures

The authors Genetic Algorithm based 2-D CAs (Bilotta et al., 2002), display similar patterns. For example, we can generalize (11.1) to produce:

$$a_{ij}\left(t+1\right)=a_{i-1,j-1}\left(t\right)\dotplus a_{i-1,j+1}\left(t\right)\dotplus a_{i+1,j-1}\left(t\right)\dotplus a_{i+1,j+1}\left(t\right) \tag{10.3}$$

The development of this CA is depicted in the first row of (Figure 2). The last column on the right provides a side-view of the 3-dimensional pattern. This is similar to the pattern produced by ECA 90. In the fractal literature, it is known as Sierpinski's arrowhead.

Let us now consider other 2-D rules we can produce by generalizing (10.3).

$$a_{ij}\left(t+1\right)=a_{i-1,j-1}\left(t\right)\dotplus a_{i-1,j+1}\left(t\right)\dotplus a_{i+1,j-1}\left(t\right)\dotplus a_{i+1,j+1}\left(t\right)\dotplus 1 \tag{10.4}$$

$$a_{ij}\left(t+1\right)=a_{i-1,j-1}\left(t\right)\dotplus a_{i-1,j}\left(t\right)\dotplus a_{i-1,j+1}\left(t\right)\dotplus a_{i,j-1}\left(t\right)\dotplus a_{i,j+1}\left(t\right)\dotplus a_{i+1,j-1}\left(t\right)\dotplus a_{i+1,j}\left(t\right)\dotplus a_{i+1,j+1}\left(t\right) \tag{10.5}$$

$$a_{ij}\left(t+1\right)=a_{i-1,j-1}\left(t\right)\dotplus a_{i-1,j}\left(t\right)\dotplus a_{i-1,j+1}\left(t\right)\dotplus a_{i,j-1}\left(t\right)\dotplus a_{i,j}\left(t\right)\dotplus a_{i,j+1}\left(t\right)\dotplus a_{i+1,j-1}\left(t\right)\dotplus a_{i+1,j}\left(t\right)\dotplus a_{i+1,j+1}\left(t\right) \tag{10.6}$$

The second, third and fourth rows of the (Figure 2) show the patterns produced by applying these rules (10.3)-(10-6). Once again, we observe nested fractal structures. These are generalizations of Sierpinski's arrowhead.

Pascal's Triangle as A Meta-Model for Other Rules

According to Wolfram, ECA rule 90 can be considered a meta-model for the block dynamics of other ECA rules (Wolfram, 2002). If we begin with a double seed (11)

Figure 2. Evolution of a CA from a single seed using the local rules described in (10.3)-(01.6). The last column provides a side-view of the 3-dimensional spatio-temporal pattern.

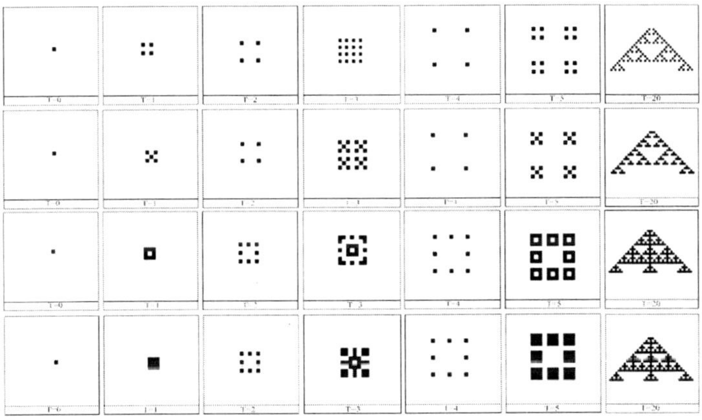

and apply rule 126, we obtain the pattern shown in (Figure 3a). If we consider the spatio-temporal block $\begin{pmatrix} 1 & 1 \\ 1 & 1 \end{pmatrix}$ in ECA 126 as equivalent to the single seed in ECA 90 (Figure 3b), (Figure 3c) and (Figure 3d),

$$\begin{pmatrix} 1 & 1 \\ 1 & 1 \end{pmatrix}_{ECA\,126} \rightarrow (1)_{ECA\,90}$$

Figure 3. Row 1: a spatio-temporal "block" produced by rule ECA126 can be interpreted as a single seed in ECA90. The patterns produced are the same. Given the right initial conditions, a single seed in ECA 90 is governed by the same dynamics as ECA 126. By performing a computational operation on the structure of ECA 126, we obtain rule ECA 90. Even if we change the initial conditions, we observe the same dynamics. In images e and f, we see the effects of a symmetric shift in the seeds. Now ECA 126 evolves from a single seed. Images g, h, .i and l show the evolution of blocks produced by ECA 90. The result is a fractal pattern. Images m and n depict the block dynamics for rule ECA 22. Images o and p show the block dynamics for rule ECA 150.

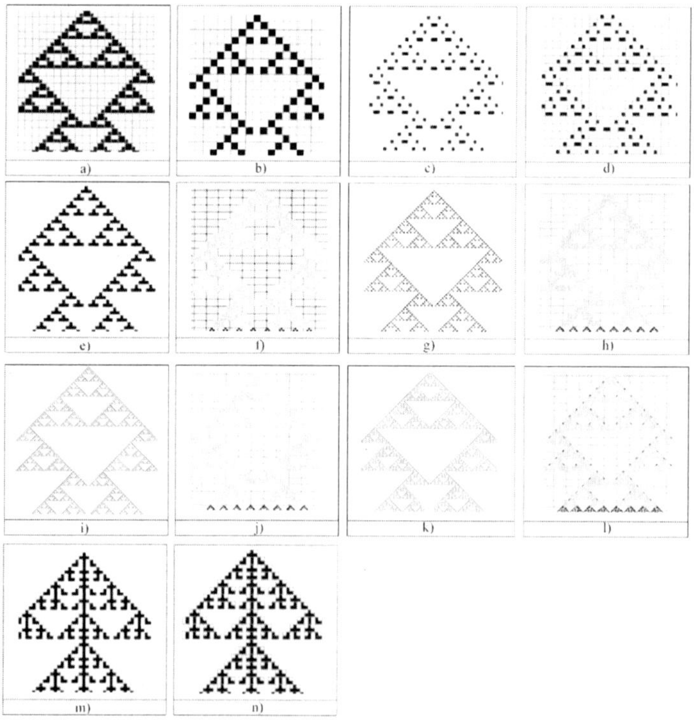

we obtain the patterns shown in (Figure 3c) and (Figure 3d). In other words, the evolution of ECA 90 provides an adequate representation for the evolution of a "block" in ECA 126. To regenerate the pattern for ECA 126, it is enough to apply the inversion:

$$(1)_{ECA\,90} \rightarrow \begin{pmatrix} 0 & 1 & 1 & 0 \\ 1 & 1 & 1 & 1 \end{pmatrix}_{ECA\,126}$$

ECA 90 also produces the pattern shown in (Figure 3a), when the initial conditions specify two seeds. In other words, ECA 90 has the ability to represent its own block dynamics:

$$(1)_{ECA\,90} \rightarrow \begin{pmatrix} 0 & 1 & 1 & 0 \\ 1 & 1 & 1 & 1 \end{pmatrix}_{ECA\,90}$$

This is an important property that depends on the fractal nature of the nested patterns produced by the CA. If we shift the blocks in ECA 90 to the right and to the left, we produce an even better match. This is what we see in (Figure 3e) and (Figure 3f). (Figure 3e) shows 40 steps in the evolution of ECA126, beginning with a single seed. In this case, we consider the following transformation:

$$\begin{pmatrix} 0 & 1 & 0 \\ 1 & 1 & 1 \end{pmatrix}_{ECA\,126} \rightarrow (1)_{ECA\,90}$$

We then allow ECA 90 to develop for a further 20 steps and superimpose the two patterns, iteratively shifting the cells in the new pattern by +1 and -1 positions, with respect to the initial seed. In this way we obtain the perfect match shown in (Figure 3f).

Once we have understood the dynamics of shifts in ECA 90 patterns, we can apply the same mechanism to larger blocks in ECA 90 itself. (Figure 3g) shows 80 steps using the transformation:

$$\begin{pmatrix} 0001000 \\ 0010100 \\ 0100010 \\ 1010101 \end{pmatrix}_{ECA\,90} \rightarrow (1)_{ECA\,90}$$

This time we iteratively apply a horizontal shift of +3 and -3 with respect to the initial seed. Once again, we obtain a perfect match (see Figure 3h). As shown in (Figure 3i) and (Figure 3j), we can apply the same procedure to even bigger blocks. These images show 160 steps in the development of the block shown below:

000000010000000
000000101000000
000001000100000
000010101010000
000100000001000
001010000010100
010001000100010
101010101010101

This time we apply a horizontal shift of +7 or -7 with respect to the initial seed.

With the right initial conditions, we can apply the same procedure to rule 22. In the pattern shown in (Figure 3k), the initial seed is $(10101)_{ECA22}$. In this, as in the previous case, we use an 8*15 block and a horizontal shift of +7 or -7 with respect to the initial seed. (Figure 3l) clearly shows the match between the two patterns.

To conclude this section, consider rule ECA 150. Here too, we observe nested patterns with a fractal structure. Just like rule ECA 90, ECA 150 can simulate its own behavior using block dynamics. This is what we see in (Figure 3o) and (Figure 3p). We use the following block:

$$\begin{pmatrix} 0 & 1 & 0 \\ 1 & 1 & 1 \end{pmatrix}_{ECA150} \rightarrow (1)_{ECA150}$$

As can be clearly seen in (Figure 3n), a shift of +1 or -1, with respect to the initial seed, produces a perfect match between the blocks. Here again, as in the previous cases, we use the mod 2 sum operator.

MAIN FOCUS

The main goals of this chapter are to show that:

1. The growth of self-replicators produces structures with fractal symmetries;
2. The patterns produced by 1-D and 2-D CAs contain precisely defined mathematical structures, such as the Sierpiński triangle and other similar objects;

3. Artificial systems can display Von Neumann's "life-like logic" and that mathematical modeling of structures such as 2-D CA self-replicators can throw light on the properties of biological organisms;
4. Growth can be a synchronized process and that synchronization between elementary models such as Sierpinski's triangle and other growth processes can produce complex or chaotic patterns, unlike those produced by known models.

The Fractal Nature of K-Totalistic Self-Replicators

We will now consider self-replicators produced by k-totalistic rules C2.2, the simplest of these self-replicators, with parameters $d = 2$, $k = 4$, $r = 1$. The first row of (Figure 4) shows the spatio-temporal pattern it produces. The self-replicator consists of a single cell in state 1. Over time, this becomes a 3x3 square of cells 1. In the next step, the self-replication concludes, with the production of four cells 1 on the corners of a 5x5 square.

$$1 \rightarrow \begin{matrix} 1 & 1 & 1 \\ 1 & 1 & 1 \\ 1 & 1 & 1 \end{matrix} \rightarrow \begin{matrix} 1 & 0 & 0 & 0 & 1 \\ 0 & 0 & 0 & 0 & 0 \\ 0 & 0 & 0 & 0 & 0 \\ 0 & 0 & 0 & 0 & 0 \\ 1 & 0 & 0 & 0 & 1 \end{matrix}$$

If we limit our analysis of the CA to this self-replicator and we consider only 2-D Boolean CAs, we can express the local rule by saying that if any cell in the neighborhood of the cell has a state of 1, the state of the cell will be 1; otherwise it will be 0. In algebraic terms:

$$a_{ij}\left(t+1\right) = \begin{cases} 1 if \sum_{i,j=-1}^{1} a_{ij}\left(t\right) = 1 \\ 0 if \sum_{i,j=-1}^{1} a_{ij}\left(t\right) \neq 1 \end{cases} \tag{10.7}$$

As can be seen in the first row of (Figure 4), the development of the self-replicator resembles that of a Sierpiński arrowhead. Self-similarity is evident (as we can see from side views of the object at different stages in the evolutionary process). The patterns are reminiscent of the Sierpinski triangle. As the following section will show in analytical detail, the Sierpinski Triangle is in fact an attractor for the system.

Figure 4. a) Spatio-temporal evolution of C2.2, an elementary self-replicator, based on a single seed. b) self-similarity in the pattern. c and d) self-similarity generated by a modified version of rule C2.2 (change of a single character).

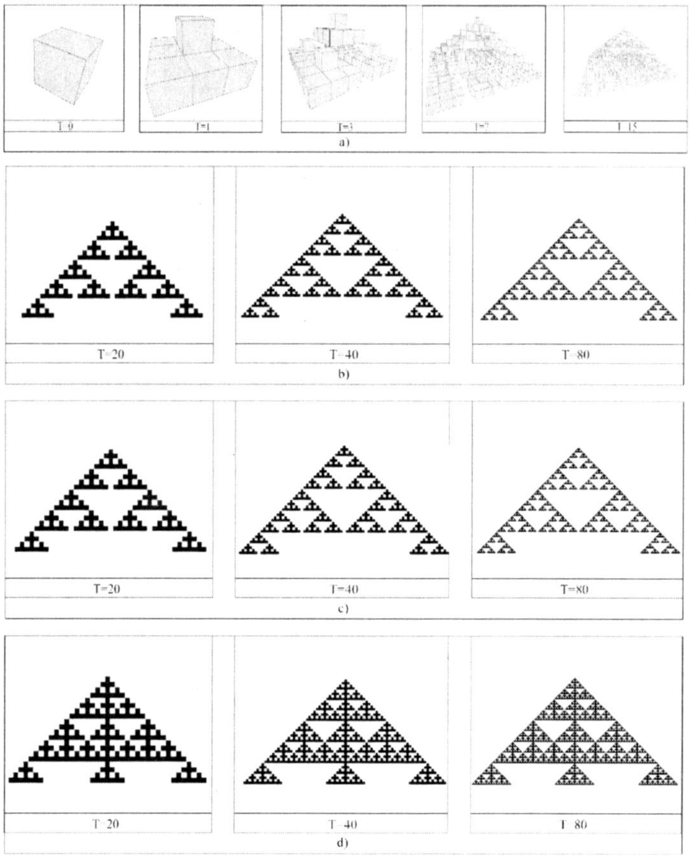

$$1 \rightarrow \begin{matrix} 1 & 1 & 1 \\ 1 & 1 & 1 \\ 1 & 1 & 1 \end{matrix} \rightarrow \begin{matrix} 1 & 0 & 0 & 0 & 1 \\ 0 & 0 & 0 & 0 & 0 \\ 0 & 0 & 1 & 0 & 0 \\ 0 & 0 & 0 & 0 & 0 \\ 1 & 0 & 0 & 0 & 1 \end{matrix} \rightarrow \ldots \rightarrow \begin{matrix} 0 & 0 & 1 & 0 & 0 & 0 & 0 & 0 & 0 & 0 & 1 & 0 & 0 \\ 0 & 0 & 0 & 0 & 0 & 0 & 0 & 0 & 0 & 0 & 0 & 0 & 0 \\ 0 & 0 & 0 & 0 & 0 & 0 & 0 & 0 & 0 & 0 & 0 & 0 & 0 \\ 0 & 0 & 0 & 0 & 0 & 0 & 0 & 0 & 0 & 0 & 0 & 0 & 0 \\ 0 & 0 & 0 & 0 & 0 & 0 & 0 & 0 & 0 & 0 & 0 & 0 & 0 \\ 0 & 0 & 0 & 0 & 0 & 0 & 0 & 0 & 0 & 0 & 0 & 0 & 0 \\ 0 & 0 & 0 & 0 & 0 & 0 & 0 & 0 & 0 & 0 & 0 & 0 & 0 \\ 0 & 0 & 0 & 0 & 0 & 0 & 0 & 0 & 0 & 0 & 0 & 0 & 0 \\ 0 & 0 & 1 & 0 & 0 & 0 & 0 & 0 & 0 & 0 & 1 & 0 & 0 \end{matrix}$$

If we modify the local rule for the self-replicator, changing the character in position 55 from 0 to 1, we obtain a new variant (Bilotta et al., 2003). As before, the self-replication process produces 4 points. However, the process now takes 4 steps instead of 2 and intermediate structures are more complex.

As can be seen clearly from the third row of figure 4, the self similar nature of the emergent structure is conserved. Once again, we observe a strong resemblance to rule ECA 90. The pattern produced by the modified version of the CA is a slightly modified version of the Sierpiński arrowhead.

A third way of generating a self-replicator is to modify the value of the 210th character from 0 to 1. Instead of the 4 copies produced by C2.2, the new self-replicator creates 8 copies of itself, again in 2 steps.

$$\begin{array}{cccccccccc}
 & & & & 0 & 0 & 1 & 0 & 0 \\
 & 1 & 1 & 1 & 0 & 0 & 0 & 0 & 0 \\
1 \rightarrow & 1 & 1 & 1 \rightarrow & 1 & 0 & 0 & 0 & 1 \\
 & 1 & 1 & 1 & 0 & 0 & 0 & 0 & 0 \\
 & & & & 0 & 0 & 1 & 0 & 0
\end{array}$$

The local rule for the self-replicator (10.7) becomes:

$$a_{ij}\left(t+1\right) = \begin{cases} 1\,if \displaystyle\sum_{i,j=-1}^{1} a_{ij}\left(t\right) = 1,3 \\ 0\,if \displaystyle\sum_{i,j=-1}^{1} a_{ij}\left(t\right) \neq 1,3 \end{cases} \tag{10.8}$$

Although the two rules are obviously not equivalent, the self-replicator produces a pattern (see row 4 in Figure 4) similar to the pattern produced by rule (10.4). From the images, it is clear that the pattern displays self-similarity.

The Dynamics of Cell Blocks In Pattern Formation

As we saw in previous sections, it can sometimes be useful to consider the behavior of blocks rather than that of individual cells. Consider a CA with local transition rule: (2.2)

$$f : S^{n} \rightarrow S$$

Now consider that the space occupied by the CA is covered by a set B of blocks, each containing M cells. Thus

$$B \subset Z^d, \bigcup B = Z^d \text{ and } B_i \cap B_j = \varphi \ \forall i, j \text{ with } i \neq j. \text{ where}$$
$$\mathrm{B} = \left\{ B \,/\, B \subset Z^d, B\_is\_a\_block \right\},$$

To each block we can assign a state s. We can then design a transition function for blocks similar to (2.2). In the new transition function, N is the neighborhood of the block. We thus obtain:

$$g : \mathrm{B}^n \rightarrow S \tag{10.9}$$

In reality, one of the key points of interest in blocks of cells is that can also incorporate a time dimension. In other words, we can collect spatial blocks for T steps and aggregate them in a single spatio-temporal block. We thus obtain a set of spatio-temporal blocks B', each composed of T spatial blocks:

$$B' = B_1 \cup B_2 \cup \ldots \cup B_T \tag{10.10}$$

Figure 5. Row 1: on the left, the basic pattern produced by the self-replicator; in the middle, the extraction of its module; on the right, the evolution of the pattern produced by ECA 90. Row 2: a side view of the evolved pattern after the subtraction of the evolved module. We can clearly see the presence of a residual single seed (two seeds if we consider the evolution of the module in 3-D.)

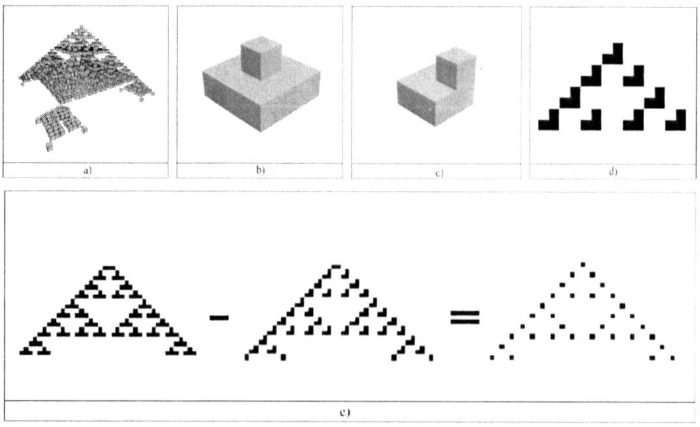

In this, as in the previous cases, we can assign a state s to each block. We can then define a transition function (10.9)

$$g': (B')^n \to S \tag{10.11}$$

Block dynamics make it possible to use rules to emulate the behavior of other rules. In reality, this is possible not only for ECAs but also for several k-totalistic rules that contain self-replicators. In what follows, we will show that this is not just a case of similar behavior but a very general feature of CA self-replicators governed by k-totalistic rules.

To understand the process, consider the self-replicator shown in Figure 5a. The pattern for the self-replicator includes a basic shape composed of two blocks (10.10) (Figure 5b):

$$B' = B_1 \cup B_2, \text{ where } B_1 = \begin{pmatrix} 0 & 0 & 0 \\ 0 & 1 & 0 \\ 0 & 0 & 0 \end{pmatrix} \text{ and } B_2 = \begin{pmatrix} 1 & 1 & 1 \\ 1 & 1 & 1 \\ 1 & 1 & 1 \end{pmatrix}$$

Now consider four subsets of this basic structure, denoting them with $C', D', E' \; and \; F'$, and defining them such that (Figure 5c):

$$C' = C_1 \cup C_2$$

$$C_1 = \begin{pmatrix} 0 & 0 \\ 0 & 1 \end{pmatrix} \text{ and } C_2 = \begin{pmatrix} 1 & 1 \\ 1 & 1 \end{pmatrix},$$

$$D' = D_1 \cup D_2$$

$$D_1 = \begin{pmatrix} 0 & 0 \\ 0 & 0 \end{pmatrix}, \text{ and } D_2 = \begin{pmatrix} 0 & 1 \\ 0 & 1 \end{pmatrix},$$

$$E' = E_1 \cup E_2$$

$$E_1 = \begin{pmatrix} 0 & 0 \\ 0 & 0 \end{pmatrix} \text{ and } E_2 = \begin{pmatrix} 0 & 0 \\ 1 & 1 \end{pmatrix}, \text{ and}$$

$$F' = F_1 \cup F_2$$

$$F_1 = \begin{pmatrix} 0 & 0 \\ 0 & 0 \end{pmatrix} \text{ and } F_2 = \begin{pmatrix} 0 & 0 \\ 0 & 1 \end{pmatrix}.$$

If we allow C' (Figure 5d), D', E' and F' to develop, using local rule (10.11), we obtain four distinct spatio-temporal patterns. If we then superimpose these patterns, we obtain the original self-replicator (see figure 5e):

$$B'(t) = C'(t) \cup D'(t) \cup E'(t)) \cup F'(t) \tag{10.12}$$

It follows that rule ECA 90, operating with a single seed, is the "generator" of the pattern produced by the C2.2 self-replicator. The self-similar pattern generated in this way can be considered as the *attractor* for the self-replicator. Figure 5 provides a schematic view of the process. Image 5a on the first row shows the basic structure. The next two images provide a 3-dimensional view of block B' (Figure 5b) and C' (Figure 5c). The last image in the first row shows side views of the pattern produced by the self-replicator for 4 steps in the evolutionary process, as well as the evolution of structure C' (Figure 5d). The two patterns are practically identical. Image 5e on the second row shows how we can subtract the pattern generated by C' from the pattern produced by C2.2. The side view is now the same as the pattern generated by applying ECA 90 to D' or E'.

Many other k-totalistic CAs display similar behavior. In each case we can identify a *module* (in general, the basic structure of the pattern or a subset of the basic structure). The growth of the module can be simulated by a 2-D Boolean CA with rules similar to those expressed by (10.3)-(10.8). In these cases, the pattern generated by applying an elementary evolutionary rule to the module, is entirely contained in the pattern produced by the k-totalistic CA. Residual elements can be interpreted as the product of other blocks. The evolution of these blocks can be described by the same Boolean rule, used to describe the development of the module. The final result is thus a composition of 4 distinct shapes all governed by a single law (11.12). In other words:

$$P_A = P_M + P_R \tag{10.13}$$

Where P_A is the final pattern, P_M is the pattern generated by the module and P_R is the residual pattern, generated by the composition of the residual elements. To summarize:

Figure 6. Generation of k-totalistic patterns from Boolean CA

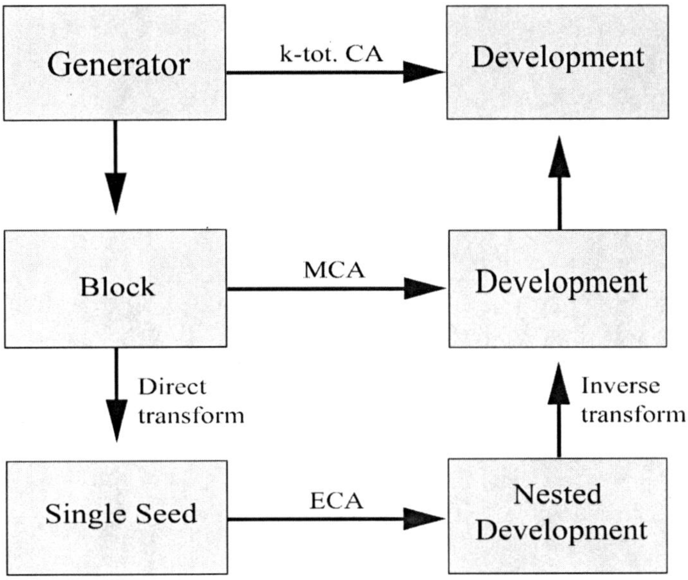

a. The final pattern is a composition of multiple patterns;
b. Each pattern is governed by an elementary CA that controls its development.

In a certain sense, we can see modules as abstract mathematical objects, whose development is governed by a Boolean CA and whose basic structure can be obtained via appropriate instantiation of a module and the residual blocks. An alternative approach would be to shift the position of the blocks at each step in the development process, as in the work described in the initial section of this chapter. This makes it possible to develop a single block with no residual elements. In these cases, what we are developing is the basic structure itself. As we will see later on, when we analyze certain specific structures, it may be necessary in some cases to shift the position of blocks not just in space but also in time.

In brief, we can summarize our procedure as follows:

1. Begin with a self-replicator;
2. Generate the spatio-temporal pattern, for a substantial number of steps;
3. Identify the basic structure of the self-replicator;
4. Identify the module in the basic structure; this can be the structure itself or part of the structure, which has evolved in a way which can be described using an ECA or a Boolean CA; this operation can be described by a function *f*;

Figure 7. Dynamics of CAs (C10.1, C10.2, C10.3, C8.3 respectively (Figure 7a), (Figure 7b), (Figure 7c), and (Figure 7d) where the number of self-replicators doubles each step. The block dynamics of these CAs are well described by ECA 90. Other CAs (C8.1, C8.2, C10.4, C10.5, respectively (Figure 7e), (Figure 7f), (Figure 7g), and (Figure 7h) that develop linearly. Patterns (C10.6, C2.2, C8.5, C8.1 respectively Figures i, j, k, m) in which each self-replicator gives rise to a square (the first three images) or a rhombus (the last image).

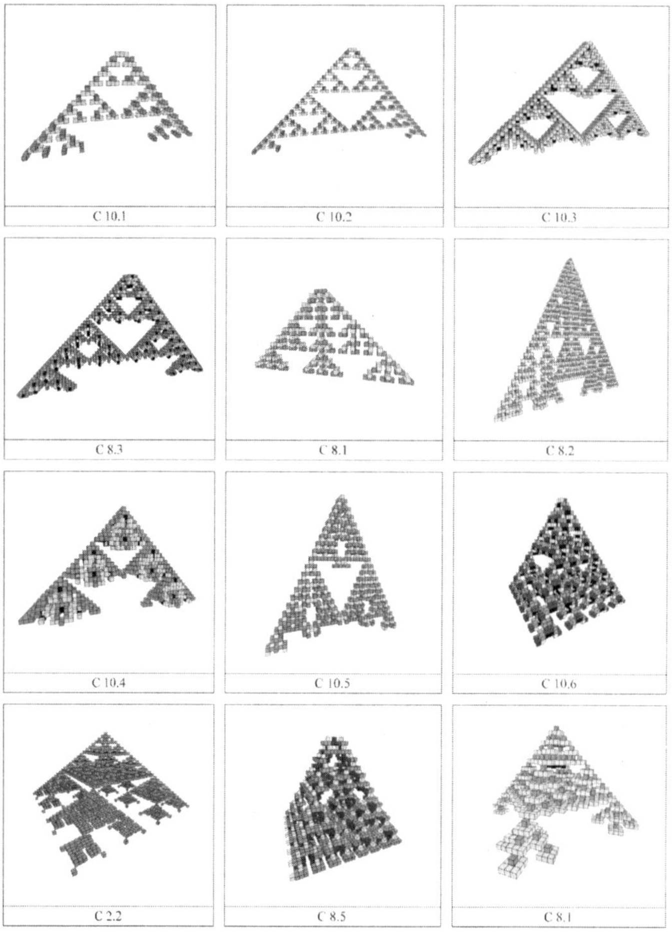

5. Use an application, g, to associate the module with a single seed, x;
6. Evolve the seed, applying a Boolean CA G^T (x);
7. Use the inverse function, f^{-1}, to replace each seed in the developed pattern with the original module;

Figure 8. The replication process. Note the linear pattern of development. Each replication cycle involves intermediate phases in which the system is "over-populated"

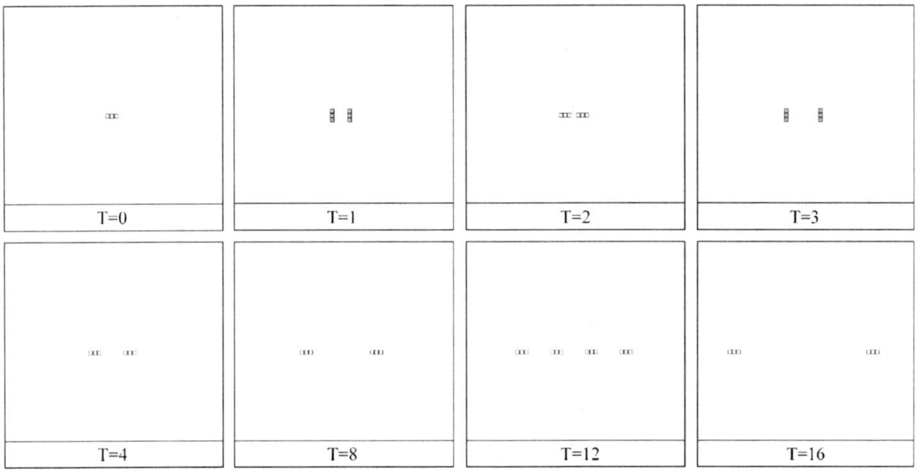

8. If necessary translate the basic structure or add the residual, obtaining the original pattern developed by the k-totalistic rule r (see Figure 6).

Symmetries: Replicators That Develop Linearly

How many different kinds of attractor exist? Is it possible to describe them in summary terms? Do they have characteristics we can use as the basis for a classification?

In the self-replicators we created with our Genetic Algorithms we found cases of:

1. Multiplication by 2,

Figure 9. Matching behavior by rule 90 and the structure generated by the evolved rule C10.1. Each step by the interpreter CA corresponds to two steps by the evolved rule

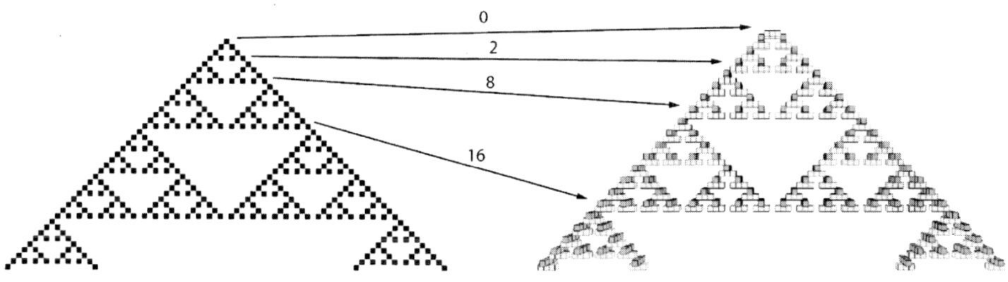

Figure 10. Match between the first 3 linearly growing patterns in (Figure 7) (first row, a, b, c) and the patterns generated by ECA 90 (at the top left of this image). The patterns clearly display the presence of the Sierpiński triangle and their general fractal nature.

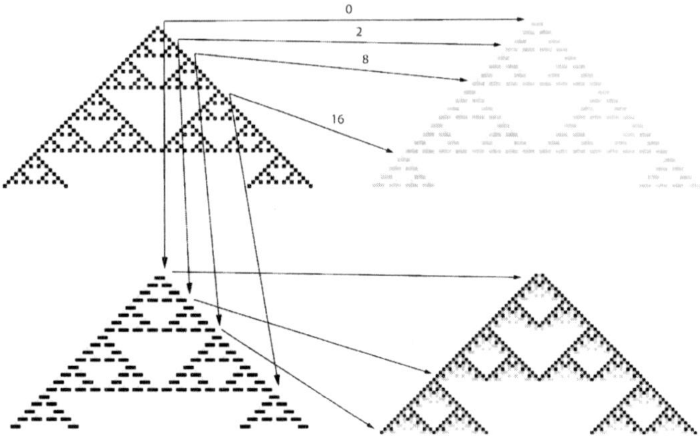

2. Multiplication by 3,
3. Multiplication by 4,
4. Multiplication by 5,
5. Multiplication by 8,
6. Multiplication by 9,
7. Multiplication by 1 (a glider).

Figure 7 shows examples of these dynamics.

Doubling (Multiplication by 2)

Consider rule (C10.1) responsible for the first pattern in Figure 7. In Figure 8, we observe the creation of a colony that develops linearly.

The initial seed is a structure (333) containing 3 elements. The basic replication process involves 8 steps. The first 4 create the conditions necessary for replication. In the next 4, the two structures move away from each other. This requires a series of intermediate steps, leading first to over-population and then to the elimination of elements not included in the new structures.

As in the block dynamics, described in the previous sections, this process can be represented using another ECA, more specifically rule 90 (Figure 9).

Many linear replication processes involve the doubling of the original structure. For instance, this is what we see in the case of rule C10.2. The initial seed is (222).

The second image in (Figure 7) shows the pattern produced by the rule. C10.3 produces similar behavior. Here the initial seed is:

$$A = \begin{pmatrix} 0220 \\ 3003 \\ 0220 \end{pmatrix}$$

The resulting pattern is shown in the third image in (Figure 7). (Figure 10) demonstrates the perfect match between the patterns generated by rule ECA 90 and those created by the 3 k-totalistic rules just described (C10.1, C10.2 and C10.3). The self-replicator generated by C8.3 displays similar behavior. The dynamics of this system are discussed in Chapter 8.

Other Forms of Replications

In other linearly developing patterns, the number of self-replicators increases threefold on each step. This is what we observe, for instance, in the case of rule C8.1, depicted in (Figure 7e.) The self-replicator produced by rule C8.2 (Figure 7f) has

Figure 11. The first 16 steps in the evolution of a structure of 4 self-replicators, governed by local rule C10.6. After 2 steps the structure doubles. After a further 4 steps it doubles again. In the following steps, the population of the colony increases and decreases rhythmically. This process is well replicated by block dynamics.

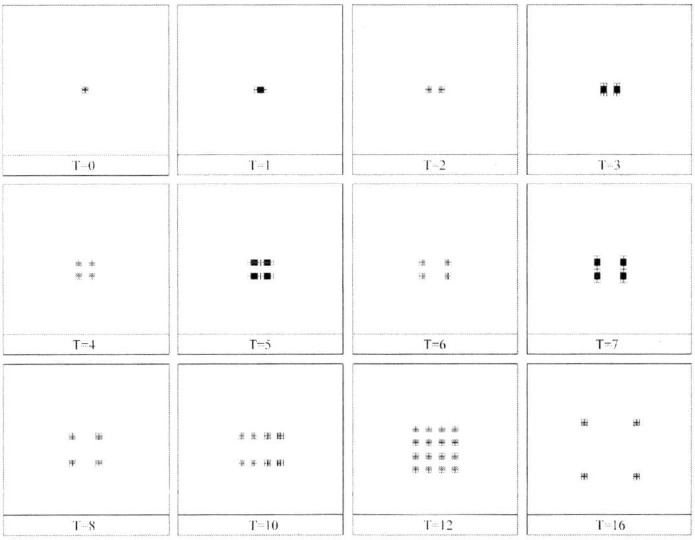

a more complex structure. Even though it too is based on a doubling process, the increased complexity of the pattern does not affect the ECA's ability to reproduce the Sierpiński pattern. This is what we see with rules C10.4 and C10.5 (Figure 7g) and (Figure 7h). Once again, we observe a doubling process. In these cases, the two self-replicating structures are A= (22) and A= (11).

Symmetries: Square-Patterned Replication

Several interesting replication processes multiply the original structure by 4. (Figure 7i) depicts the pattern generated by one such process (C10.6). The self replicating structure is given by:

$$A = \begin{pmatrix} 44 \\ 66 \end{pmatrix}$$

Figure 11 shows 16 steps in the evolution of the structure.

After two steps, the structure doubles. One of the new structures rotates by 90°, the other by – 90°. Now the self-replicators are in opposite left-right phase. After 4 more steps the structures double again, this time rotating by 180° and – 180°, respectively. Now they are in opposite top-down phases. If we consider the self-replicators as a single block we observe a perfect square symmetry. The block dynamics are similar to those of the Sierpiński arrowhead, described in an earlier

Figure 12. Row 1: a side view of the pattern produced by C10.7, respectively at 20 (a), 40 (b) and 80 steps of simulation. Row 2: a pattern generated using block dynamics (d); the residual e) and the part of the residual interpreted by a Boolean rule with r= 2 (f).

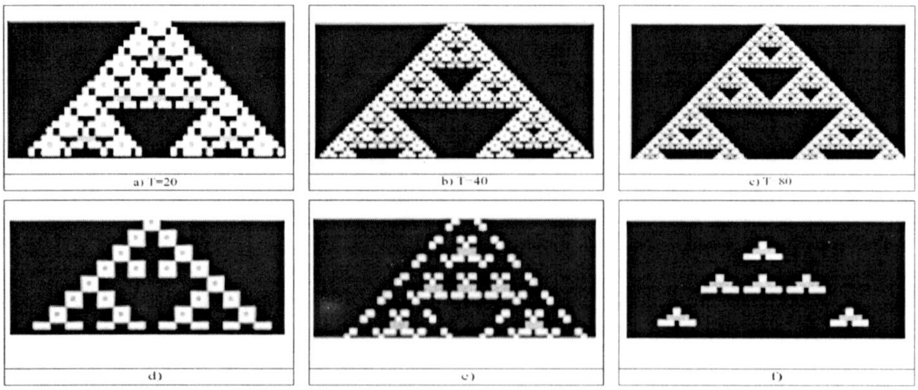

section (Figure 7i). In previous chapters, we have described other self-replicators with a square patterned replication process. One example is the self-replicator created by rule C8.5 and discussed in Chapter 8 (see Figure 7k).

Another way of multiplying the number of self-replicators by 4 is through replication processes with rhombic symmetry. This is what we see in the first self-replicator (C8.1) described in Chapter 8 (Figure 7l). As described in the section on block dynamics, we can also find systems in which the number of replicators multiplies 5, 6, 7, 8 and 9 times on each step.

In Chapter 12, we discuss the symmetries of other basic structures in Boolean CAs with the ability to act as self-replicators.

Replicators Described By Non-Elementary Rules

Not all our experimental data can be explained in terms of the universals just described. All the patterns we have analyzed so far can be generated by Elementary Cellular Automata. In some cases, however, we find emergent shapes that can only be explained in terms of more complex systems, requiring the combined action of multiple attractors.

Consider the (C10.7) self- replicator

$$A = \begin{pmatrix} 111 \\ 111 \end{pmatrix}$$

shown on the first row of (Figure 12) (a),(b), and (c) at different steps. The self-replicator grows linearly under the control of ECA 90, with block:

$$B = \begin{pmatrix} 111 \\ 141 \\ 313 \end{pmatrix}$$

In linearly growing systems, the third dimension (depth) is irrelevant. To analyze the behavior of the system it is enough to consider the side-views. Once again, we observe the fractal nature of rule 90.

If we subtract from this pattern the pattern generated by the block (Figure 12d), we obtain the residual pattern shown on the second row of (Figure 12e). Analyzing this pattern, we realize that it conserves an additional structure which we can describe in terms of a Boolean CA. As in the case described in the previous section the white dots have no importance; they can be described by ECA 90. In the new structure, we can identify the module shown below (see Figure 12f):

$$B = \begin{pmatrix} 000100 \\ 004440 \\ 033033 \\ 000000 \\ 000000 \\ 000000 \end{pmatrix}$$

The evolution of the module can be described by a Boolean CA of radius 2 with the following rules table:
00000001000101100001011101111100

FUTURE TRENDS

During the replication process, self-replicators produce colonies. The growth of these colonies is rhythmic and can be effectively described using block dynamics, governed by CAs whose rules are simpler than those of k-totalistic systems. These findings raise a number of questions:

1. Comparing ECA 90 to other rules governing block dynamics, can we say that it plays a special role in colony development?
2. Is it possible to interpret all possible replication processes in terms of simpler rules?
3. When the ECA regulates the dynamics of a spatio-temporal block, does this block represent some basic feature of the self-replicator?
4. What is the relationship between the period, phases and rhythm of the self-replication process for the original structure and the equivalent features of the block dynamics?
5. When the interpretation of block dynamics requires multiple rules, what are the relationships between them?

To the first question, for the moment we can offer no answer. All we have are empirical findings showing a very high number of replicators whose block dynamics are regulated by ECA 90. In our experiments, the number of replicators based on ECA 90 is higher than that of all the other replicators we discovered. Another very popular rule is the rule governing local rules (10.4) and (10.5): we have found several pieces of evidence suggesting that these rules play a central role in block dynamics. In Chapter 12, we will see that these two rules share a number of important properties with the Sierpinski arrowhead. These include additivity.

All replication processes leading to the formation of colonies display symmetries that we can interpret in terms of elementary rules. We can ask: is this an absolute truth, or does it depend on the cases we have chosen to examine? Is there some kind of relationship between the elementary rules and the symmetries we observe in colonies? These questions remain open. In our work, we have discovered more complex replication processes that do not seem to lend themselves to descriptions in terms of elementary rules. This is a problem that requires further research.

Coming to question c) it seems evident that blocks play an essential role in the replication process. At least to some extent, they have their own spatio-temporal dynamics. As a result, they have their own period, phases and rhythm of growth. In reply to questions c) and d), we could say that replication is a *dynamic process* that takes place in time and not just in space. This point will become clearer in Chapter 11, when we discuss dynamic networks, and changes in the replication process.

Finally, it is interesting to note that we can apply block dynamics, not just to the evolution of the main module, but also as a tool to study residual patterns. It is obvious that the two blocks can grow at different rates and follow different rhythms. In the end, however, they will have to match. In other words, they should have the same period, or at least periods that are proportional to each other. In some cases, changes in the growth of one of the blocks will halt the whole replication process. The same replication processes can display a range of different periods, some driven by the fractal structure of the patterns, while others show signs of synchrony worthy of further investigation. As we will see in Chapters 15, these features play an important role when we use CAs to produce music.

CONCLUSION

We opened this chapter with a brief introduction to fractal geometry as applied to pattern formation, going on to analyze the fractal geometry of k-totalistic self-replicators. In our discussion, we showed how some of the phenomena we originally found in ECA rules reappear in 2-D CAs, where they take the form of general laws of development. Although growth in k-totalistic self-replicators follows different patterns, it can always be interpreted in terms of the growth model for Sierpinski's Triangle and the variant known as Sierpinski's arrowhead. The final structures produced by evolutionary self-replicators are extremely complex. In our opinion, however, they contain layers that display the life-like logic discussed by von Neumann. The logic underlying the growth of 2-D CAs ensures that the final pattern is determined by a module, representing the basic laws of development for the self-replicator. In the majority of the self-replicators we found with our Genetic Algorithms, this general logic resembles Sierpinski's. However, we cannot exclude the possibility that self-replicators with periods and functions closer to those of human beings may obey different rules (Bilotta & Pantano, 2008b). Growth models are complicated by residual elements that contribute to the final pattern, modifying the basic Sierpiński model. If we extract these patterns from the self-replicator, they continue to follow a Sierpiński model of development. Their growth is synchronous, leading to spatial geometries that create secondary symmetries around the basic logic. This ensures that the final pattern will be fractal in nature, and that it will display multiple forms of emergent synchrony. All this becomes very clear when we examine block dynamics. This analysis shows the key role of Sierpinski's Triangle and Sierpinski's Arrowhead as well as the presence of other self-replication processes that do not seem to be based on the Sierpiński model. In Chapter 11, we will return to examine these growth processes in greater detail.

REFERENCES

Alexander, C., Ishikawa, S., & Silverstein, M. (1977). *A pattern language: towns, buildings, construction*. New York: Oxford University Press.

Bilotta, E., Lafusa, A., & Pantano, P. (2002). Is self-replication an embedded characteristic of artificial/living matter? In Proceedings of Artificial Life VIII, Sydney, Australia, (pp. 38-48).

Bilotta, E., Lafusa, A., & Pantano, P. (2003). Life-like self-reproducers. *Complexity, 9*(1), 38–55. doi:10.1002/cplx.10118

Bilotta, E., & Pantano, P. (2006). Structural and Functional Growth in Self-Reproducers. *Complexity, 11*(6), 12–29. doi:10.1002/cplx.20131

Bilotta, E., & Pantano, P. (2008a). *A Gallery of Chua Attractors (with DVD-ROM)*. Singapore: World Scientific Publishing.

Bilotta, E., & Pantano, P. (2008b). Biological traits in artificial self-reproducing systems . In Orsucci, F., & Sala, N. (Eds.), *Complexity Science in Information Communication Technologies: Interaction Between Living Systems and Information Technologies* (pp. 109–123). Hershey, PA: IGI Global.

Chua, L. O. (Ed.). (2007). *A Nonlinear Dynamics Perspective of Wolfram's New Kind of Science* (*Vol. 2*). Hackensack, NJ: World Scientific Publishing Company.

Chua, L. O. (Ed.). (2009). *A Nonlinear Dynamics Perspective of Wolfram's New Kind of Science* (*Vol. 3*). Hackensack, NJ: World Scientific Publishing Company.

Crutchfield, J. P., & Schuster, P. (2003). *Evolutionary Dynamics: Exploring the Interplay of Selection, Accident, Neutrality, and Function*. New York: Oxford University Press.

Flake, G. W. (1998). *The computational beauty of nature: Computer explorations of fractals, chaos, complex systems, and adaptation*. Cambridge, MA: MIT Press.

Frontier, S. (1987). Application of fractal theory to ecology . In Legendre, P., & Legendre, L. (Eds.), *Development in Numerical Ecology* (pp. 335–378). Berlin: Springer.

Hilborn, R. C. (2000). *Chaos and Nonlinear Dynamics: An Introduction for Scientists and Engineers*. New York: Oxford University Press.

Hofstadter, D. R. (1999). *Gödel, Escher, Bach: an Eternal Golden Braid*. New York: Basic Books.

Koehler, W. (1947). *Gestalt Psychology: An Introduction to New Concepts in Modern Psychology*. New York: Liveright Publishing Corporation.

Mandelbrot, B. B. (1982). *The fractal geometry of nature*. San Francisco: W. H. Freeman.

Murray, J. D. (1989). *Mathematical Biology*. Berlin: Springer-Verlag.

Normant, F., & Tricot, C. (1993). Fractal simplification of lines using convex hulls. *Geographical Analysis*, *25*, 118–129.

Piaget, J. (1960). Equilibration and development of logical structures . In Tanner, J. M., & Inhelder, B. (Eds.), *Discussions on child development* (*Vol. 4*, pp. 98–115). New York: International Universities Press.

Piaget, J. (1970). Piaget's theory . In Mussen, P. H., & Kessen, W. (Eds.), *Handbook of child psychology* (4th ed., *Vol. 1*, pp. 103–128). New York: John Wiley & Sons.

Schroeder, M. (1991). *Fractals, Chaos, Power Laws: Minutes from an Infinite Paradise*. New York: W. H. Freeman and Company.

Shepard, R., & Meltzer, J. (1971). Mental rotations of three-dimensional objects. *Science*, *171*, 701–703. doi:10.1126/science.171.3972.701

Stewart, I. (2000). Mathematics: The Lorenz attractor exists. *Nature*, *406*, 948–949. doi:10.1038/35023206

Stewart, I. (2007). *Why Beauty Is Truth: A History of Symmetry*. New York: Basic Books.

Strogatz, S. H. (2000). *Nonlinear dynamics and chaos: With applications to physics, biology, chemistry, and engineering*. New York: Perseus Books Group.

Turcotte, D. L. (1992). *Fractals and chaos in geology and geophysics*. Cambridge, UK: Cambridge University Press.

Wolfram, S. (1984). Universality and Complexity in Cellular Automata. *Physica, 10D*, 1–35.

Wolfram, S. (Ed.). (1994). *Cellular Automata and Complexity: collected papers*. Reading, MA: Addison-Wesley.

Wolfram, S. (2002). *A New Kind of Science*. Champaign, IL: Wolfram Media.

APPENDIX

Rule Codes

C10.1:

0000000000021003322123000000000003101002000003000000100000003002
0033300000013002100000000100000000100000200000030210000030002200002
3200000000020031100000002300000000100011000030200010003001000020323
000020233302000000202000

C10.2:

3020000001211131133231300231022022233223310230230232311222223221
1313123010012000133112300113222230022311133000220000002011332222 12
2122311223233222222331322132221323112230000002113323333211333221 11
0320210000020212020011 00

C10.3:

1012000033020001203000000100200000000000012000020300120000232120
0110020013213100121203320112310233003020000031112100220032000000000
011000102200011331002303000002200000302111320000201310000000001300
00000000000000000320330000

C10.4:

1012000033020003003200200330010000100000120000233001210002321 20
0120000003213100100303320132310233003020000031113100220032000000000
0110001022000113310023030000022000003021213203200013000000000001331
000000001020000032033000 0

C10.5:

000200000000000000004000002040020000400040004000000000200400000 30
0120100400400000200000000010001000001100000000003000000000000000034
0004100000000000402344303411243142300001000001112243212003431430040
40133200231342232321311320000040002023444003000340310043020021131
0000034410302300024441114443331200000000300000000000300000001200 3
0002441000400010003040314000001100000000341200040041200400020000 0
3010440201000000001003003430000003300110000403200034013300002020 1
2003001230011000444130002343230303101000120004440004002000010100 0
0102000200100130300440001030100200304020100030000301300000200022 0
0000002400000000000010000030004220004000000100003430120010003000 1
0000000000410323300020001002004000000000440030003000100200303203 00

C10.6:

0600000050000000615200050000000000000000000000001000001000500000000
6600330070061505000741753030000070040000063000002430000000500000000
0000543127200000000100024130050000000570000003000004000400000000005
0700050100005007002040664406000230076003105000740006000200000600000
0700000024000005000000000000000000000002030000000000000502400200000000
0000500000570001600370023000403002000200570400010614200070000000000
0100000006000140003400000000400000000000135060000035104306000300010
0071000000000000031000030066727300000100000000000300000000120050003
0000000060000000000060000000000000600000000600000000000000100301000000
000000000000000000000000000000001000000000010700700000000000000000007
5753001000000000206000000000000000000000000100445000007500000000300000
0000005000004550000000073000050000200107000020057650000000020000000
7000000000000600000000000000000330000000000007000000060000060000000000
0000000000054000600330303000015000000000000000700500010570000700000
0000000000006000060600000520000000000500300000624300000000000000000
00000000000030050000060000006000000200200000662130664000100000145000
0455117611500300000007070000000200410000000000000000000500040000000000
0020000055040050001200002200002540077600000000500000562200000004044
0002000300000000000000400032100040020047220000000072000000050000
0000000000000000000000000000000000070000600025353000000000000600000000
405100000000000500000006000010000000006000005100000607000400030700
0007000400032020020040000000030001004030000100030000260000020600000
2002460000000230725006600000000000000000013300050030010060010463000
0000074004400400010002000000000000000007000005021001000650400000065
0344071002005000000000004005000260567406000030006500042570000430622
0020200000000000222000700000000000000060050000100000000000400500030000
00060000000002000000003000005020750100000014152000000000307002600020
200020003502020000000000000000500000000160600000000005000300002000630
640000000000000000600510007000000700020000000000000000000000000000000
000000650040000000000000000400210300005022207100050002000040201450555
6170000000000020000300000000040320000000000001000000007100001000000030
00000000000000000005074007042000003300110000000506000000000500000200
0000000700230000000000040003005004076020000100070010024000600000000
0000400000000400000000400000000042514000001000000000500000302000000044
2000600000000000000100000071540705000004720007000005000000000000655
1700006736000000006000060000000205000004400360600760000000000000300
0000137000000141326506364070000700500000000300000000002000000032000
0000040000500000006200000406000070000000000000000002000010103050000
000000000000000000000000500000600000000007077660000000000062030627

03075003000000000000000001000660000070003507000040000000100300407
00000000000001003600000460001051400000000000000000000000030000000
04510252010022100000753000000000000000000000000000000002000000000
00000000031740000000000000000055000000700070016000000022000560075
00070601000040300700005000000000000000000000004005077300000000034
70000200000600000000006000022000000700000000127100000300300402700
00003140600051004007400000000000000000005400500002000006000000700
50
005000420000077030000026306000074007761605000000604100010100200
70
00707732300073505401510012420000007400000000600000000104500040001
20000060402006000004270000022500000000550207040030000000003000301
0
04001000000700020160000000000000000060353610030100000000042000030
11
00000000002000000000000500600000000200000070002000000000000000000
070
00000040000500000000000000007000600030570460000013507060000206003
00540000006204000700000000004000023700000000206000000300000000000
0
00000000000607600000200040000000050000000000000000500010000001041
00000031404653000000060000002000000510420006000000050000420000004
00000420000060000000000610000530700334562440000000050055430000000
00001000000000040000000000000200000303000000000200400050000100620
0
00700007200000100263000360400000160000000020001702001006070000030
04000000000000000730000000002100000000030000000000000000001000050
00000000000302400006000210700100000000407700603003000050007010300
67400070040000035100046077000003704000067001050000000000000002030
0
20000500000000033000200000000000000000000560000006060000100000100
00000207700000000001040000000011000003000000000000410005000437245
5
37502000000000200002606070004303700010066502000630040700510006000
00000000003502000556300001022000506000066000070362100007430040500
00000553066160054000001300060306000000000500040000020050000015000
00000000200270000100000400000040070000033020470010000000026250000
00000000000000000000000000000000000000044200000220200000000000000
3
40040000000703035300005000020001050000620056000033060270000000044
00005000004300534500030060030060760100000000020000000000000000000
00000000004000000000000000000000000000000000000000000000000000000
00000000050100603606000500003000033225001730307051007000000000000
00000020000000000000000000000040000000000000670000060000000060000
02000040400250510000500000040000000000365000040000000000000000000
00000000000000200000000020700071000370005000000400000030000300001
05000000000600700000000066000003000000500600000000000002037200200
000
05000000000000070002000020402350050700720017000007206006005643000
00106000700004300050003030000000000000630070030143000000601606072
04000000000000000000000000000000000000000700043000025020000000003

000000001000000000500706050000003000020000003007070001007540062102
200000070002000003600004212154000000000000000000000000000000000000
000000000000000000000000605000000065200000061000000000405000000 2
0040670000000020050004000000000000500000000000000000000001005000046
6000000060000206002020200010040000000000000000000000000020000004000
0000066050200644350003000011006007500601300000020000010070420040 5
6000010007000000000000400000007200000000600000000000043002000000000
0040200020000050000000000000000200206000410000005000430000000001
0000004000061000006000776000000600000000000000000000004510007500000
0000000000000000000500002500033400743200000653610000000000000100
0000000000000000000040005313000000460000200600010000000000000000
0000000000004040210762601000065010363300523713537000003000016620
5200000000000000032000000200000000007026575400000575032600300030 1
54000000700006005000000070002000450217050000010630000000000045320
50007000000000070200070000000004000000276701003000000000000007000600
00040036200713507000771200001543000000000000000000000202502000720 0
2000070003062500700205002300004000060000500000000002000060415000 0
0004000300200000006444736000002000000100000000660000350500000051 0
0030000004003420000600000000600007620000500000010003505001000000 0
00000400000000000306400034464022752437000000006500003505000000000
5600000000000000000020000000002000000000000000000000000000000000 0
00000002205110200020044010000000000700200400000000030050050000000 0
000000000066600051200600030310000000000040070006000000002700561100
0000005000000000000000000065005000533002027730007003140500000000 0
00000640200200000070400002010000000000000000000115075000003000300 0
0034000004040001000450005300131000600000000000000000000050600004000
2000030513400000000000000000000070000040324000031700000072000006 07
50500000000610700715000400002130050407060000070003700045750000005 1
0036035040206000000000050000500003006000070000200040050300000340 7
70000000000006000000000000004000000200044000005000000010000005006
00000020430000040000000000000000000700020700001000100000067000040 0
00200000000200000515040000560004000000000000011461000000030000000 00
510000010000307004000700700001000060000300000000000101150400006006 0
00070200000000306500500045000100010004006000002000700000000600767
0000707470300200402003000000065025006331070000003500000000060000 0
0000000000006000000000035650000350570000000000000000000000000000 2
31000000762061701000015674000030500030650300000060050030006000000
000000001000000000000000000000000000000000000000007004000000031000
0000000000010002175030000000177003000000560001100000000500000000 0
00000000000000002000000000050000000000000000100650000000000400500

730000000000170030716024006230000000000100000004000000000000007060000
000304070006000000000000000050000000070040060000000607003007000010
000000000020000020000000000000307713000000000000000000000007000200054
070000000000020077600000000000000000000000000000000000000004000000
000035100061600020000000000300005651300100000000000013003700000600 2
055004620000000000720300000000000700006000000000000700100000000000000
000000000000002000001000700000000000071000020000027000170000040040 0
651073000100264000050000000040004400000000000400000200000000007710000
000040602665055000000000003040366000000000200000000000000034000400000
000001010000003000005007004555061030000002000000000000000000000000
000000000000000000000006070071000000001040006620006200000000000000
000000704000040000010304204000000000000000042000001000000000000002000
064020070000000000001450130000001600000105000050050003120000000000
000000000005006000000001000000000204006002000000000000000000000004360
000000003000100000000642000060000050000306305230000204000050000003
001000000000300207000050000000000225041300000000000000434006050010020
000000000007000042020001400000100000000700000600000241007030000000
600000004300000600010076000700000000000000000060045077000257000000000
000050701000042000000500004000400000000060000040000000000000000000
003000000001043400000000000000000000000000000001000000000000000000000
000200060000000000000000000000460123000000500003000060000000072400000
000000000002000004000002700001053400014200000000060070070000000020
004000001300000000000000001040170300000110200000004073700555070100000
000000000000000002000000001000007727001000000000000005073704005000
060000010001400000000200000400003070040004000050107300000032066050
000000701003000000410005003201003004000000000000000370040003000003000
031004000000300000000000000000000010200040000030000000645004000000
300000300022000100074200000000000000700000000000000000000000000000
000000000000006036000400000400000000060002200000004050700010203300 0
000000000000000002620100050100737473600003007000000070000000500200
002520030000000020000000000603000000000000000000400070006251006006 0
007010100107000064000004500000000050000000007240300000073000003070 7
003170320040300200000000000000000000001031600001020000000100000000000
025000500000000323071000004000000000000300204000000000000000000000
200000000002000000000000530600000000000000010020650162760002010001 40
060000000000004000100030073300600670015060000070100123506700100770
000700040000170005000000000700000700000000700000540340033003100040 0
106717400065005073013000000003003000050300200000006100077070703020 0
070300057000770005000000000000500000000000460000000000000000600500001 1
030000004000000000002000000400600000300000500000000000000000400000000

000007000000000000000000000300000001000000030030000000003000000040
000000000005000000050600700000000000000003600000300000070007020000
0002000500710050100000004535650200000000000420000000060000052304070
400300060310001104000000000027005000050000400001620000000065000700
070002000063050000000507470303600000100006004004100000000000260450
005000030001000003400300050030000000777030000000052162200006240070
110600770004005730402300000605062463000000500206000006641000050000
00000000000000000010000500000014000000000000000000003000700020000000
500000000000300000000000000000000007000000012200000000000002700060000
000700000000304410600441075007340460000000000000120030000000000071
300000200342000000000020000002000000001063600060070000000030451032
031034105000070000051662000051007000000000000000000000000700500330
006000020106440060000000000000007000000001360700305310740635000000
460700400600000300000370464730000006000100312303000000000000000000
0010100000000000005000000000007000000000005040000000000040637200002
000006670000000000000000000300060000000000105013005000030000305004
060000660000000000000000720000000300006000000000000000

C10.7:

00030000000000000012000022001300000000040444000200400000001020000
00000004000142130000000023000042130000300000000000400042000140000000
12001030000000000000200100100000100000104014300000000000004003000010
0040100400302004000000300000000000040000000000000021000000400000004
100010310300000000000000431000210000200230040042001000000000000000
00000010001000000244201000000042201000000000001000000000001300000010
00000000000000001000000002000000003200004100303000003103343300024
0041400004214100130030044110124002000000000000024044010201040020043
000200200004310000000000001003003400000130341300000000004000113000
00043000003013000001003000020001340020020001000001000040001000000
100030404010000000004014000423004000300000440000000000201000130000

Chapter 11
A Genetic Approach to the Study of Self-Replication

INTRODUCTION

The basic mechanisms underlying development have long been a focus of attention for biological research. Development - or morphogenesis - involves a special sequence of transformations determined by a wide range of elementary processes. It is these processes - genetic regulation, changes in cell morphology, self-replication and proliferation of cells, cooperation among cells (Gierer and Meinhardt, 1972) - that allow organisms to adapt to their environment and to survive.

One of the processes most commonly studied is cellular replication. The results of these studies pose interesting questions. Is it possible to identify the basic mechanisms of biological self-replication, and apply them to artificial organisms? If so, is it possible to use such organisms as experimental models? Can models of their behavior provide new insights into biological processes?

The first authors to investigate this interplay between biological behavior and mathematical models of mechanism were d'Arcy Thompson (1917) and Alan Turing (1952). D'Arcy Thompson hypothesized that the creation of new structures is determined by the mathematical and physical properties of living matter, in this way laying the foundations for theoretical analyses of morphogenesis. Alain Turing, for his part, proposed a reaction-diffusion model of molecular dynamics, showing

DOI: 10.4018/978-1-61520-787-9.ch011

how patterns could develop from an initial homogeneous state. The model makes no explicit mention of growth. Rather it describes chemical reactions between substances that diffuse through an organism. According to Turing's main hypothesis, pattern formation is determined by reaction-diffusion reactions that take place when certain specific chemicals (*morphogens)* are present at specific densities. Turing's model provides a paradigm for the study of morphogenesis. Examples of Turing's reaction-diffusion reactions have been found both in biological organisms (Kondo and Asay, 1995; Meinhardt, 1982; Meinhardt et al. 2003; Murray, 1989; 2003) and in artificial systems systems (Kauffmann, 1993; 1995; Adamatzky & Komosinski, 2005; Miyashita & Murata, 2005; Goras et al., 1995a; 1995b; Goras & Chua, 1995; Arena et al. ; 1999; Bedau et al., 2000).

The study of biological behavior using artificial systems, (inspired by classical work in cybernetics, systems theory and dynamic systems by authors such as Ashby (1952), Rosen (1991), Hacken (1977), Kelso (1995) involves a wide range of disciplines including mathematics, biology, psychology and philosophy. The majority of research is based on a conceptual framework, involving the steps listed below

a. Collection of evidence (from observation and experiment), providing a (partial and incomplete) picture of the behavior of biological systems.
b. Construction and validation of simplified models of biological behavior, based either on observational and experimental evidence or on engineering approaches with no direct link to experimental data.
c. Construction and validation of *artificial laboratories,* dedicated to the development of algorithms, simulations or embodied artificial agents that can be validated using mathematical analysis and benchmark problems. The construction of these *laboratories* requires new computational metaphors, providing appropriate representations for system components, effective methods for evaluating interactions among components and processes to control and modify components. The *laboratory* should also supply descriptions of the characteristics of biological systems it seeks to represent and interfaces allowing the visualization, manipulation or physical implementation of these characteristics. Artifiical laboratories may also define principles for the design and analysis of bio-inspired algorithms. Some of these algorithms may be suitable for application to non-biological problems; others are specifically designed to address issues in a specific domain of knowledge (Bonabeau et al., 1999; Dorigo & Stützle, 2004). But even in these cases, the end-results is new data that can be fed back into biological research. At this point, the cycle repeats, starting from step 1.

The kind of artificial laboratories we have just described raise a number of fundamental issues.

a. Openness. Computational systems never stop. What they are simulating is continuous evolution, growth, and incorporation of new resources in a continuous time domain. In other words, the computational systems used in artificial laboratories are similar to open systems, far from dynamic equilibrium. How is this openness controlled by the structure of the system and its interaction with the environment and with other systems? How can systems maintain their integrity in the presence of this openness?
b. Diversity. Diversity is present in the behavior, structure and interactions of all natural biological systems. How much diversity is necessary at each level of the structure? Is it reasonable to talk about the average diversity of a system?
c. Interaction. We can think of systems as agents that interact with their environment and among themselves (this is true above all for bio-inspired systems - whether simulated or embodied in physical systems). What interactions occur within individual levels? Between individual levels? Among multiple layers?
d. Structure. Even though the layers needs not be clearly defined, biological systems are organized as hierarchically organized structures. Should the layers in the hierarchy act as filters for the methods we use in our models or for the kind of systems we build? How can we recognize layers? In what conditions is it appropriate to model them as a hierarchy? Is it possible for the structure to influence the interaction? What is the relationship between physical structure and the structure of information? What is the relationship to functional specialization or to localized functionality?
e. Scale. Biological systems operate on many different scales and contain huge numbers of components. What are the key transitions (critical points, phase transitions)? What is their function? How small does a system have to be to be considered as an emergent system? When is a system too big? How important is multiscale modeling? What are the relations between scale and diversity? We should also consider that behavior too can be analyzed on many different scales (locally) as well as globally (interactions between levels).
f. The role of organisms or artificial agents in the environment. How can an organism emerge? How does it grow? How does it interact with the world and with other organisms (populations of agents)? What are its mechanisms of communication?

These are the issues we will address in this chapter. We will begin by briefly explaining the main mechanisms underlying the functioning of DNA. In this way, we will clarify our modeling procedure. We will then go on to provide a formal definition of the genetic networks underlying the self-replicators generated by 2-D CAs and show how these networks act as basic building blocks for the emergence of patterns. We will then analyze various forms of mutation, describing sequences of mutations in simple self-replicators and showing ways in which it is possible to represent these evolutionary processes. Finally we will describe the changes in the shape of self-replicators that these processes determine.

BACKGROUND

With the exception of RNA viruses, all living systems, from bacteria to plants, from animals to human beings, have the same basic DNA,. James Watson and Francis Crick, a biologist and a physicist, discovered the functioning of DNA in the 1950s. Although they never performed a single experiment with DNA, they collected all the information available and used it to construct models of the molecule, leading, in 1953, to their successful identification of its structure. In 1962, their work was rewarded with the Nobel Prize, which they shared with Maurice Wilkins, who had provided them with X-Ray diffraction images of DNA.

DNA has the ability to carry genetic information and make copies of itself, allowing the transmission of this information, both during cell division and from one generation to the next. In brief, DNA is life's information technology. The molecule is composed of 4 bases (or nucleotides) A (Adenine), T (Thymine), C (Cytosine), and G (Guanine). Base-pairing rules ensure that the bases join in complementary pairs: A always joins to T; C always joins to G. The famous double helix consists of two strands of complementary bases, for example.

strand 1: C = A = T = G = T = C = T = A = G
| | | | | | | | |
strand 2: G = T = A = C = A = G = A = T = C

It is this complementarity that allows DNA to copy itself. The sequence of nucleotides in the first base is arbitrary. However, the base-pairing rules guarantee that the sequence in the second strand is complementary to the sequence in the first. Both strands are coupled to the backbone of the molecule via strong covalent sugar phosphate bonds, held together by hydrogen bonds between the bases (2 in the cases of A-T pairs, 3 for C-G pairs). The two anti-parallel strands wrap round each other to create the double helix. In standard, so-called B-DNA, the 3-dimensional structure of the molecule is right-handed. The order, or sequence of the bases, carries the genetic information governing the functioning of the organism. During cellular replication, the two DNA chains separate.

The two ends of a strand are called the "3' end" and the "5' end.", Beginning from a specific location on one of the DNA chains (the so-called *origin),* DNA synthesizes a short sequence of complementary DNA that acts as a primer for subsequent DNA replication. The replication process creates a new sequence of DNA, complementary to the template provided by the first sequence. To achieve this, it continuously adds new bases to the 3' end of the primer, proceeding towards the 5' end. This process is repeated for both chains.

- Eukaryotic (multi-cellular) organisms contain several classes of DNA sequence:
    - Single-copy, protein-coding genes.
    - Genes present in multiple copies.
- Sequences of known function:
    - Coding sequences,
    - Non-coding sequences,
- Sequences of unknown function:
    - Repeats (dispersed or in tandem),
    - Transposons,
    - Spacer DNA.

Spacer DNA contains numerous repeated sequences. These are sequences that appear repeatedly in different locations and especially in special regions known as centromeres and telomeres. Repeats can be of differing lengths, and have differing patterns of distribution across the genome. This makes them ideal candidates as molecular markers (Flavell & Moore, 1996).

Mutational processes give rise to variation or polymorphism. There are many different kinds of event that can give rise to changes in DNA sequences.

- Point mutations. These are changes in a single base-pair. Point mutations do not change the length of the sequence.
- Insertions and deletions. These mutations involve the insertion or deletions of sequences of base-pairs, producing a change in the length of the sequence.

- Reorganization of the chromosome, via genetic recombination or the insertion of new elements between existing ones. This kind of change may or may not involve changes in sequence length. Even when the length of the sequence does not change the change in composition may be sufficient to create a new polymorphism.

Many authors have used models of DNA to model the behavior of cellular automata (Chua, 2004; Crutchfield and Schuster, 2003; Bilotta & Pantano, 2005; Mitchell, 2006). In our *artificial laboratory,* we have developed a metaphor in which we consider self-replicators as artificial proto-organisms (agents) inhabiting an abstract environment and displaying a range of replication mechanisms and behaviors.

MAIN FOCUS

The main focus of this chapter is on the methodology used to model the behavior of a 2-D CA as a life-like agent. In the rest of this Chapter, we present the genetic approach we have used in our studies of self-replication. In this work, we have developed a genetic code for these systems, along with computational tools allowing us to identify the code actually used at specific stages of the evolutionary process. Self-replicators control their growth through dynamics that simulate the operation of DNA and RNA processes in cells. These dynamics give rise to emergent, meaningful patterns, reflecting an "internal or innate logic of life" (von Neumann, 1966). They behave as if they were executing a well-defined genetic program, described by an interconnected set of points in the rule space for 2-D CAs. It is these sets, we believe, that underlie the life-like properties of self-replicators and their ability to self-replicate and grow. In other words, our artificial agents have a genetic code (a genotype) which can give rise, via an evolutionary process, to numerous artificial agents (phenotypes). When we change the structure of the genotype, creating polymorphisms, we not only change the position of the CA in rule space; we also change its functionality and shape. Sometimes the changes can be small but sometimes they are large enough to destroy the self-replication capabilities of the system. Figure 1a shows 20 steps in the evolution of a k-totalistic CA (C11.1), with $D = 2, k = 4, r = 1$

The CA generates the self-replicator A=(22) (see Figure 1b).

Using (2.18), we obtain the following string:

l=(1012000033020002203000000000020000003000012000023300120000232120
01100200132131001212033202323102330030200000311121002200320000000
00110001022000113310023030000021000003021213200002013100000002013
01000000010000000320330000) (11.1)

Figure 1. a) Development of rule C11.1 after 20 steps of simulation, starting from random initial data. b) Evolution of self-replicator (22) for 40 steps

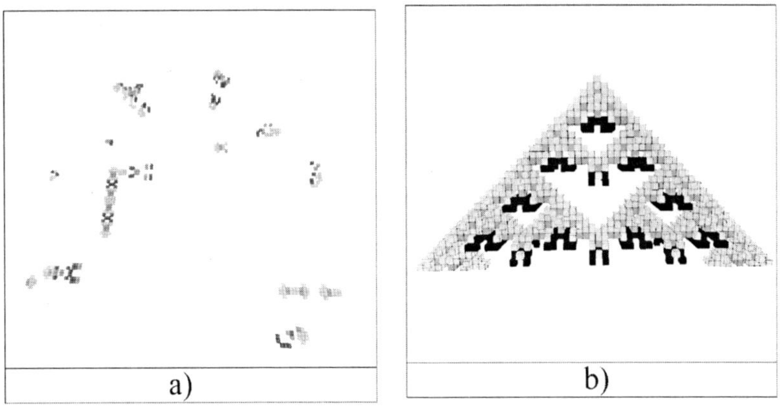

We then proceed as follows.

1. We represent the rules table as a string of characters in which each rule is defined by a number between 1 and 220 (R1, R2, R3, R4, . . ., R220) (see Figure 2).

2. We proceed to identify the elementary rules used by the self-replicator during its evolution. The rules for the 40 evolutionary steps shown in the Figure 1b are reported in Table 1.

Figure 2. Representation of the rules tables as a sequence of characters

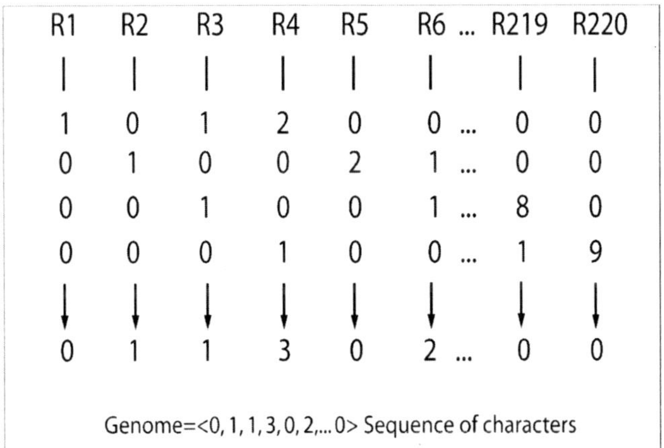

This pool of rules represents an initial core of artificial DNA (Figure 3)

3. We map the string to a sequence of DNA bases, using the procedure described in Chapter 3. Thus:
   $0 \rightarrow A$ (Adenine)
   $1 \rightarrow T$ (Thymine)
   $2 \rightarrow C$ (Cytosine)
   $3 \rightarrow G$ (Guanine)

In this way, we obtain the following sequence of *Artificial DNA*:
TATCAAAAGGACAAACCAGAAAAAAAAACAAAAAAGAAAAT-CAAAACGGAATCAAAACGCTCAATTAACAATGCTGTAATCTCAGGCAC-GCGTACGGAAGACAAAAAGTTTCTAACCAAGCAAAAAAAAATTA-AATACCAAATTGGTAACGAGAAAAACTAAAAAGACTCTGCAAAACAT-GTAAAAAAACATGATAAAAAAATAAAAAAAAGCAGGAAAA

4. We create the dynamic genetic networks, the building blocks for patterns (see Figure 4).

During the simulation, the self-replicator uses specific genes (rules) from its genome (the rules table).

Figure 3. The genome of a 2-D CA C11.1, with k=4, r=1. The rules are shown on the left side of the image. For the sake of clarity, the mapping between the rules and the artificial DNA sequence is not shown.

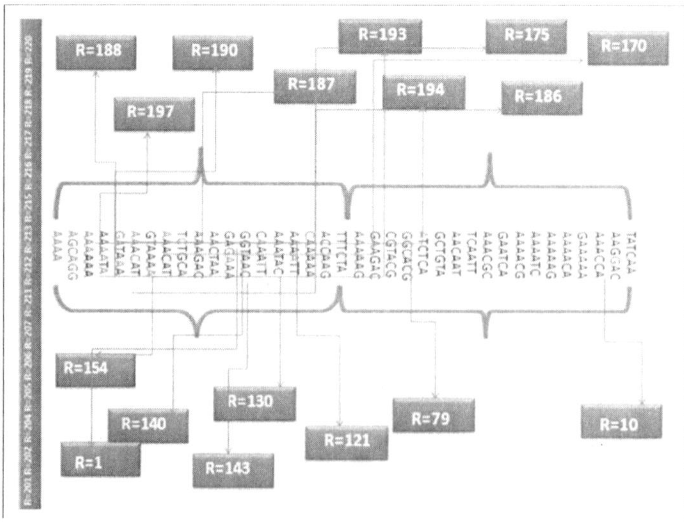

Figure 4. The first eight steps in the development of a single self-replicator, showing the genes responsible for the shape of the system

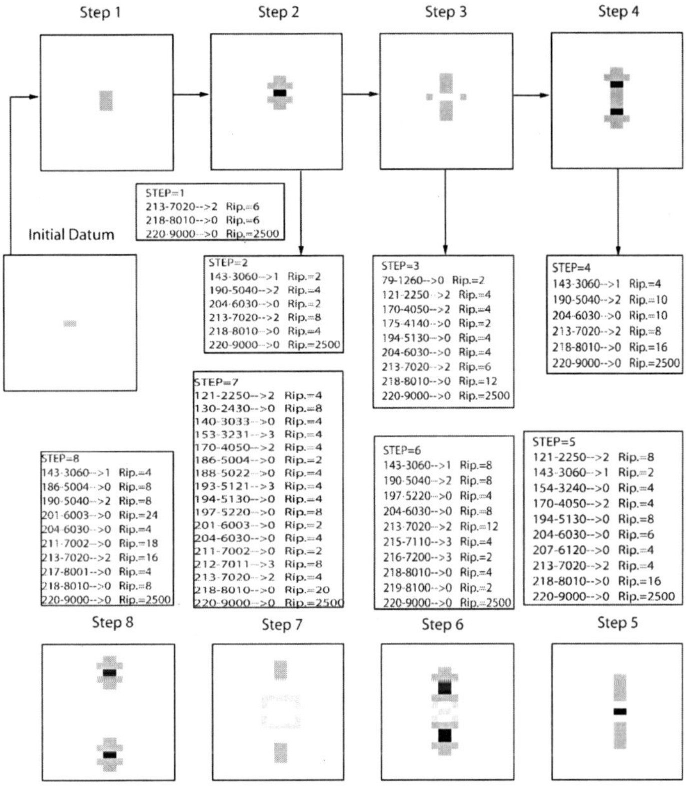

We can thus exploit our computational tool not only to identify the genes the self-replicator uses to generate the overall pattern, but to see how the components of this pattern are realized step by step and which genes are involved in this construction process.

5. We display the resulting dynamic networks as graphs or glyphs or small world networks (Strogatz, 1994). In our model, we define the rules used by the self-replicator at different stages of its development as a dynamic network in which the nodes are the rules (Figure 5a). Figure 5b restricts the representation to rules the system actually uses. Figure 5c represents the network as a glyph. Use of R220 (R=220) depends on the size of the network and is thus far more frequent than use of the other rules. To avoid "squashing" the data for the other rules, we omit R220 from our diagrams,

Another way of representing the development of the self-replicator is to use the representation presented in Chapter 3. Briefly speaking, we can depict the development process on a graph in which the vertical axis represents time and the horizontal axis shows the rules used at different stages in the process. In this way we can visualize the sequences of genes used as the system develops (Figure 5k). When we stop the simulation, we can collect these components into a unified representation that allows us to detect genes shared between different members of the same family of self-replicators. This in turn, allows us to investigate the possibility of general development models, shared among family members.

6. We introduce polymorphism. This allows us to introduce changes (changes in the genome of a single self-replicator; changes affecting a population of systems) that provide us with information about the influence of genes on the behavior (self-replication mechanism) and structure of the self-replicator or the population of systems. This work has enabled us to identify sequences whose functions are immediately visible in our artificial system and others which have no apparent effect - the equivalent of repeats and spacer DNA. In brief, we have created an artificial sequencing laboratory in which we can study the genetics underlying the local and overall behavior of the system and identify the polymorphisms responsible for the patterns described in previous work.

Figure 5. Row 1: different views of the genetic network. Row 2 and first two images in row 3: snapshots of the network from steps 1 to 6. The snapshots show periodicity in the use of elementary rules Row 3 next to last and last image: snapshots of the consultation networks, in different representations.

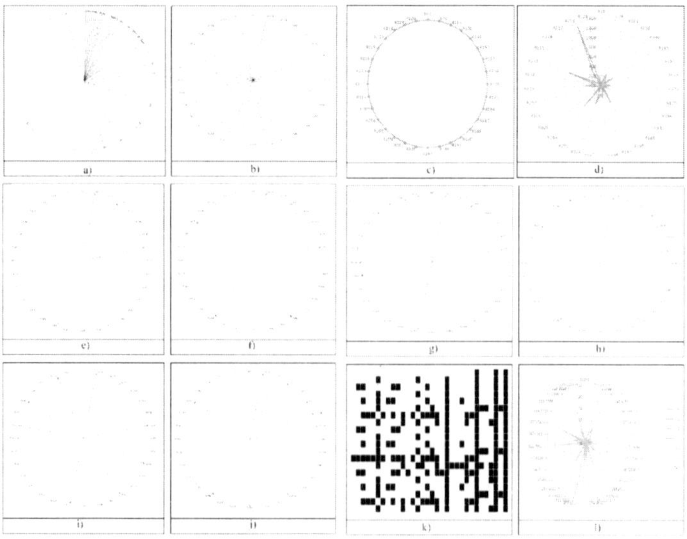

7. We allow the system to evolve for a long period, creating representations of the evolutionary process and resulting polymorphisms.
8. We use Artificial Genetic Engineering t techniques, already tested with 1-D CA (Chapter 9), to create new organisms.

In the following paragraphs we describe an experiment implementing this procedure.

Self-Replicators and Networks

Consider a structure belonging to a 2-D CA. The rules table for the CA can be described by a string (2.18) l:

$$l = \left(s_1, s_2, \ldots\ldots, s_{(k)^{N_T}} \right)$$

If we simulate this system, we observe that it uses only some of the elementary rules in the rules table. In other words, it uses only some of the characters in string (2.18). We can create a new string l' of length $N_{T,}$ (the *consultation string*), representing the rules actually used during the evolutionary process. In this new string, we represent rules that are not used with a neutral symbol y. We thus obtain:

$$l' = (y, y, s_i, y.., s_j, y...s_k, y...) \tag{11.2}$$

Let us consider L as the set of all possible strings l of length $N_{T,}$, and L' as the set of all possible consultation strings. Sets L and L' constitute two languages, defined over alphabets $S = \{0,1,2,....,k-1\}$ and $S' = \{0,1,2,....,k-1,x\}$ respectively. Assume that, in the initial state, $c(0)$, relatively few cells in the CA have a state different from 0. If we define a function $|l'|$, representing the number of characters in l' with values different from y, the value of the function will grow over time. We can now use this scheme to visualize the behavior of self-replicator C11.1. The pattern produced by the system is shown in the first image in Figure 2b. In the first 40 steps of its development, the self-replicator uses just 32 of its 220 elementary rules. The string for the system (11.1) is thus:

$l' =$ (yxxxxxxxx3xxxxxxxxxxxxxxxxxxxxxxxxxxxxxxxxxxxxxxxxxxxxxx
xxxxxxxxxxxxxxxxx0xxxxxxxxxxxxxxxxxxxxxxxxxxxxxxxxxxxxxx2xxxx
xxxx0xxxxxxxxx0xx1xxxxxxxx30xxxxxxxxxxxxxx2xxxx0xxxxxxxxxxx000x2
xx30xx0xxx00x0000xxx032x330000

Imagine a diagram in which the characters of the string are arranged clockwise on a circle, with the first character at the top and with lines linking the rules used to the center of the diagram. Image a in Figure 5 shows the diagram for C11.1. If we restrict the diagram to the rules actually used, we obtain Image b. We can now join and label the nodes. In this way, we obtain image c. Using the glyph shown in image d, we can depict how often each rule is used (see the values in Table 1). In the image, we have artificially left out the last rule. In this way, we do not "squash" the other rules.

In some cases, the value of the function stabilizes so that from time T onwards,

$$m_s^a : L \to L$$
$$l \in L \mapsto Ra\_s \qquad (11.3)$$

C11.1 is not the only rule displaying this behavior. In reality, all self-replicators conserve the symmetries described in Chapters 8 and 10. If (11.3) is satisfied, we can represent the stable version of l' in a graph. We call this graph, the *consultation network* for the CA (see Figure 5a-d).

As can be seen in Table 2, the CA uses only a subset of the elementary rules in any given step. In some cases, the value of the function stabilizes so that from time T onwards,

Let us define the *instantaneous consultation string* $l''(t)$, as the string l' generated by a single step in the development process, starting with $c(t)$. Then L'' is the set of all instantaneous consultation strings and L'' is a language over S'.

In the case of (11.3), the instantaneous consultation strings realize a cycle. Thus $\exists \overline{T} \in N$:

$$l''(t) = l''(t + \overline{T}) \qquad (11.4)$$

We call $\overline{T}$ the *period of the consultation network.* If we continue the simulation, we observe that the period of the network is T=6 T=2. This allows us to construct a new graph in which the nodes are the nodes in the consultation network. Nodes that are consulted consecutively in the sequence of instantaneous consultation net-

Table 1. Frequency of use of elementary rules over 40 steps, by rule C11.1. (self-replicator 22)

Elementary Rule	Frequency
R10	12
R79	62
R121	264
R130	120
R140	60
R143	284
R153	60
R154	76
R170	188
R175	62
R186	150
R187	12
R188	60
R190	398
R193	72
R194	264
R197	180
R201	402
R202	36
R204	586
R205	36
R206	24
R207	88
R211	360
R212	144
R213	740
R215	84
R216	30
R217	312
R218	1314
R219	30
R220	…..

Table 2. Frequency of use of elementary rules during the first 9 steps in the development of C11.1

Rule\Time	T=0	T=1	T=2	T=3	T=4	T=5	T=6	T=7
10-0090	0	0	0	0	0	0	0	0
79-1260	0	0	2	0	0	0	0	0
121-2250	0	0	4	0	8	0	4	0
130-2430	0	0	0	0	0	0	8	0
140-3033	0	0	0	0	0	0	4	0
143-3060	0	2	0	4	2	8	0	4
153-3231	0	0	0	0	0	0	4	0
154-3240	0	0	0	0	4	0	0	0
170-4050	0	0	4	0	4	0	4	0
175-4140	0	0	2	0	0	0	0	0
186-5004	0	0	0	0	0	0	2	8
187-5013	0	0	0	0	0	0	0	0
188-5022	0	0	0	0	0	0	4	0
190-5040	0	4	0	10	0	8	0	8
193-5121	0	0	0	0	0	0	4	0
194-5130	0	0	4	0	8	0	4	0
197-5220	0	0	0	0	0	4	8	0
201-6003	0	0	0	0	0	0	2	24
202-6012	0	0	0	0	0	0	0	0
204-6030	0	2	4	10	6	8	4	4
205-6102	0	0	0	0	0	0	0	0
206-6111	0	0	0	0	0	0	0	0
207-6120	0	0	0	0	4	0	0	0
211-7002	0	0	0	0	0	0	2	18
212-7011	0	0	0	0	0	0	8	0
213-7020	6	8	6	8	4	12	4	16
215-7110	0	0	0	0	0	4	0	0
216-7200	0	0	0	0	0	2	0	0
217-8001	0	0	0	0	0	0	0	4
218-8010	6	4	12	16	16	4	20	8
219-8100	0	0	0	0	0	2	0	0
220-9000	2500	2500	2500	2500	2500	2500	2500	2500

works are shown as connected. The images in Figure 5e-j show the instantaneous consultation networks for the first eight steps in the development process. The node at the top of the circle represents the rest state, which appears in every step. To avoid complicating the graph, the links to the other nodes are not shown. The sequence provides clear evidence of periodicity in the way the rules are used. This is even more evident in Figure 5k. Here we see an $n * t$ diagram, where *n* is the number of nodes in the network and *t* is time. Whenever a rule is used, the corresponding node is colored in black. We call this diagram the *decoding scheme* for the CA. The last image in Figure 5l shows a glyph representing instantaneous consultation frequencies for all rules in the network. In the case of T_220, the value is artificially left out to avoid squashing the values of the other nodes.

The Growth of Patterns

Now consider the case of a new self-replicator, generated by rule C8.5 - a 2-D CA, with k=5.

$$A=\begin{pmatrix} 3 & 3 & 3 \\ 3 & 3 & 3 \\ 2 & 2 & 2 \end{pmatrix}$$

The description of this self-replicator is in Chapter 8, where we saw that it gives rise to a colony that grows in a pattern with a square symmetry. If we reflect the initial condition, the resulting pattern does not change. As we have seen, the logic underlying the growth of this apparently complex pattern can be interpreted as a Sierpinski Arrowhead. Below we list the number of cells in the state 3, sampled at 2-step intervals (on the odd steps the state 3 disappears).

6,12,24,24, **24**,48,96,48, **24**,48,96,96, **96**,192,384,96, **24**,48,96,96, **96**,192,384,192, **96**,192,384,384, **384**,768,1536,192,.......

We observe that the first value in each group of 4 is the value of the *fundamental sequence* multiplied by 6.

$$1,4,4,16,4,16,16,64 \ldots\ldots \rightarrow 6,24,24,24,96,24,96,96,384\ldots \tag{11.5}$$

The fundamental sequence is obtained by counting the number of cells in state 1, in each time step, during the development of the pattern. This is regulated by rule

(10.3) - the same rule that produces the Sierpinski arrowhead. This rule represents the *basic logic* of the self-replicator. As we see in Figure 6, this basic logic recreates A, once every 8 steps. This finding is confirmed by an analysis of the side-views for the pattern.

The consultation network is composed of 19 elementary rules:

R 34, R 512, R 523,R 527, R 533, R 650, R 657, R 660, R 684, R 687, R 690, R 703, R 704, R 705, R 706, R 711, R 712, R 713, R 715.

To identify symmetries, we can analyze neighborhoods (corresponding to elementary rules) and the frequency with which they appear during the simulation. For the sake of simplicity, we will limit our analysis to the first 9 steps. The consultation network for the self-replicator is composed of 19 nodes. The instantaneous consultation networks for the individual steps contain 15,4,15,4,15,4,15, 4 and 9 nodes respectively. All networks with 4 nodes are identical. The networks with 15 nodes come in pairs (see Figure 7). As in the previous figure, links to the last node (representing the rest state) are not shown.

The networks with 15 nodes all have 13 nodes in common. Networks 5 and 8 have two nodes which are different from those in networks 11 and 8. The networks with

Figure 6. The Sierpinski Triangle and the colony generated by self-replicator A

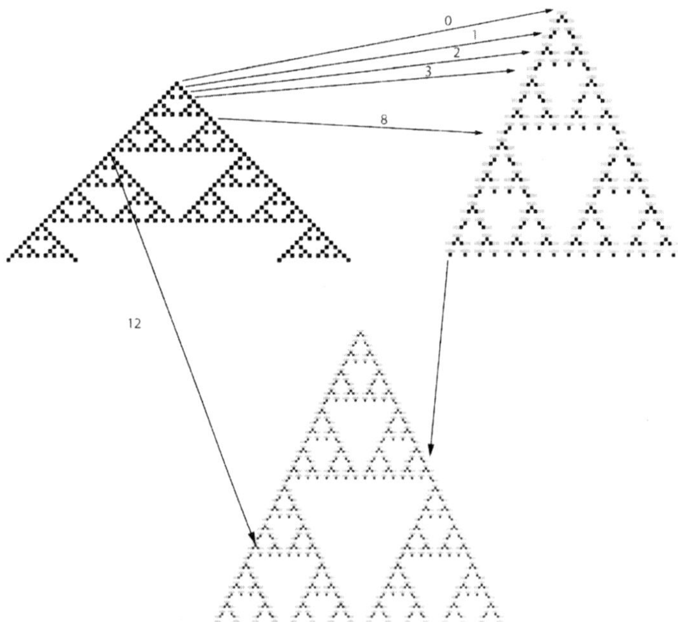

4 nodes contain 2 nodes which are never consulted by the networks with 15 nodes. The decoding schemes, shown on the last row of Figure 7, provide clear evidence of periodicity in the instantaneous consultation networks. In a certain sense, the decoding schemes contain all the information the self-replicator needs. The rules, for each step shown in the scheme, generate the phenotype for the following step. It is important to observe the unbreakable association between the decoding scheme and the phenotype. The following figure (Figure 8) shows the growth pattern produced by the CA and the way in which it uses the elementary rules.

Figure 7. Instantaneous consultation networks for the first 9 steps. All the 4-node networks are identical. The 15-node networks appear in pairs. The first image in the last row shows a representation of the decoding scheme. The last image shows data for IE.

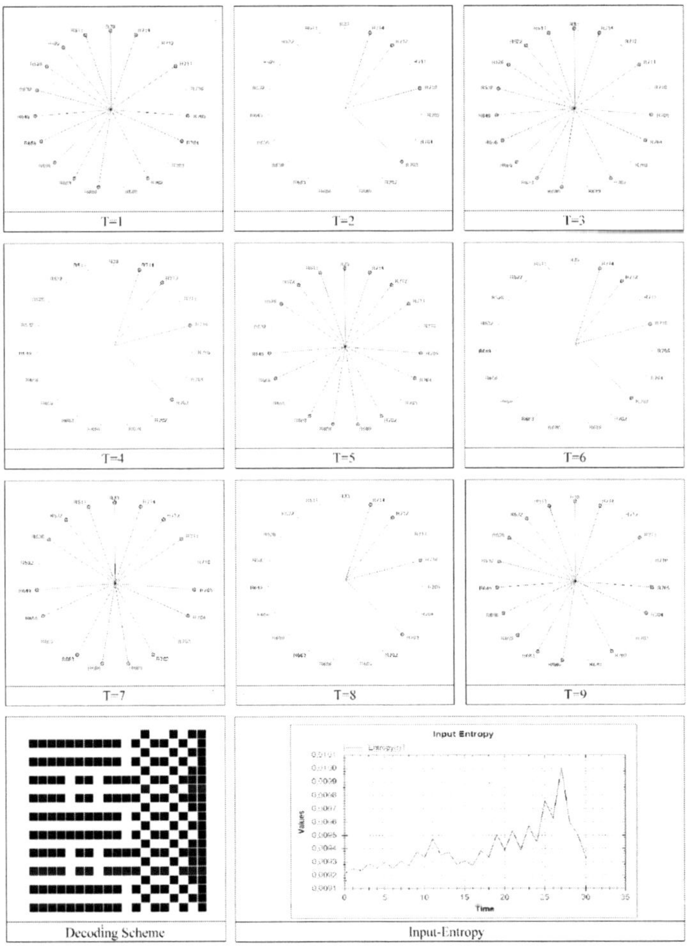

This process (see Figure 9), brings to mind the process determining the form of biological cells

a. Specific regions of the genome are selected for expression.
b. The selected regions give rise to new strings (in biology, Messenger RNA, in our computational model, dynamic networks).
c. These strings end up inside the ribosomes where they translate into proteins; these in turn determine the cell phenotype (defining the spatial organization of the cell).
d. At the end of the process, the cell divides and the process starts over again.

In terms of the fundamental sequence (11.5), we observe an 8-step cycle. To obtain further information, we can examine the frequency with which the nodes are consulted on the first step of each cycle, as shown in Table (3)

Figure 8. The decoding scheme controls the evolution of the phenotype, synchronizing the development process

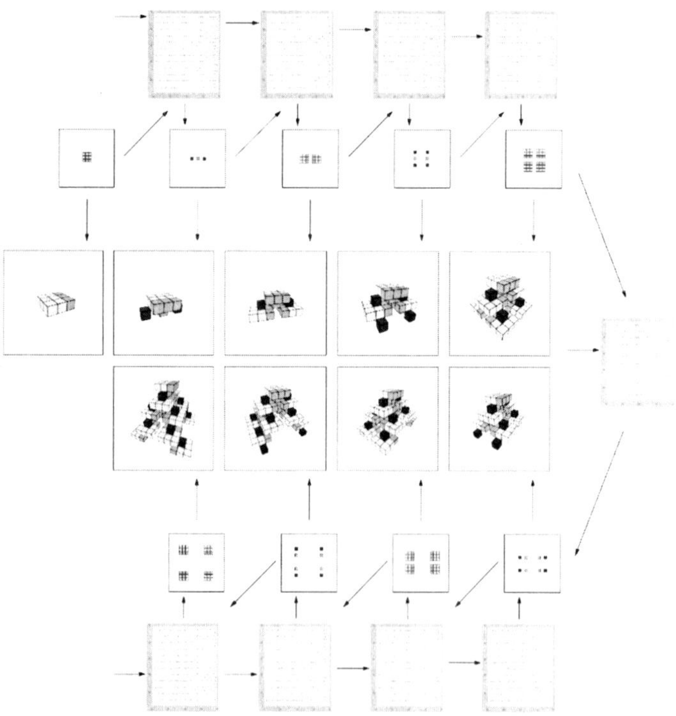

Figure 9. Process of construction of artificial self-replicating organism, modeled on the DNA-RNA method

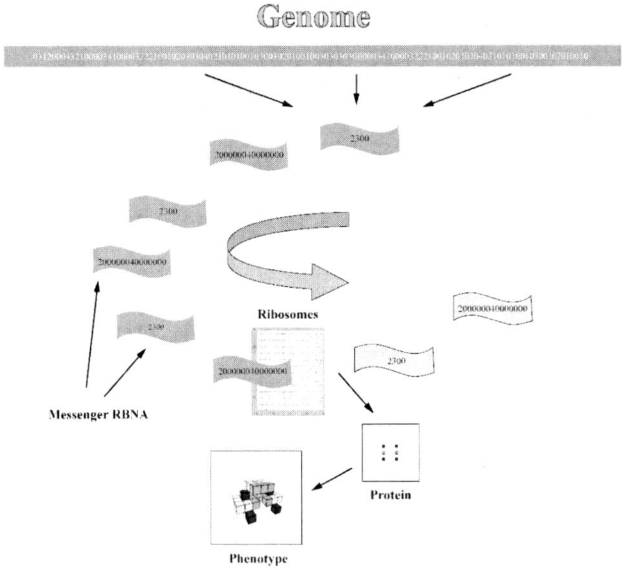

From the table, we see that the system uses the same nodes at the beginning of each cycle. The frequency with which they are used are multiples of the coefficients for the fundamental sequence for the Sierpinski triangle (11.5):

1,4,4,16,4,16, 16, 64….

To gain a better understanding of this process, let RD_1 be the dynamic network for the first cycle and RF_1^n the dynamic network for cycle n. Both networks can be represented by 14x1column matrices in which each of the elements represents the frequency with which a specific node is used (the 15th node, representing the rest state is omitted). We therefore have the following relationship:

$$RF_1^n = \alpha_n * RF_1^1 \qquad (11.6)$$

where α_n is the nth element in the Sierpinski sequence. The same relation is satisfied by the other 7 dynamic networks. This finding confirms the results described in the

Table 3. Frequency of consultation of nodes on the first step of each cycle. Nodes that are not used are not shown. The last node is excluded

Nodes/Cycles	1	2	3	4	5	6
1	1	4	4	16	4	16
2	1	4	4	16	4	16
3	2	8	8	32	8	32
4	1	4	4	16	4	16
6	2	8	8	32	8	32
7	2	8	8	32	8	32
9	1	4	4	16	4	16
10	2	8	8	32	8	32
11	1	4	4	16	4	16
12	4	16	16	64	16	64
14	2	8	8	32	8	32
15	2	8	8	32	8	32
17	2	8	8	32	8	32
18	2	8	8	32	8	32

previous chapter and allows us to explain the growth of the phenotype in terms of the properties of the genotype, as seen in the consultation networks.

We can summarize these results as follows.

- During its 8-step development cycle, the original self-replicator produces 4 copies of itself.
- Each of these 4 structures begins a development cycle of its own. This explains the quadrupling of frequencies of use on the second cycle.
- Any replication process requires space. The replicator itself can grow until it occupies all available space or until growth is halted by regulatory processes. However, space for the replication of substructures is obviously limited. The growth process is symmetrical in all four directions but the space available to the replicator is not. This implies that the occupation of internal space is regulated. The regulation creates a crystal-like structure reminiscent of Sierpinski's arrowhead. At this point, however, the elementary unit is no longer the single cell, but the complete module undergoing replication. The way in which these modules assemble into larger systems is regulated by the Sierpinski arrowhead logic. The number of modules that can be assembled in this way is given by the fundamental sequence (11.5).

Dynamic networks and consultation frequencies provide very detailed information. If we require higher level information, we can use statistical measures. Several such measures can be computed directly from dynamic networks.

Populations

If $p_i(t)$ is the number of cells in state s_i at time *t*, we obtain:

$$p_i(t) = \sum_{h=1}^{v} \beta_h(i) * RF_h^{\ t} = \sum_{h=1}^{v} \beta_h(i) * \alpha_{\mathrm{int}(1+t/8)} * RF_h^{\ t \bmod 8} = \alpha_{\mathrm{int}(1+t/8)} * \sum_{h=1}^{v} \beta_h(i) * RF_h^{\ t \bmod 8} \qquad (11.7)$$

where $v+1$ is the number of nodes in the dynamic network (excluding the node representing the rest state), $\beta_h(i)$ equals 1 when the value of node h equals i, otherwise 0; $\mathrm{int}(1 + t / 8)$ is the whole part of the argument.

Population Entropy

Given a CA, with states between 0 and k-1, we can define *population entropy* by: (5.2)

$$H^t(X) = -\sum_{i=0}^{k-1} \frac{h_{ii}(t)}{n*n} \log \frac{h_{ii}(t)}{n*n}$$

Where $n * n$ is the total number of sites.

Neighborhood Entropy or Input-Entropy

Neighborhood Entropy, also known as Input Entropy is given by:

$$H^t(X) = -\sum_{h=1}^{v(t)} \frac{RF_h^{\ t}}{n*n} \log \frac{RF_h^{\ t}}{n*n} = -\alpha_{\mathrm{int}(1+t/8)} * \sum_{h=1}^{v(t)} \frac{RF_h^{\ t \bmod 8}}{n*n} \log \frac{\alpha_{\mathrm{int}(1+t/8)} * RF_h^{\ t \bmod 8}}{n*n} =$$

$$= \alpha_{\mathrm{int}(1+t/8)} * \sum_{h=1}^{v(t)} \frac{RF_h^{\ t \bmod 8}}{n*n} (\log \frac{RF_h^{\ t \bmod 8}}{n*n} + \log \frac{\alpha_{\mathrm{int}(1+t/8)}}{n*n}) \qquad (11.8)$$

where $v(t)+1$ is the number of nodes in the dynamic network at time t. These entropy curves are fractal in nature: in each 8-step cycle the self-replicator undergoes a process of translation and expansion. The coefficients for these non-linear transformations are expressed in the fundamental sequence for the Sierpinski Arrowhead.

In Chapter 5, we depicted the time series for (11.8) as points in a two-dimensional phase space, where each point represented the values of the pair $H^t - H^{t+1}$ The trajectory of these points provides a signature for the self-replicator. The trajectory depends on two factors:

- The precise period of the basic replication module,
- The typical behavior of the Sierpinski Arrowhead, describing the higher level logic governing the development of the pattern.

It is important to observe that these characteristics are not specific to this particular self-replicator. As discussed at length in the previous chapter, they are shared by many. Replication periods and basic modules can change. However the trajectories and the basic Sierpinski logic are conserved. In a certain sense, we can consider the trajectories as a special kind of attractor in phase space. Their repetitive behavior brings to mind limit cycles. At the same time, however, their fractal behavior and ability to always stay in a well-defined region of space is reminiscent of strange attractors. We should not take this property too literally: in empty space, the time series would grow indefinitely. However, the regions in which the self-replicator has space to grow can be described by non-linear transformations based on rules similar to (11.8). In reality, therefore, they are predictable. We will see later that the ability to reason in terms of trajectories and phase space can be useful even in cases in which we are unable to directly derive the logic underlying a specific pattern of growth.

Mutations

In previous sections, we have seen that consultation networks for self-replicators are limited in size and that instantaneous consultation networks are cyclical with period T (11.4). In other words:

$$l''(t) = l''(t+\overline{T})$$

where $\overline{T}$ is the period of the network.

If L'' is the set of all instantaneous consultation strings and L' is the set of all possible consultation strings, $L'' \subset L'$. We can thus define:

$$l'(t) = l''(0) \oplus l''(1) \oplus .... \oplus l''(t) \tag{11.9}$$

where $\oplus$ is the string sum operator:

$$\oplus : L'' x L'' \to L'$$
$$(l''_1, l''_2) \mapsto l' = l''_1 \oplus l''_2 .$$

Elements $c_i \in l'$ are: $c_i = x$ if $s_{2i} = s_{1i} = x$ and $c_i = s_{1i}$ or $c_i = s_{2i}$ if any one of the corresponding elements in strings l''_1, l''_2 is different from x. For instantaneous consultation networks described by the same string l, if both elements are different from x, $s_{2i} = s_{1i}$.

We define a *mutation* as an application associated with $l \in L$:

$$l \in L \mapsto Ra\_s \tag{11.10}$$

$Ra\_s \in L$, is a new string made up of the same characters as l, except for the character in position a, which has value s. If $Ra\_s = l$, the mutation is the *identity* operation. String l' with $|l'| = z$, satisfies (11.4). In other words, the consultation network for the network is constant over time and contains only z elements with values different from x. It follows that (11.10) defines $k * z$ mutations, in which z elements are *identities*. Strings $Ra\_s \in L$ are realized by substituting characters $0,1,...,k-1$ in position *a* in string l, in such a way that all characters in l' have values different from y.

Set $M^l_{l'}$ of strings $Ra\_s \in L$ represents the space of all possible mutations:

$$M^l_{l'} = \{m^a_s(l) = Ra\_s \in L \,/\, \forall a, \forall s = 0,1...k-1\} \tag{11.11}$$

Mutations in a Simple Self-Replicator

To see the phenotypical effects of changes in the genotype, consider the simple self-replicator, $A = (1)$ generated by rule C2.2. The consultation network for the structure consists of 7 nodes corresponding to elementary rules R55, R164, R200, R210, R216, R219 and R220. Given that the network displays the temporal symmetry described in (11.4), it is stable over time. Figure 11 depicts the way the self-replicator changes its behavior when we mutate an elementary rule. If we observe the changes produced by the mutations shown in the first image in Figure 11, we see that two out three (R55_1 and R55_2) leave the system intact. However, mutation R55_3 makes it chaotic and destroys its functionality. The process becomes even clearer if we consider the formation of colonies based on the Sierpinski Arrowhead. In this case too, the first two mutations leave the structure intact, while the third destroys it. By contrast, mutations in the second elementary rule (R164) have a catastrophic effect, destroying the self replicator (see the second image in Figure 10).

If we analyze all variants for all rules (except R220), we discover that 3 variants (R55_1; R55_2 and R210_1) conserve the system's self-replication capabilities. 2 (R219_0; R219_2) lead to the null state; 13 (R55_3, R164_1,R164_2, R164_3, R200_1, R200_2, R200_3, R210_2, R210_3, R216_1, R216_2, R216_3, R 219_3) generate chaos, destroying the self-replicator.

Variants of R55 have no effect on the shape or function of the self-replicator. With respect to the original rule, R55_1 uses a different space-filling technique and R55_2 contains a parasite that exploits the basic replication mechanism to replicate itself. All three rules are governed by the logic defined in ECA 90. By contrast, variant R210_1 shows an important change in the way the self-replicator fills its internal space. For example, we observe the emergence of new shapes. In addition, replication is no longer governed by rule (10.4).but by rule (10.6).

We can now consider variants to the variants. In this case, we will restrict our analysis to rules R55_1, R55_2 and R210_1: the three variants that conserve the self-replication process.

Mutations in Rule R55_1

The consultation network for R55_1 consists of 9 nodes: R55, R100, R164, R185, R200, R210, R216, R219 and R220. Compared to the original network, it contains two new nodes: R100 and R185. Mutations to these rules lead to the following results:

a. 9 variants (R55_0, R55_2, R100_1, R164_1, R185_1, R185_2, R200_1, R210_1, R216_1) conserve the replication process or produce gliders;

Figure 11. Instantaneous consultation networks for the self-replicator generated by rule C11.2. The behavior of the network is cyclical with period 6.

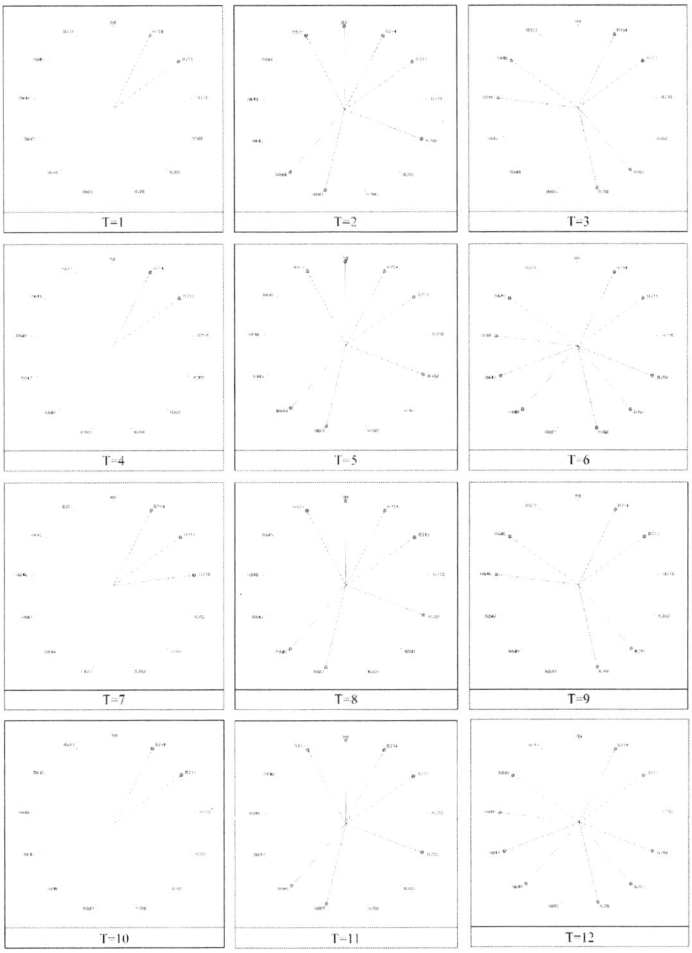

b. 13 variants (R55_3, R100_2, 100_3, R164_2, R164_3, R185_3, R200_2, R200_3, R210_2, R210_3, R216_2, R216_3, R219_3) lead to chaos, destroying the replication process
c. 2 rules (R219_0, R219_2) lead to the null state

The second image in Figure 10 depicts some of these mutations.

Figure 10. Row 1, behavior of the self-replicator when the rule is mutated. Row 2, the first image shows the conservation of self-replication in a self-replicator that has survived a mutation. The second image shows variations in the first self-replicator to undergo mutation. The third and fourth images show emergent self-replicators following a double mutation.

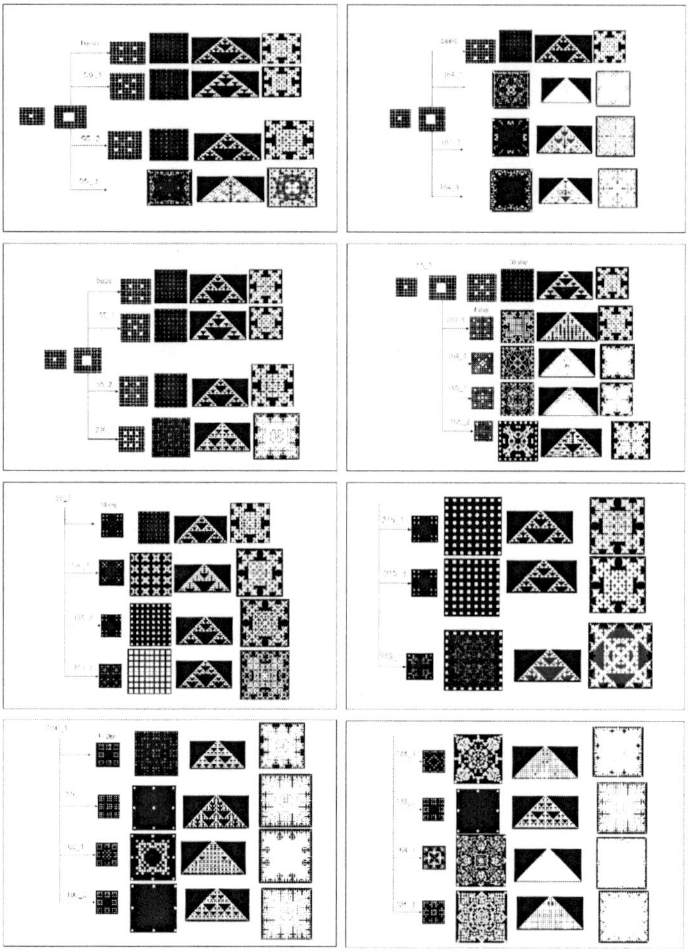

Mutations in Rule 55_2

The consultation network for rule 55_2 contains 11 nodes: R_55, R_100, R_164, R_185, R_200, R_210, R_215, R_216, R_218, R_219 and R_220. Compared to the original rule, there are two new nodes: R_215 and R_218.

The changed rules give the following result.

a. 8 variants (R55_0, R55_1, R100_1, R185_2, R185_2, R210_1, R215_1, R215_3, R218_1) conserve the replication process or produce gliders.
b. 20 variants (R55_3, R100_2, R100_3, R164_1, R164_2, R164_3, R185_1, R185_3, R200_1, R200_2, R200_3, R210_1, R210_3, R215_2, R216_1, R216_2, R216_3, R218_2, R218_3, R219_3) lead to chaos, destroying the replication process.
c. 2 rules (R219_0, R219_2) lead to the null state.

Row 3 in Figure 10 depicts six variants (R100_1, R185_2, R185_2, R210_1, R215_1, R215_3, R218_1) that conserve the replication process. The other two mutations with this property were depicted earlier. These mutations confirm the key role of the replication logic in ECA 90. In addition, R100_1 and R218_1 both lead to interesting changes in the behavior of the CA. In the first case, we observe an expansion of the basic block, in time and in space. The second case is marked by an expanded role for cells in state 2. The main changes in the other variants involve parasitic structures.

Mutations in Rule R210_1

The consultation network for rule R210_1 contains 10 nodes: R55, R100, R136, R164, R185, R200, R_210, R216, R219, R220. Compared to the original rule, there is 1 new node (R136). Two nodes (R215 and R218) are no longer present.

Mutating these rules we obtain the following results

a. 8 variants (R55_1, R100_1, R100_2, R136_1, R136_2, R164_1, R185_1, R200_0, R210_0, R216_1;) conserve the replication process or produce gliders
b. 15 variants (R55_2, R55_3, R100_3, R136_3, R164_2, R164_3, R185_2, R185_3, R200_2, R200_3, R210_2, R210_3, R216_2, R216_3, R219_3) lead to chaos, destroying the replication process
c. 2 rules (R219_0, R219_2) lead to the null state

The last row of Figure 10 shows the patterns produced by some of the mutations in which the replication capability of the CA is conserved. We observe important changes both in the basic module and in the replication logic. Two variants (R100_2 and R136_2) conserve the original logic; rule R55_1 uses logic reminiscent of ECA 90.

The Complex Nature of the Mutation Process

Consider a self-replicator with $A = (3)$, (C11.2, in the Appendix of this Chapter). During its evolution, the self-replicator uses 13 elementary rules (see Table 4):

If we analyze the dynamic behavior of the instantaneous consultation networks, we observe that after a brief transient, the network settles into a stable of cyclical behavior with a period of 6 (see Figure 11)

The behaviors of the individual dynamic networks for the self-replicator generated by rule C11.2 are described in the Appendix of this Chapter. As in the previous case, the logic of the colony is the same as the logic for the Sierpinski Arrowhead.

Now consider string (11.2), representing a CA with $D = 2, k = 5 and\ r = 1$.

The string contains 715 characters. However, the dynamic consultation network which controls the CA's behavior has only 13 (see Table 4). The other 702 have no effect, even if they mutate. Let us now examine all possible mutations to the 13 rules that do have an effect. The number of these mutations is given by:

$$\mathrm{M} = k * r \qquad (11.12)$$

where M is the number of possible mutations, *k* is the number of possible states and *r* is the number of nodes in the dynamic network. If we analyze these changes, we discover that r mutations have no effect. Mutations in the last character either leave the network unchanged or lead to radical changes.

The number of valid mutations M' is thus:

$$\mathrm{M}' = (k-1) * (r-1) \qquad (11.13)$$

For rule C11.2, $\mathrm{M} = 65 and \mathrm{M}' = 48$. Figure 12 depicts the patterns produced by each of the 65 including those that leave the network unchanged and those that

Table 4. Elementary rules used during the evolution of the self-replicator and their frequency of use. It is clear that the last rule controlling the rest state is the rule most used

R10	R512	R646	R647	R648	R650	R684
29	116	289	464	112	126	116
R701	R702	R703	R711	R712	R715	
60	240	248	234	565	99801	

modify the rest state. The first image shows the pattern produced by the basic, unmutated rule.

We observe that, in many cases, the CA's self-replication capability is conserved. In others, it is replaced by chaos or by the null state. In one case (R702_4), the self-replicator generates two gliders traveling in opposite directions.

To obtain a general picture of these mutations (see image a in Figure 13), we order the rules from top to bottom and used color to identify those that have undergone valid mutations. We call this image the rule spectrum for the self-replicator. In cases where a single rule generates more than 1 self-replicator, the rule spectrum changes when we change self-replicator. The spectrum displays significant regularities. As

Figure 12. Patterns corresponding to all possible mutations in rule C11.2

we can see in the second image b), valid mutations are concentrated towards the end of the string: characters with positions lower than 650 are hardly affected. If we analyze the rules used by variants of the basic CA, the results are even more interesting. Here too, the rules used are concentrated in the right hand region of the image. If we highlight these rules, we discover that in some cases the consultation network includes relatively few nodes. In most cases, however, the number of nodes is much larger than in the original version and involves regions that were not directly affected by the mutation. In other words, mutations lead to profound changes in dynamic networks, involving the whole genome and modifying the self-replication process. This is true even for mutations affecting a single locus. These findings are additional evidence for the complexity of the self-replication process. However, we also observe rules that are never used. In the spectrum, we represent these regions by white stripes (see for example image 4 in Figure 13d) and call them "prohibited regions". Even after the original self-replicator has mutated, they play no role in its behavior. Although new mutations may restore their functionality, at this stage in the evolutionary process, with this particular self-replicator, they are "junk."

Let us now analyze mutated rules that leave the self-replication process intact. As we can see in Table 5, there are 23 such rules. The figures in the row beneath the name of the rule show the number of nodes in the consultation network for the rule. This varies between a minimum of 13 for rule R711_3 (the same as for the unmutated rule) and a maximum of 25 for rule R647_3.

As we can see in Figure 14a, the number of nodes in the consultation networks for the mutated networks varies widely. However, the values for mutations that conserve the self-replication process lie within a much narrower range. When the number of nodes falls below a certain threshold, the system tends towards ordered behavior; when it is higher, the resulting behavior is chaotic. If we consider longer periods of evolution, the number of nodes used by chaotic systems increases dramatically. However, relation (15b) guarantees that the number of nodes used by self-replicators remains constant. So once again, we discover a border region, separating order from chaos. It is important to observe that the existence of such a region is not just a characteristic of one particular example. Rather, it is a general feature of replicated processes displaying the symmetries described in Chapter 10 (Bilotta & Pantano, 2005; 2006).

Let us now briefly analyze the patterns produced by the 23 mutated self-replicators. All except two have growth models similar to those of the un-mutated system (see Figure 16c). As we have already seen, this logic is well described by the Sierpinski Arrowhead. The two variants whose growth model changes are R702_1 and R711_3. The model for R702_1 has a period of 6, just like the basic self-replicator (Figure 15d). However, the pattern it produces is completely different. Although both systems follow the logic of ECA 90, the first self-replicator in the mutated

Figure 13. a), the spectrum for a self-replicator generated by C11.2 starting with a single seed in state 3. b): regions especially sensitive to mutations. c): frequency of rule use in the consultation networks, d): "closed regions" of the genome, never used during the mutation process.

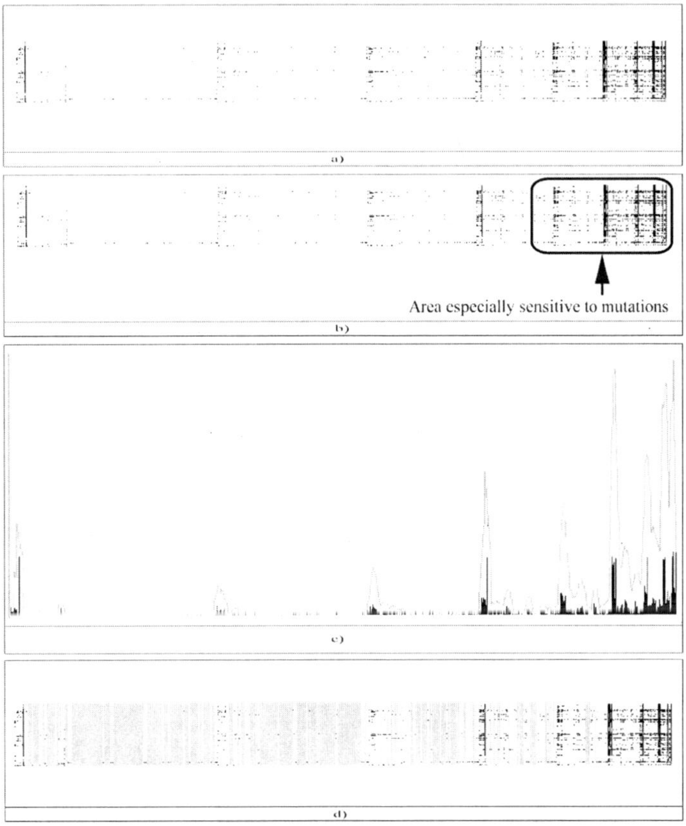

Table 5. Rules that conserve the self-replication process. The row below the name of the rules shows the number of nodes in its consultation network

Basic	R10_1	R10_2	R10_3	10_4	R512_4	R646_2	R646_4
13	15	23	15	15	18	17	15
R647_3	R648_1	R648_2	R648_3	R648_4	R650_0	R650_1	R650_2
25	17	18	19	15	13	22	16
R684_4	R701_2	R701_4	R702_1	R703_1	R703_2	R711_2	R711_3
19	14	13	24	22	16	19	13

Figure 14. a): distribution of the number of nodes in the consultation network. b) Self-replicators are located in a border region separating null and chaotic rules. c): side views of the patterns produced by the self-replicators.

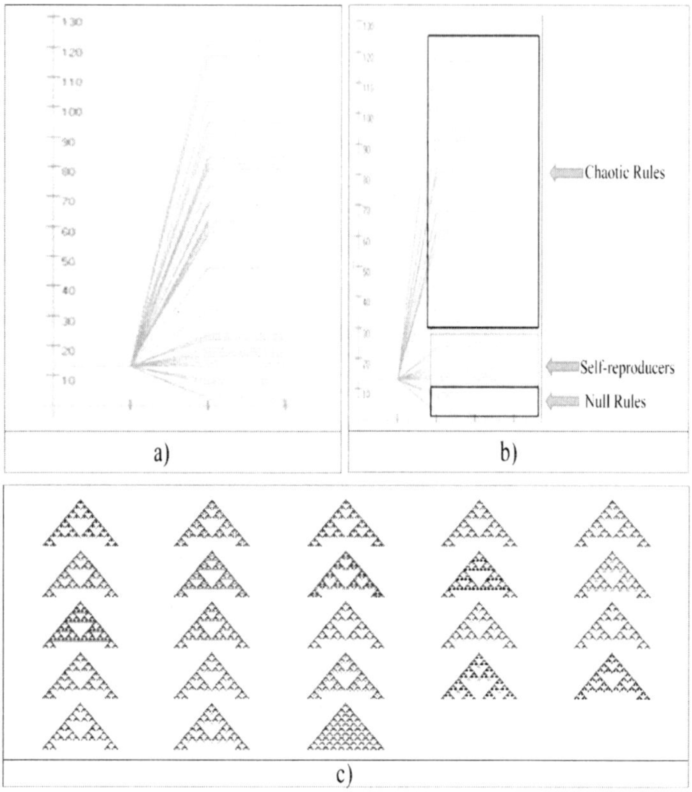

variant appears after 9 instead of 6 steps. In the case of R711_3, the self-replicator is completely different from the original. The replication period for the mutated system is no longer 6 but 3.

These findings are confirmed by these systems' trajectories in phase space (Figure 15e). In a sense they are cyclical. During the evolutionary process, however, they tend to expand, like a fire that is out of control. With two exceptions, we find similar behavior in all of the models regulated by ECA90. Different mutations produce slight differences in the trajectories. However, the overall behavior remains the same. In the first of the two exceptions, the structures take longer to form and this prevents closure of the replication cycle. In the second, the period is much shorter and each cycle contains just 3 points. In this case the expansion is very clear.

We can sum up the behavior of these trajectories as follows:

Figure 15. d): the decoding scheme for the self-replicators .e): dynamic models of the growth process. Input Entropy values are displayed in the phase plane.

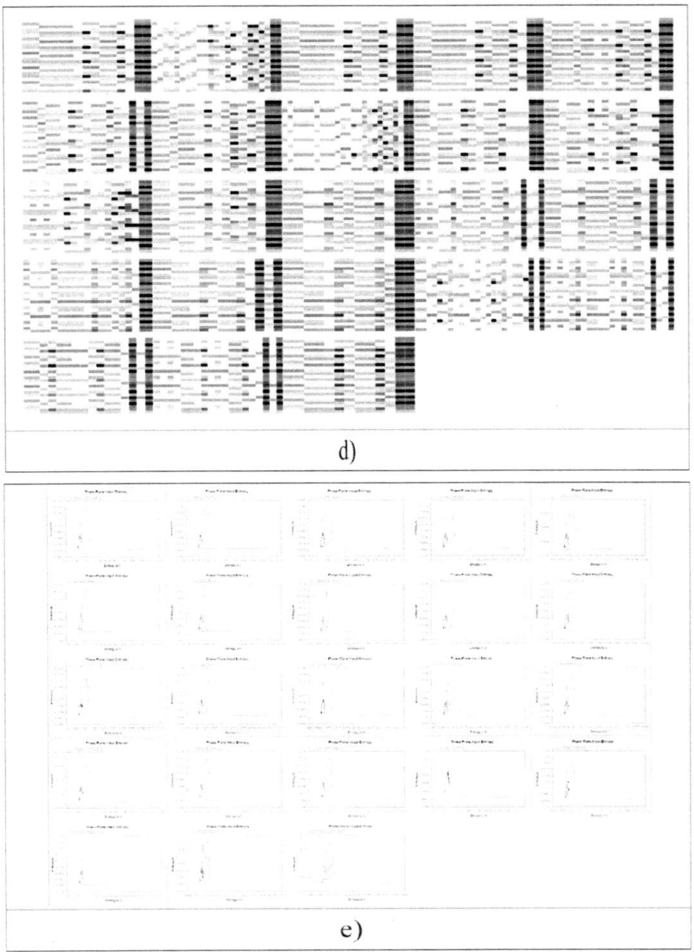

a. The trajectories for self-replicators are regular; the main variations concern the period of the replication cycle and the growth model for the self-replicator.
b. Even when the curve does not occupy a well-defined region of phase space, the trajectories follow well-defined cycles.
c. Chaotic behavior produces trajectories that occupy ever larger regions of phase space, with no regularity in their behavior.
d. Ordered behavior (limit cycles, fixed points etc.) produces regular trajectories.

We observe that the self-replicators display the same regularities in behavior as ordered systems. However, the way they grow is more typical of a chaotic system.

The behaviors of self-replicators in phase space constitute what we call *growth attractors*.

Evolutionary Dynamics and the Problem of Representation

It is easy to imagine that the genetic networks governing replication change during the evolutionary process. Some of these changes are conservative or adaptive, leaving replication intact. Others are destructive, leading to extinction or chaos. In some cases, we observe the emergence of second order structures or parasites. In these ideal spaces, self-replicators behave as if they were regulated by well-defined symmetry groups and as if the transformations they undergo represented fixed sequences, determined by the growth process. From a mathematical point of view, a group of symmetries is a set of transformations that conserves structural relations between members of the same group. If we change the genetic code for the self-replicator, we break these symmetries, creating multiple sub-sets of symmetries. These can be seen as representing a layered model of development. We have been able to sketch two representations of this model, starting from a single self-replicator. We have also identified ancestors of the self-replicator that share its basic shape and function. We have thus proved that ontogenesis recapitulates phylogenesis.

The evolutionary history of a self-replicator follows a trajectory that stays away from the chaos typical of other branches of the evolutionary tree. As evolution proceeds, each self-replicator produces a growing number of new self-replicators. The number of these systems grows exponentially. An exhaustive analysis of all possible evolutionary trajectories is probably impossible. All we can do is examine a few steps in the process. There are two different ways in which we can represent the process (Figures 16a and b). We observe that some apparently chaotic structures are actually transients towards stable structures such as gliders or other kinds of self-replicator. Can we interpret these phenomena as equivalent to the birth of new species?

This approach matches several features of biological evolution:

a. The process begins with an agent that already exists.
b. The evolutionary process is historically determined.
c. Many mutations have catastrophic effects.
d. New agents present structural (morphological) similarities to their ancestors.
e. The process allows complexity to increase spontaneously.
f. It allows the emergence of new species.
g. Exhaustive analysis of all possible evolutionary trajectories is impossible.
h. Different species may share common ancestors.

Figure 16. a) and b): Two representations of the mutation-based evolutionary process. c) and d) The difference between the self-replication process for the basic rule and the rule for the last generation.

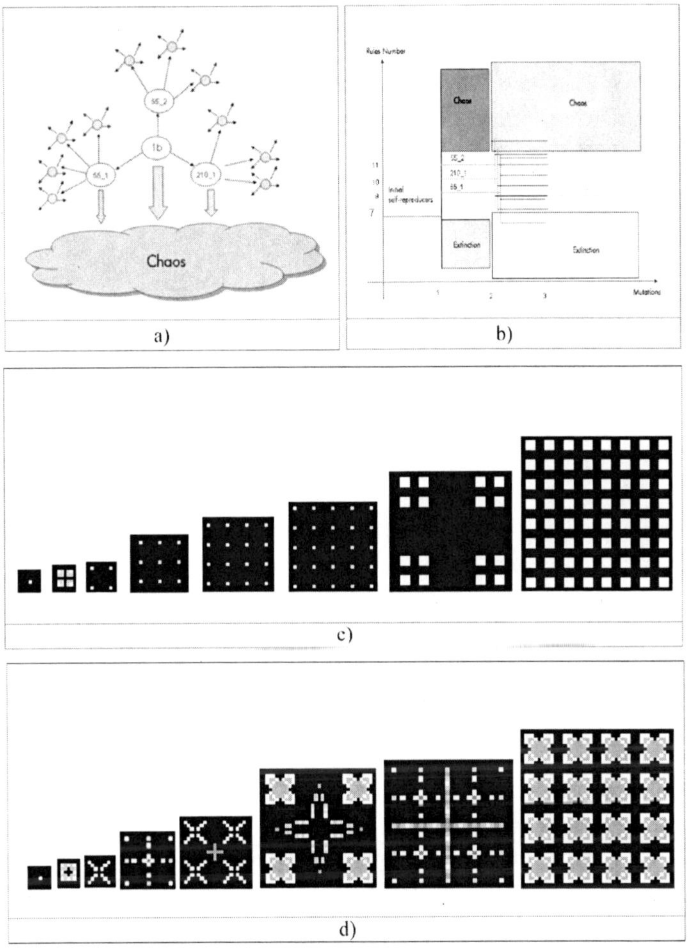

The diversity of phenotypes increases during the evolutionary process even though many phenotypes maintain the same shape and function as their ancestors. Mutation space can be partitioned into three regions: the first produces self-replicators that continue to function (albeit with many variations); the second region (the largest of the three) produces chaotic systems; the third leads to extinction. The border region between the first and the second region is clearly fractal in nature. It is this region where self-replicators and chaotic systems compete for space. In the figure we also see an "edge of chaos" - a thin border region separating order from chaos (Figure 16b).

The differences between the behavior of the mutated and the original self-replicators can be grouped into 5 major categories (Bilotta & Pantano, 2006):

a. Deletions or additions of elements that do not change the high level or internal logic of the original structure;
b. Replacement of elements in the high level or internal logic;
c. Merging of structures to create new structures with a replication cycle whose period is a multiple of the period for the original structure;
d. Changes in the reference model;
e. Changes in the period of the replication cycle.

Changes in categories a) and b) should be considered as minor, non-catastrophic perturbations in the original behavior, based on redundant elements available in the network. Changes in category c) represent a change in the scale of the structures making up the system that nonetheless conserves their basic behavior. These kinds of change are very frequent. In most cases, they are the consequences of changes on the left hand side of the Hamming distribution. Changes in classes d) and e) are less frequent. In the case discussed, we observe only one case, characterized by a large Hamming distance between the original and the new rules and major changes in behavior. Such changes are similar to the changes described by Catastrophe Theory. The points at which they occur represent catastrophic bifurcations in system behavior.

Patterning Phenomena

It is useful to illustrate the arguments of the previous section with an example. In what follows, therefore, we will describe a series of mutations to the consultation network for the basic self-replicator in rule C11.2. Each of these mutations creates new self-replicators with new consultation networks. However it is computationally very expensive to track long sequences of mutations and the vast range of systems to which they give rise. To avoid this difficulty, the best strategy is to limit the analysis to strings presenting specific characteristics. In each generation, therefore, we focus our analysis on the variant whose consultation network contains the highest number of nodes. In this way, we generate the following sequence of mutations:

R647_3 →R684_3 →R 648_2 →R 508_1 →R 594_1 →R1_2 →R668_3 →R628_1 →R688_4 →R621_1→R634_2 →R 185_2 →R600_1 →R 443_4 →R 680_2 →R 625_2 →R 700_3 →R681_2→R 513_2 →R 11_2 →R 675_3 →R 7_3 →R 680_3 →R 570_4 →R 686_4 →R 580_3 →R 571_4→R 686_2 →R 550_4 →R 696_2

Figures 16c and d picture the behavior of the basic self-replicator and of its final variant (C11.3) after 2, 3, 6, 9, 12, 16 and 22 steps. During the evolutionary process, we observe an increase in the complexity of the self-replicator. This is especially apparent if we examine side-views of the patterns (Figure 17e). The increase in complexity becomes even more evident if we analyze the number of nodes in the Consultation Networks for systems at different stages in their development (see Table 6). Figures 17f, 18g and 18h depict their respective decoding schemes, consultation networks and phase plane behaviors.

Figure 17. e) Side-views of the patterns for each generation. f) Decoding schemes and consultation networks for the various phases of the process

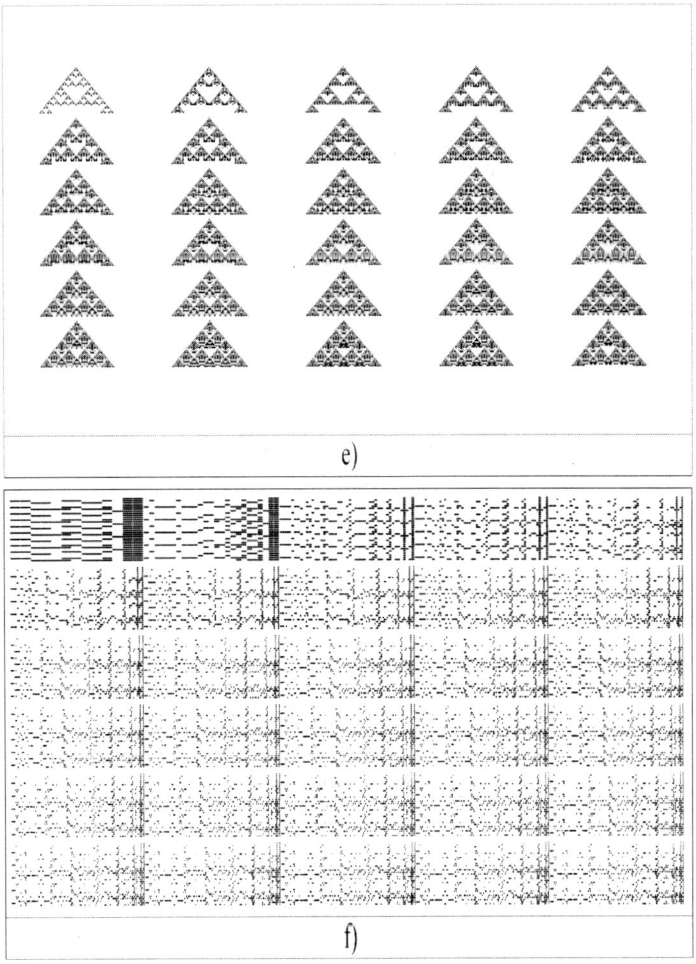

Table 6. Number of elementary rules by generation

Generation	Elementary Rules
1	13
2	25
3	47
4	55
5	73
6	83
7	92
8	96
9	102
10	109
11	112
12	117
13	119
14	121
15	123
16	124
17	125
18	126
19	128
20	129
21	130
22	131
23	132
24	133
25	134
26	138
27	138
28	138
29	138
30	138

Initial Stages (Generation 0 to Generation 4)

At the beginning of the evolutionary process, the basic self-replicator uses just 13 elementary rules. However, the very first mutation almost doubles this number. The basic replication logic does not change. The Sierpinski Arrowhead plays a key role

Figure 18. g) Decoding schemes and consultation networks for the various phases of the process. h) Trajectories in phase plane for each generation.

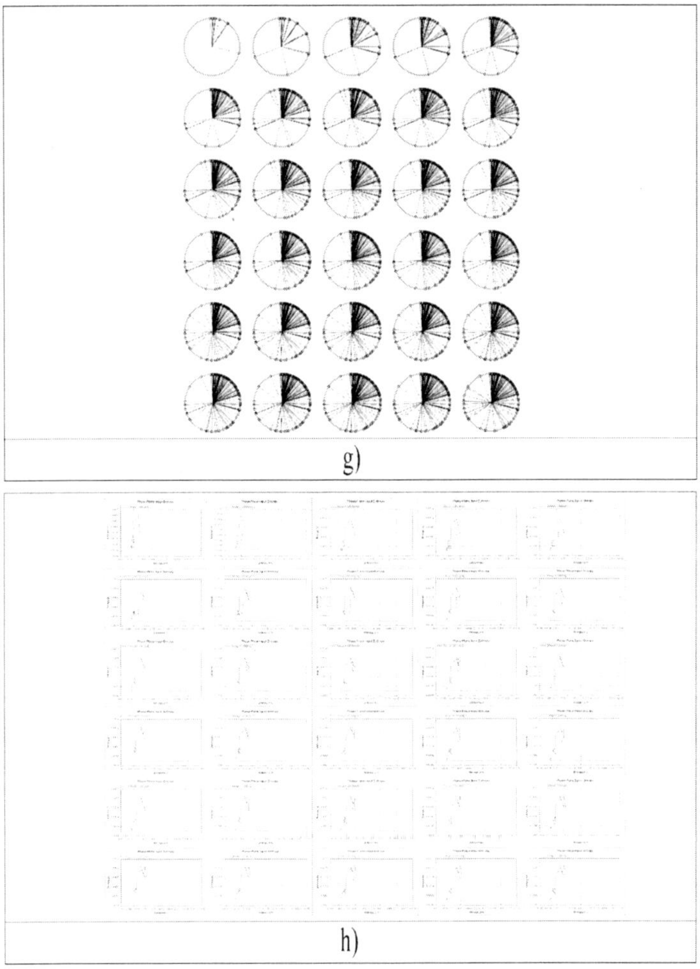

from the beginning to the end of the process. Nonetheless we observe a substantial change in the basic module. The next major change occurs in the next generation, by which time the number of nodes in the network has already reached 47, almost double the previous figure. The nodes in the old network are incorporated in the new one. However, the links between nodes are completely different. The process that began in the previous generation continues, further filling the available space in the basic module. Generation 3 is marked by a small increase in the number of nodes, which is followed by a larger increase in generation 4, when the number of nodes rises to 73. The increase produces increased pattern complexity and further filling of the available space. It is important to observe that the pattern produced in generation 2 is entirely contained in the pattern for generation 4; the length of the replication cycle doubles, to 12. Similarly the network for generation 4 completely incorporates the network for generation 2.

Constant Growth (Generation 5 to Generation 11)

In generation 5, the evolutionary process discovered a new module, allowing it to fill space more efficiently. Now, the process stabilizes, continuing to fill up the internal space. To achieve this result without changing its basic logic, the system uses an ever growing number of elementary rules. The number of nodes in the consultation network increases gradually until it reaches 117 (in generation 11).

Slower Growth (Generation 12 to Generation 25)

This phase is marked by a slowing down in the growth of the number of nodes. Between generation 12 and generation 25, this passes from 117 to 134. The growth in the complexity of the system also slows. The space-filling process produces an ever denser pattern.

Maximal Complexity (Generation 26 to 30)

By generation 26, the system has reached the maximum complexity possible. The features of the system include a large number of parasitic structures and very complex basic modules. At this stage, the effects of individual mutations are almost negligible. The only changes are replacements of specific elements, deletions and addition of new elements. These changes belong to categories a) as defined in the previous section.

It is very interesting to observe the trajectories in phase space (see Figure 18h). These provide clear evidence of the way the module repeats. The fact that the trajectories for the last generations in the series are not closed is due exclusively to the lack of time for the evolutionary process to finish its work. Similarly to the processes described in the previous section, the ECA 90-based replication process involves the repetition of a module that expands several times during the process. As the number of nodes in the network increases, the module expands still further. The change in the length of the replication cycle leads to a further increase in complexity. The process continues to be dominated by the persistent symmetries of ECA 90.

FUTURE TRENDS

Our future work will focus on the design and prototyping of the artificial laboratory. Although many components have been completed, we will continue work on our tools for the visualization of evolutionary processes, individual self-replicators and populations of self-replicators. This will allow us to generate an enormous quantity of data which we will make available to other researchers. It will also allow other teams to conduct their own evolutionary experiments, currently feasible only for specialists. The simple simulation environment provided by the artificial laboratory will allow researchers to analyze different evolutionary trajectories, investigate the function and morphology of phenotypes, study interactions within competing and collaborative gene pools and gain insights into very large scale population dynamics. It is likely that this work will cast new light on the biological processes simulated and visualized in the artificial laboratory.

CONCLUSION

We began this chapter by providing a brief description of scientific approaches to pattern formation in living systems and to key genetic processes within the cell. We went on to describe data from experiments in synthetic biology using artificial systems such as 2-D CAs, in an *artificial computational laboratory.* The artificial organisms metaphor allowed us to reproduce a series of processes that normally require genetic experiments. We began by identifying the genetic sequence for our artificial organism. We then identified the genes determining the system's shape and function and went on to map the sequence to DNA bases, using an appropriate encoding scheme. The methods described in Chapter 9, allowed us to treat the sequence as a code or language. At each step in the simulation, we identified the genes responsible for the behavior of the system. Given that these genes belong to a dynamic process, we represented them as part of what we called dynamic genetic networks. We then created representations of the way the networks function: network models and symbolic representations of the rules consulted at each phase of the development process. These studies allowed us to identify development models shared by large numbers of systems. We found that modifications to these models produce multiple changes in the structure and function of the organisms. In an alternative representation, we used Input Entropy to map the discrete growth process into a continuous dynamic system with a limit cycle. We observed how changes in an artificial system's genome are reflected in changes in its growth model the associated developmental trajectory.

Our computational models have allowed us to sequence artificial DNA, identifying coding and non-coding sequences as well as repeats and other features present in biological DNA. The experiments provide us with an enormous volume of data that we can compare with data for self-replication in biological systems. The laboratory allows us to visualize the development of individual organisms and the evolution of populations. Last but not least, it makes it possible to perform *digital genetic engineering,* using trial and error, or series of rules to construct new self-replicators.

REFERENCES

Adamatzky, A., & Komosinski, M. (Eds.). (2005). *Artificial Life Models in Software*. New York: Springer. doi:10.1007/1-84628-214-4

Arena, P., Fortuna, L., & Branciforte, M. (1999). Reaction-diffusion CNN algorithms to generate and control artificial locomotion. *IEEE Transactions on Circuits and Systems. I, Fundamental Theory and Applications*, *46*(2), 253–260. doi:10.1109/81.747195

Ashby, W. R. (1952). *Design for a Brain*. London: Chapman & Hall.

Bedau, M. A., McCaskill, J. S., Packard, N. H., Rasmussen, S., Adami, C., & Green, D. G. (2000). Open problems in artificial life. *Artificial Life*, *6*(4), 363–376. doi:10.1162/106454600300103683

Bilotta, E., & Pantano, P. (2005). Emergent Patterning phenomena in 2D Cellular automata. *Artificial Life Journal*, *11*(3), 339–362. doi:10.1162/1064546054407167

Bilotta, E., & Pantano, P. (2006). Structural and Functional Growth in Self-Reproducers. *Complexity*, *11*(6), 12–29. doi:10.1002/cplx.20131

Bilotta, E., & Pantano, P. (2008). Biological traits in artificial self-reproducing systems . In Orsucci, F., & Sala, N. (Eds.), *Complexity Science in Information Communication Technologies: Interaction Between Living Systems and Information Technologies* (pp. 109–123). Hershey, PA: IGI Global.

Bonabeau, E., Dorigo, M., & Theraulaz, G. (1999). *Swarm Intelligence: From Natural to Artificial System*. New York: Oxford University Press.

Crutchfield, J. P., & Schuster, P. (2003). *Evolutionary Dynamics: Exploring the Interplay of Selection, Accident, Neutrality, and Function*. New York: Oxford University Press.

Dorigo, M., & Stützle, T. (2004). *Ant Colony Optimization*. Cambridge, MA: MIT Press.

Flavell, R. B., & Moore, G. (1996). Plant genome constituents and their organization . In Foster, G. D., & Twell, D. (Eds.), *Plant gene isolation: principles and practice* (pp. 1–25). New York: John Wiley & Sons.

Gierer, A., & Meinhardt, H. (1972). A theory of biological pattern formation. *Kybernetik*, *12*, 30–39. doi:10.1007/BF00289234

Goras, L., & Chua, L. O. (1995). Turing patterns in CNNs. II. Equations and behaviors. *IEEE Transactions on Circuits and Systems. I, Fundamental Theory and Applications*, *42*(10), 612–626. doi:10.1109/81.473568

Goras, L., Chua, L. O., & Leenaerts, D. M. W. (1995b). Turing patterns in CNNs. I. Once over lightly. *IEEE Transactions on Circuits and Systems. I, Fundamental Theory and Applications*, *42*(10), 602–611. doi:10.1109/81.473567

Goras, L., Chua, L. O., & Pivka, L. (1995a). Turing patterns in CNNs. III. Computer simulation results. *IEEE Transactions on Circuits and Systems. I, Fundamental Theory and Applications*, *42*(10), 627–637. doi:10.1109/81.473569

Haken, H. (1977). *Synergetics; An Introduction Non-equilibrium Phase Transitions and Self-Organization in Physics, Chemistry and Biology*. Berlin: Springer-Verlag.

Kauffman, S. A. (1993). *The Origins of Order: Self-Organization and Selection in Evolution*. New York: Oxford University Press.

Kauffman, S. A. (1995). *At Home in the Universe: The Search for Laws of Self-Organization and Complexity*. New York: Oxford University Press.

Kelso, J. A. S. (1995). *Dynamic patterns: The self-organization of brain and behavior*. Cambridge, MA: MIT Press.

Kondo, S., & Asai, R. (1995). A reaction-diffusion wave on the marine angelfish Pomacanthus. *Nature*, *376*(6543), 765–768. doi:10.1038/376765a0

Meinhardt, H. (1982). *Models of Biological Pattern Formation*. London: Academic Press.

Meinhardt, H., Prusinkiewicz, P., & Fowler, D. R. (2003). *The Algorithmic Beauty of Sea Shells*. New York: Springer.

Miyashita, S., & Murata, S. (2005). Topology Changes Enable Reaction-Diffusion to Generate Forms. Paper presented at 8th European Conference on Artificial Life, ECAL 2005. In J. Kelemen, & P. Sosik (Eds.), *Lecture Notes in Computer Science*. Heidelberg, Germany: Springer-Verlag.

Murray, J. D. (1989). *Mathematica Biology*. Berlin: Springer-Verlag.

Murray, J. D. (2003). *Mathematical Biology: II Spatial Models and Biomedical Applications* (3rd ed.). New York: Springer-Verlag.

Rosen, R. (1991). *Life Itself: A Comprehensive Inquiry into the Nature, Origin, and Fabrication of Life*. New York: Columbia University Press.

Thompson, D. W. (1992). *On Growth and Form* (2nd ed.). New York: Dover Publications Edition.

Turing, A. M. (1952). The chemical basis of morphogenesis. *Philosophical Transactions of the Royal Society of London, Series B, 237*, 37–72. doi:10.1098/rstb.1952.0012

von Neumann, J. (1966). *Theory of Self-Reproducing Automata* (Burks, A. W., Ed.). Urbana, IL: University of Illinois Press.

APPENDIX

Instantaneous Consultation Networks for the Self-Replicator Generated By Rule C11.2

Network 1

The first step in the simulation reads the following neighborhoods:

R712-80010>>3 frequency=9

R715-90000>>0 frequency=9991

The dynamic consultation network now contains 2 nodes of which 1 is in the rest state.

Network 2

The second step in the simulation reads the following neighborhoods:

R10-00090>>0 frequency=1

R512-30060>>0 frequency=4

R650-50040>>4 frequency=4

R684-60030>>0 frequency=4

R703-70020>>4 frequency=8

R712-80010>>3 frequency=4

R715-90000>>0 frequency=9975

The dynamic consultation network now contains 6 nodes plus one in the rest state.

Network 3

The third step in the simulation reads the following neighborhoods:

R646-50004>>0 frequency=9

R647-50013>>0 frequency=16

R701-70002>>0 frequency=4

R702-70011>>0 frequency=16

R712-80010>>3 frequency=4

R715-90000>>0 frequency=9951

The dynamic consultation network now contains 5 nodes plus one in the rest state.

Network 4

The fourth step in the simulation reads the following neighborhoods:

R712-80010>>3 frequency=36

R715-90000>>0 frequency=9964

The dynamic consultation network now contains 2 nodes of which 1 is in the rest state

Network 5

The fifth step in the simulation reads the following neighborhoods:
R10-00090>>0 frequency=4
R512-30060>>0 frequency=16
R650-50040>>4 frequency=16
R684-60030>>0 frequency=16
R703-70020>>4 frequency=32
R712-80010>>3 frequency=16
R715-90000>>0 frequency=9900
The dynamic consultation network now contains 6 nodes plus one in the rest state.

Network 6

The sixth step in the simulation reads the following neighborhoods:
R646-50004>>0 frequency=40
R647-50013>>0 frequency=64
R648-50022>>0 frequency=16
R650-50040>>4 frequency=1
R701-70002>>0 frequency=8
R702-70011>>0 frequency=32
R703-70020>>4 frequency=4
R712-80010>>3 frequency=4
R715-90000>>0 frequency=9831
The dynamic consultation network now contains 8 nodes plus one in the rest state.

Rule Codes

C11.1: $D = 2, k = 4, r = 1$

1012000033020002203000000000020000003000012000023300120000232120
0110020013213100121203320232310233003020000031112100220032000000000
0110001022000113310023030000021000003021213200002013100000000201301
00000001000000000320330000

C11.2: $D = 2, k = 5, r = 1$

0100010300000000000000022000003000000000020001133000100000000400
0004000000000000000000000400041223000000000000000000000330003100000002
4010120020022000000000001000000000300030102020002000000100021204300
00000000300133340004000000203000000010000000000000000003401011000
40000000000000000000400000000000010003000000000001000000110203003000
00000021000030000030000004000000000200000100000000000340004000010000

003000000100000020000000001000012000010000001000402000003000000000
000034400000300000000000000010010000032301000000000000340000200001
300000000020000000004004400102000233202002400103030100400400042000
002000010004020400100003000000000000002201000010000000030000000004
000200000030000040400000000000000000000003001000000000440000100302 0

C11.3: $D = 2, k = 5, r = 1$

21000133002000000000000220000030000000000200011330001000000000400
000400000000000000000004000412230000000000000000000033000310000000 2
401012002002200000000000100000000300030102020002000000102021204300 0
000000003001333400040000002030000000100000000000000000003401011000
400000000000000000040000000000001000300000000000100000011020300300 0
000000210000300000300000040000000002000001000000000034000400001000
003000000100000020000000001000012000010000001000402040003000000000
000034400000300000000000000010010000032301000000000001340020200001
300000000020000000004004400102400233202002400103034400400400342000
002000011004021400100003000000000000010222101001002000003000000320 4
000200000030000043400000300003200302040030010200030044000010030200

Chapter 12
The Universal Constructor

INTRODUCTION

At the beginning of the 1950s, John von Neumann (1966) asked himself whether it is possible to design a machine with the ability to create exact copies of itself which would themselves have the ability to produce new copies. Such a machine would have reproductive capabilities comparable to those we find in biological organisms. In this setting, von Neumann's goal was to design a *Universal Constructor* capable of reading the instructions for, and assembling, any machine the designer might seek to build. If the instructions specified a Universal Constructor, the machine would have the ability to build a copy of itself. In other words, it would be able to reproduce. If we wanted the copies of the machine to share this ability, all we would have to do would be to copy the instructions and incorporate them in the new machines.

Von Neumann's model showed that in principle it is possible to create a self-replicating artificial machine. In the early stages of his work, he proposed a mechanical design. However, the design was not convincing and the mechanical details were so complicated they obscured the replication process. After discussions with Stanley Ulam, the great physicist, von Neumann proposed a model based on Cellular Automata. The model - derived from Turing's universal computer and including several feedback mechanisms - was still very complex. However, Langton (1984)

DOI: 10.4018/978-1-61520-787-9.ch012

was to show that CAs can generate self-replicating structures that are not universal constructors. This proved that the presence of a Universal Constructor is not a necessary condition for self-replication.

From this point on, the real problem was how to build systems that could do more than self-replicate. In 1995, Tempesti (1995) proposed a loop, similar to Langton's, enhanced to allow the incorporation of an executable program that would be duplicated and incorporated in every new copy. The program is conserved inside the loop where it is connected to the self-replicating segment of the code. Tempesti considered CAs with 10 states and a neighborhood of 9 cells. His work was followed by studies from Perrier, Sipper and Zahnd (1996), who implemented a loop with the ability to implement any possible program, so long as it was written in a very simple language. A further contribution came from Mange et al. et al. (2004), who built a loop with the ability to act as a Universal Constructor. Julian & Chua (2002) demonstrated special self-replication capabilities in CAs based on 1-dimensional, additive rules. Wolfram (2002) made a similar proposal, disregarding earlier comments by Langton [1984] who had considered this kind of structure as trivial. In 2002, Bilotta et al. demonstrated the existence of CA's with self-replicators that join together to form complex structures. The way in which they join is far from arbitrary - each combination obeys specific conditions. Their observed behavior brings to mind, on the one hand, the typical behaviors of additive rules, on the other, the properties of a Universal Constructor. This suggests that the class of self-replicators is far larger than previously thought. In the previous chapter, we also showed how it is possible to increase the complexity of initially simple self-replicators. This suggests that the division between trivial and non-trivial replicators is arbitrary, both being based on the same underlying processes. In this chapter, we return briefly to our discussion of additive rules, showing how we can extend our treatment to other rules (such as ECA90), which can act as templates for complex self-replicators. During the discussion we will see how self-replicating structures can couple to produce more complex systems which are themselves self-replicators. We will see, furthermore that this ability is a property, not just of the rule presented in (Bilotta et al., 2002), but of many rules producing self-replicating structures. These observations bring to mind a form of additivity.

BACKGROUND

Additivity

Certain CAs display *additivity*. One of these is ECA 90, a CA whose fundamental role we have already discussed in earlier chapters. Given two initial configurations $P_1(0)$ and $P_2(0)$, a CA is said to be additive if developing a superimposition of the two configurations for a certain number of steps produces a state identical to the state generated when the two configurations, develop independently for the same number of steps. In algebraic terms:

$$\varnothing_n(P_1 \oplus P_2) = \varnothing_n(P_1) \oplus \varnothing_n(P_2) \tag{12.1}$$

where $\varnothing_n(P)$ represents the state of configuration P after n steps and is an operator that sums the two states, module 2. If we follow Wolfram (2002) and analyze the rules of the CA rather than its configuration, we can say that an elementary CA is linear if it meets the additivity condition:

$$f(N_1 \dot{+} N_2) = f(N_1) \dot{+} f(N_2) \tag{12.2}$$

where are two neighborhoods and $\dot{+}$ is an operator that sums the two neighborhoods, module 2. This condition is met by ECAs 0, 60, 90, 102, 150,170, 204, 240. The rules for the evolution of these ECAs are summarized in Table 1 (Chaudhuri et al., 1997).

It is easy to generalize (12.2) to rules with any kind of neighborhood and with $k > 2$. In this case the total number of possible rules is $k^{k^{2r+1}}$ of which k^{2r+1} are additive.

2-D Additive Rules

The previous case can be easily extended to include 2-D rules. Figures 1 and 2 shows a number of additive rules for 2-D Boolean CAs. The left hand column gives the code for the rule; the center column defines the rule (always a sum module 2); the right hand column defines the neighborhood for the rule.

The first 3 rules have square neighborhoods. The neighborhood for the first two contains 4 cells; the neighborhood for the third has 5. These rules develop symmetrically, creating a pattern similar to that of a square (see Figure 3). The right hand column of Figure 3 provides a side-view of these patterns. The pattern created by the first rule is similar to the pattern for ECA 90. The fourth and fifth rules have cross-shaped neighborhoods containing 4 and 5 cells respectively. The evolution of these rules follows a rhomboid pattern. The sixth and seventh rules have neighborhoods of 8 and 9 cells respectively and evolve following a square pattern. The eighth and ninth rules have neighborhoods of 2 and 3 cells, and develop linearly. Side-views of the patterns created by these rules are reminiscent of the patterns

Table 1. Additive rules for ECAs. The left column shows the code for the rule; the right column gives the rule (always a sum module 2).

Rules	Expressions
ECA 60	$a_i(t+1) = a_{i-1}(t) \dot{+} a_i(t)$
ECA 90	$a_i(t+1) = a_{i-1}(t) \dot{+} a_{i+1}(t)$
ECA 102	$a_i(t+1) = a_i(t) \dot{+} a_{i+1}(t)$
ECA 150	$a_i(t+1) = a_{i-1}(t) \dot{+} a_i(t) \dot{+} a_{i+1}(t)$
ECA 170	$a_i(t+1) = a_{i+1}(t)$
ECA 204	$a_i(t+1) = a_i(t)$
ECA 240	$a_i(t+1) = a_{i-1}(t)$

produced by ECAs 90 and 150. Rules 10 and 11 have neighborhoods of 4 and 5 cells respectively and evolve in an asymmetric rhomboid pattern. Rules 12 and 13 both have a neighborhood of 6 cells and evolve in a rhomboid pattern.

Self-Replication

All the rules in Figures 1 and 2 produce nested patterns. For instance, rule 8 produces the well known Sierpinski Triangle. It is this nesting that produces the fractal structure of the patterns. We also observe processes of self-replication. In these cases, the initial seed replicates via a series of intermediate phases. Table 2 shows the number of replicas after two steps in the evolutionary process. As is apparent from Table 3, the evolution of the population has a clearly defined rhythm.

The data in the table confirms the tight relationships between rule 8 and ECA 90 and between rule 9 and ECA 150, already observed in Figure 3. The strength of these relationships is confirmed by their identical populations.

Let us now examine the relationship between rule 1 and rule ECA 90. In this case, we have the relation:

$$P_{Rule1} = (P_{ECA90})^2 \tag{12.3}$$

Figure 1. Additive module 2 rules for CAs with $k = 2, D = 2$. Left: the code for the rule; center: the code; right: the neighborhood

Rule 1	$a_{ij}(t+1)=a_{i-1,j-1}(t)\dotplus a_{i-1,j+1}(t)\dotplus a_{i+1,j-1}(t)\dotplus a_{i+1,j+1}(t)$	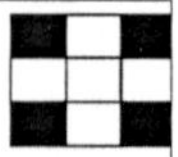
Rule 2	$a_{ij}(t+1)=a_{i-1,j-1}(t)\dotplus a_{i-1,j+1}(t)\dotplus a_{i+1,j-1}(t)\dotplus a_{i+1,j+1}(t)\dotplus 1$	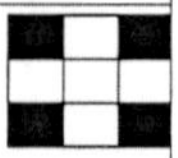
Rule 3	$a_{ij}(t+1)=a_{i-1,j-1}(t)\dotplus a_{i-1,j+1}(t)\dotplus a_{i,j}(t)\dotplus a_{i+1,j-1}(t)\dotplus a_{i+1,j+1}(t)$	
Rule 4	$a_{ij}(t+1)=a_{i-1,j}(t)\dotplus a_{i+1,j}(t)\dotplus a_{i,j-1}(t)\dotplus a_{i,j+1}(t)$	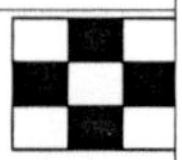
Rule 5	$a_{ij}(t+1)=a_{i-1,j}(t)\dotplus a_{i+1,j}(t)\dotplus a_{i,j}(t)\dotplus a_{i,j-1}(t)\dotplus a_{i,j+1}(t)$	
Rule 6	$a_{ij}(t+1)=a_{i-1,j-1}(t)\dotplus a_{i-1,j}(t)\dotplus a_{i-1,j+1}(t)\dotplus a_{i,j-1}(t)\dotplus a_{i,j+1}(t)\dotplus a_{i+1,j-1}(t)\dotplus$	
Rule 7	$a_{ij}(t+1)=a_{i-1,j-1}(t)\dotplus a_{i-1,j}(t)\dotplus a_{i-1,j+1}(t)\dotplus a_{i,j-1}(t)\dotplus a_{i,j}(t)\dotplus a_{i,j+1}(t)\dotplus a_{i,}$	

Figure 2. Additive module 2 rules for CAs with $k = 2, D = 2$. Left: the code for the rule; center: the code; right: the neighborhood

Rule 8	$a_{ij}(t+1)=a_{i,j-1}(t)\dotplus a_{i,j+1}(t)$	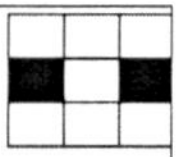
Rule 9	$a_{ij}(t+1)=a_{i,j-1}(t)\dotplus a_{i,j}(t)\dotplus a_{i,j+1}(t)$	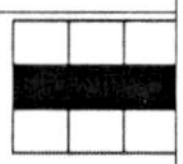
Rule 10	$a_{ij}(t+1)=a_{i-1,j}(t)\dotplus a_{i-1,j+1}(t)\dotplus a_{i,j}(t)\dotplus a_{i+1,j-1}(t)\dotplus a_{i+1,j}(t)$	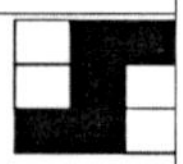
Rule 11	$a_{ij}(t+1)=a_{i-1,j}(t)\dotplus a_{i-1,j+1}(t)\dotplus a_{i+1,j-1}(t)\dotplus a_{i+1,j}(t)$	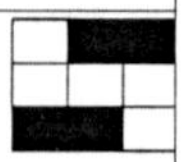
Rule 12	$a_{ij}(t+1)=a_{i-1,j-1}(t)\dotplus a_{i-1,j+1}(t)\dotplus a_{i,j-1}(t)\dotplus a_{i,j+1}(t)\dotplus a_{i+1,j-1}(t)\dotplus a_{i+1,j+1}(t)$	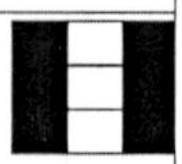
Rule 13	$a_{ij}(t+1)=a_{i-1,j-1}(t)\dotplus a_{i-1,j}(t)\dotplus a_{i-1,j+1}(t)\dotplus a_{i+1,j-1}(t)\dotplus a_{i+1,j}(t)\dotplus a_{i+1,j+1}(t)$	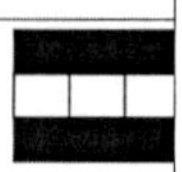

Figure 3. 5 steps in the evolution of the 13 rules in Figures 1 and 2, starting from a single seed. The right hand column provides a side view of the pattern after 20 steps. Rule 1 is clearly related to rule ECA 90; rule 9 is clearly related with rule ECA 150.

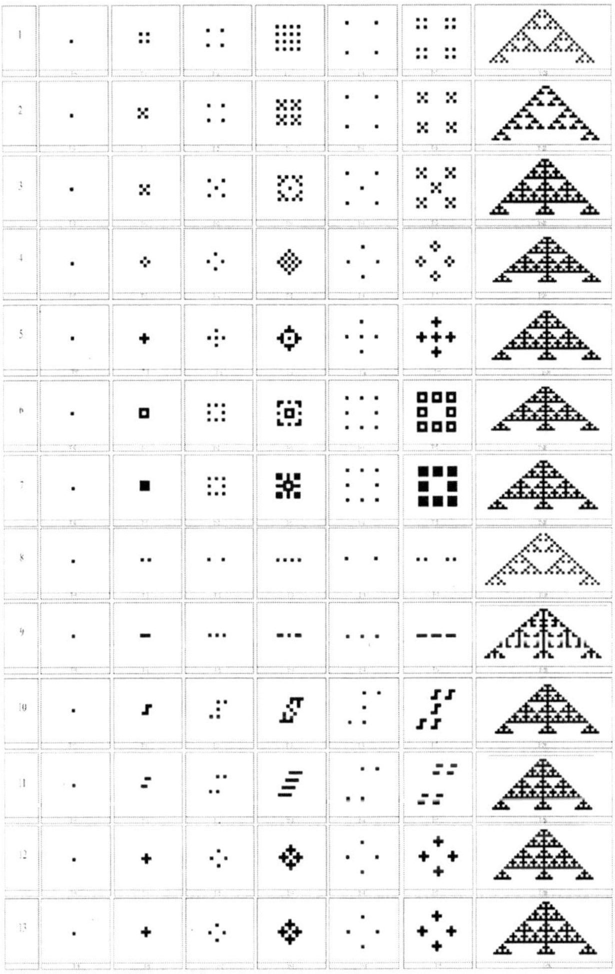

Table 2. Number of replicas after 2 steps in the evolutionary process

Rule no.	1	2	3	4	5	6	7	8	9	10	11	12	13
No. of replicas.	4	4	5	4	5	8	8	2	3	5	4	4	4

Table 3. Size of the population generated from a single seed. The column in gray shows the number of replicas after 2 steps in the evolutionary process

Time	T=1	T=2	T=3	T=4	T=5	T=6	T=7	T=8	T=9	T=10	T=11	T=12	T=13	T=14	T=15	T=16	T=17	T=18	T=19	T=20
Rule 1	1	4	4	16	4	16	16	64	4	16	16	64	16	64	64	256	4	16	16	64
Rule 2	1	5	4	20	4	20	16	80	4	20	16	80	16	80	64	320	4	20	16	80
Rule 3	1	5	5	17	5	25	17	61	5	25	25	85	17	85	61	217	5	25	25	85
Rule 4	1	4	4	16	4	16	16	64	4	16	16	64	16	64	64	256	4	16	16	64
Rule 5	1	5	5	17	5	25	17	61	5	25	25	85	17	85	61	217	5	25	25	85
Rule 6	1	8	8	24	8	64	24	112	8	64	64	192	24	192	112	416	8	64	64	192
Rule 7	1	9	8	32	8	72	24	136	8	72	64	256	24	216	112	528	8	72	64	256
Rule 8	1	2	2	4	2	4	4	8	2	4	4	8	4	8	8	16	2	4	4	8
Rule 9	1	3	3	5	3	9	5	11	3	9	9	15	5	15	11	21	3	9	9	15
Rule 10	1	5	5	17	5	25	17	61	5	25	25	85	17	85	61	217	5	25	25	85
Rule 11	1	4	4	16	4	16	16	64	4	16	16	64	16	64	64	256	4	16	16	64
Rule 12	1	5	4	20	4	20	16	80	4	20	16	80	16	80	64	320	4	20	16	80
Rule 13	1	5	4	20	4	20	16	80	4	20	16	80	16	80	64	320	4	20	16	80

where P_{Rule1} is the population generated by rule 1 and P_{ECA90} is the population generated by rule ECA 90. The close relationship between the two rules is confirmed by the first row of Figure 4, in which we see the growth of their respective populations.

Figure 4. a): size of the populations generated by rule 1 and rule ECA 90. The b) shows the size of the population on a log scale base 2. c): superimposition of patterns produced by different rules with evident additivity. d): replication of complex initial data in specific CAs; e): a complex replication process based on row 3. In this case the replication period is 16 instead of 2. The rate of growth is the same as the rate for the individual seed.

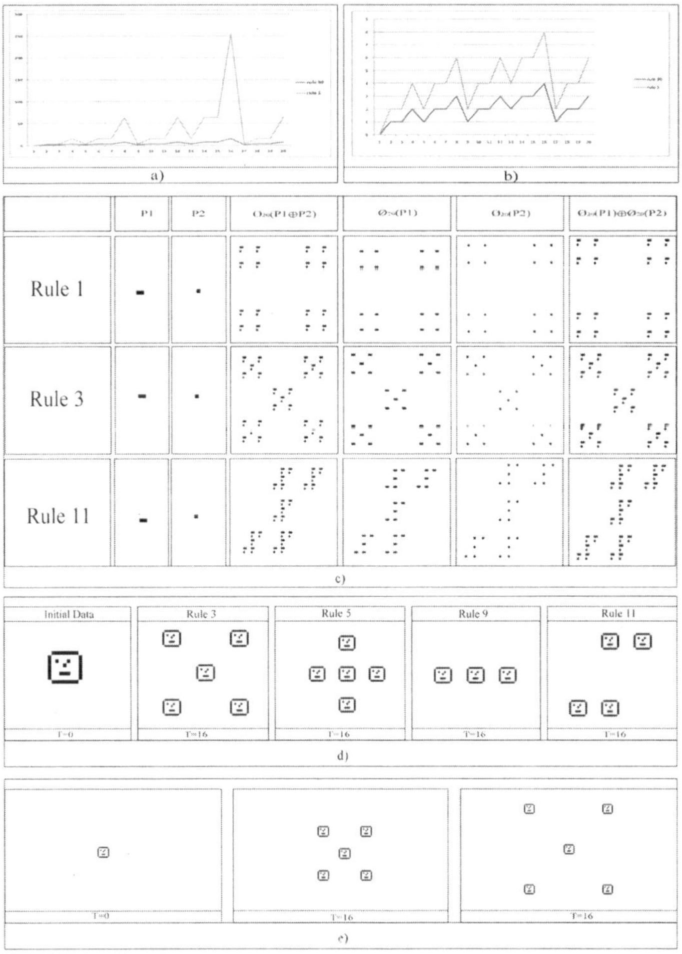

As far as concerns other relationships, Table 3 shows that the populations produced by rules 2, 12 and 13 all have similar trends. The same is true for rules 3, 5 and 10. As always, additivity between patterns is based on the sum of the patterns module 2. One example can be seen in Figure 4c, where we see the superimposition of two patterns with different initial data.

Some of the rules in Figures 1 and 2 have the ability to reproduce complex initial data. Figure 4c shows this capability for rules 3, 5, 9 and 11. The other rules are limited to the replication of initial data that meets specific conditions. The spectacular ability to replicate any initial data brings to mind some of the characteristics of a *Universal Constructor.* It is interesting to observe that the rules shown in the Figure 4d require 16 steps to complete the replication process, as opposed to 2 steps in the case for a single seed.

As we have seen, the self-replication mechanism for additive rules is anything but trivial and can in fact replicate extremely complex rules. As in the examples described in Chapter 11, colony formation is guided by very simple basic modules in the CA. The only difference is the speed at which the systems replicate.

Some of the self-replication mechanisms we have seen are very similar; some very different. One way of bringing them under a common roof is to examine their symmetries in the plane. As we already saw in Chapter 10, we can identify the following categories of self-replication process:

a. Multiplication by 2,
b. Multiplication by 3,
c. Multiplication by 4,
d. Multiplication by 5,
e. Multiplication by 8,
f. Multiplication by 9,
g. Multiplication by 1 (a glider).

This last case is degenerate, in the sense that it replicates the initial seed without alteration. A glider that does not move conserves all its symmetries. If it moves it conserves its spatial symmetries and one temporal symmetry. Rule 8 is an example of multiplication by 2. In this case, the self-replication process leads to a break in symmetry: the only symmetries conserved are the mirror symmetry and the 180° rotation symmetry. The growth process for this rule is similar to the process for ECA 90. The case of rule 9 (multiplication by 3) is similar but this time the growth process follows rule ECA 150 (see the first image in Figure 5).

Multiplication by 4 produces 2 different breaks in symmetry (see the second image in Figure 4). Rules 1 and 2 conserve the square symmetry. Rule 4 conserves the rhomboid symmetry. It is interesting to observe that both rules share the same

Figure 5. Image 1, replicators with linear symmetries and gliders. We observe several breaks in symmetry. Image 2, replication processes that conserve square or rhomboid symmetries.

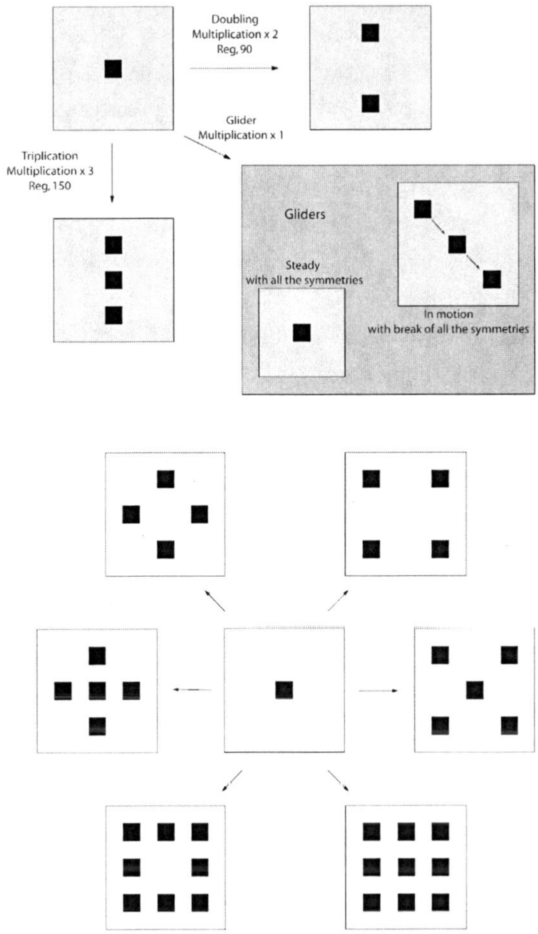

pattern of growth. This is similar to the pattern we observe in ECA 90. The main difference with respect to rule 1 is in the intermediate phases of the replication process. During the process, we observe the activation of cells that are de-activated when the process concludes (see Table 3). Rule 11 appears as a degenerate case of rule 1. With this rule, the axes of the symmetry are no longer perpendicular: one is slanted with respect to the other. However, both rules display exactly the same rate of growth. Rules 12 and 13 evolve with exactly the same symmetries as rule 4. However the rate of growth is the same as for rule 2. As a result, the two rules have different transients, even though they produce the same final result. Finally, rules 12 and 13 both produce the same result, even if they use different neighborhoods. Multiplication by 5 displays the same breaks in square and rhomboid symmetry as multiplication by 4. Rule 3 creates a square symmetry, rule 5 a rhomboid with an additional element in the center of the structure. Once again, the behavior of these rules is reminiscent of the behavior of rules 150 and 90. However, there is no obvious sign of a relationship similar to (12. 2) which connects ECA 90 to rule 1. Rules 3 and 5 share the same rate of growth, which is similar to the rate of growth for rule 10. Rule 10, like rule 9, is a degenerate version of rule 3. The last two cases involve rules 6 and 7 which only conserve their square symmetries. As usual, the difference between rule 6 and rule 7 is that the latter has an extra element in the center of the structures it produces.

To conclude, we can say that self-replication processes display:

a. A break in some symmetries and the conservation of others,
b. A characteristic period,
c. A well-defined rhythm of growth during colony formation,
d. Well-defined, frequently repeated, replication mechanisms.

It is obvious that the simple additive rules discussed in this section can give rise to more complex reproductive behaviors. As discussed in Chapter 10, blocks of complex self-replicators can evolve following the same local laws listed in Figures 1 and 2. In a certain sense, these rules act as templates for the rules governing the formation of colonies of self-replicators. .

MAIN FOCUS

We have already seen that some of the rules in Figures 1 and 2 have two very interesting properties:

a. They can replicate complex initial configurations;
b. They can act as templates for complex replication processes based on the block dynamics of certain specific self-replicators.

In this rest of this chapter, we will extend this property to complex replicators whose block dynamics are governed by the rules in Figures 1 and 2.

Universals and Laws of Composition

Consider C10.6, a replicator with a square pattern of development, already analyzed in Chapter 10.

$$A = \begin{pmatrix} 44 \\ 66 \end{pmatrix}$$

So long as the system meets certain conditions concerning phase and distance, it has the ability to join multiple objects into a more complex composite object. In other words, the system behaves just like rules 3, 5 and 11 in Figure 4. Observe, for instance, how we can replicate complex initial data by merging several examples of A (see Figure 6). The first composition event occurs after 64 steps, the second after 128. The result is a self-replication process based on multiplication by 4. The surprising aspect is that the process obeys the same logic as the Sierpinski Arrowhead. This is true not only for A but for compositions involving multiple instances of A.

It would be entirely possible for the composition process to continue further, creating further compositions of the same basic structure. This capability of rule C10.6 is evident in experimental observations of the system's behavior when it is seeded with random initial data. Multiple structures similar to A can give rise to large-scale structures which are themselves self-replicators. This process is not random but obeys precise rules:

a. The structures involved in the composition must have the same orientation; alternatively one should be rotated by 180° with respect to the other;
b. The distance between the two objects should be, where n is an integer.

Where these conditions are not met, the large scale structure is destroyed, sometimes together with the original structures. In other cases, the evolutionary process successfully handles the problem, synchronizing the two structures. In reality, this is not the only example of a self-replicator with the ability to replicate arbitrary structures. As will see in the following section, this is a capability shared by many self-replicators.

Other Examples

Consider the structure $A = (333)$. This structure, which develops linearly, is a product of rule C10.1, already described in Chapter 10. The basic logic is the same

Figure 6. Two replication processes. Image 1 represents a composition of the basic structure which is replicated after 64 steps. The second structure is replicated at step 128.

as the logic for ECA 90. The first row of Figure 7 (a and b) offers evidence of additivity. Additivity is maintained so long as the distance between the structures is exactly *4n* cells, where *n* is an integer. With distances of 2, 6 etc. the structures disintegrate. So long as the relations between the structures meet these conditions, the system can replicate complex structures, as in the previous case. The perpendicular distance between the structures involved in the replication process must always be ≥ 3. The introduction of new structures in a direction perpendicular to the original structures can lead to modifications in the replication process (see Figure 7c).

Figure 7. Row 1, linear development of C10.1 (b) with respect to a specific attractor (a). Row 2(c): production of complex structures from self-replicators. In some cases we observe minor shifts in the position of specific elements in the self-replicator (rule C10.1). Row 3 (d and e): production of complex structures from self-replicators (rule C10.2).

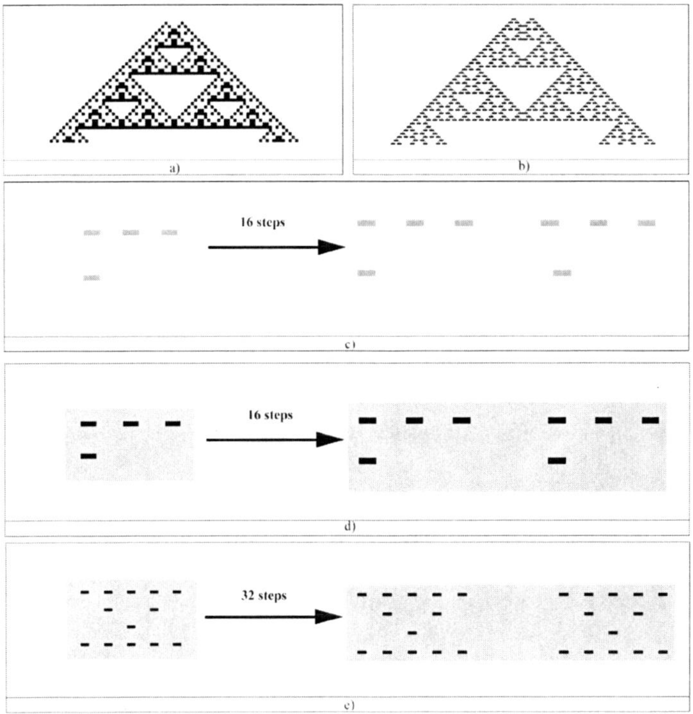

Similar behavior can be observed in another linearly developing self-replicator ($A = (222)$) governed by rule C10.2. Like the examples already discussed, the replicator has the ability to reproduce complex structures based on compositions of the basic structure (see Row 3 in Figure 7d and 7e). Here too, it is necessary to maintain a precise distance between the structures involved in the composition: in this case 5 or multiples of 5.

Now consider the self-replicator

$$A = \begin{pmatrix} 0220 \\ 3003 \\ 0220 \end{pmatrix}$$

governed by rule C10.3. This time the behavior is more interesting. The development of the replicator is linear leading to a duplication of the original structure. Composite structures are also replicated (see row 1 in Figure 8). This time, however, there is a parasitic residual: a strip of 3 cells in state 3. The residual has no effect on replication and is itself replicated during the process.

Replication occurs when the gap between the two structures contains 7 consecutive zeros. For example:

022000000000220
300300000003003
022000000000220.

Figure 8. a): composite structures derived from rule C10.3 replicate linearly. During the process, however, they create parasitic structures which are themselves subject to replication. b): composite structures produced by rule C9.5 follow a square pattern of development. In some stages of the process we observe inversions of phase. In the case shown, replication requires 32 steps.

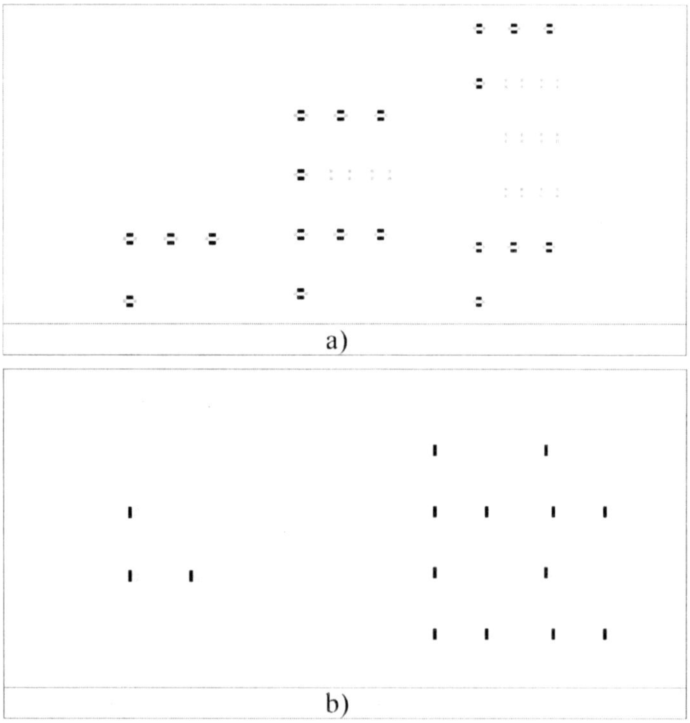

Finally, consider the self-replicator

$$A = \begin{pmatrix} 233 \\ 233 \\ 233 \end{pmatrix}$$

governed by rule C8.5. When the structures involved in the initial composition are separated by precisely 13 cells, the replicator creates complex structures that cyclically multiply by four. During the process, some components of the structures undergo an inversion of phase.

FUTURE TRENDS

In this chapter, we have analyzed additive rules as a key factor for the emergence of complexity. We have seen that self-replicating structures with additive proprieties can couple to produce more complex structures which are themselves self-replicators and that some rules are *weak-additive*. In other words, the development of complex structures in simple self-replicators is the consequence of a composition of the self-replicators. This is the result of level 2 logic, usually governed by additive rules.

The existence of weak additivity raises several questions for future research:

a. Is it possible to find expressions for weak additivity similar to those in (13.1) or (13.2)?
b. What is the true significance of weak additivity?
c. What is its relationship to self-replication?
d. Why do self-replicators join together in some specific ways and not in others?
e. Does there exist a relationship between level 2 logic and weak additivity?
f. How can we describe the basic principles involved in empirically observed weak additivity, in formal mathematical terms?

CONCLUSION

Additive rules have the ability to create complex structures as compositions of simpler ones. Sometimes the product is a complex self-replicator. This is important, demonstrating how elementary self-replicators can combine to produce more complex ones. In this chapter, we have seen that these capabilities are sometimes shared even by rules that are non-additive. In other words, we have discovered rules that produce self-replicators with the ability to join together in new complex self-replicating structures. This implies that there exist forms of additivity that cannot be described in terms of (12.1)-(12.2). We have called this phenomenon weak additivity. This however is only a working definition. We are well aware that much future work is required on this theme. If the underlying rules are additive, it is likely that the structures they govern will be able to join together in more complex self-replicators. Once again, we observe the central role of second order logic in colony formation. It is this logic that governs the behavior of modules. Given that modules represent dynamics in time and space, it is essential they should be compatible. This is why the coupling process requires such fine tuning or synchronization.

In Chapters 14 and 15 we will see how we can exploit these properties to generate interesting rhythms and sounds.

REFERENCES

Bilotta, E., Lafusa, A., & Pantano, P. (2002). Is self replication an embedded characteristic of Artificial/Living matter? In Standish, R. K., Bedau, M. A., & Abbass, H. A. (Eds.), *Artificial Life VIII* (pp. 38–48). Cambridge, MA: The MIT Press.

Julian, P., & Chua, L. O. (2002). Replication properties of parity cellular automata. *International Journal of Bifurcation and Chaos in Applied Sciences and Engineering*, *12*(3), 477–494. doi:10.1142/S0218127402004632

Langton, C. G. (1984). Self-Reproduction in Cellular Automata. *Physica*, *10D*, 135–144.

Mange, D., Stauffer, A., Petraglio, E., & Tempesti, G. P. (2004). Self-replicating loop with universal construction. *Physica D. Nonlinear Phenomena*, *191*(1-2), 178–192. doi:10.1016/j.physd.2003.11.003

Perrier, J. Y., Sipper, M., & Zahnd, J. (1996). Toward a viable, self-reproducing universal computer. *Physica D. Nonlinear Phenomena*, *97*, 335–352. doi:10.1016/0167-2789(96)00091-7

Tempesti, G. (1995). A new self-reproducing cellular automaton capable of construction and computation. In F. Morán, A. Moreno, J. J. Merelo, & P. Chacón (Eds.), *ECAL'95: Third European Conference on Artificial Life*, Vol. 929 of Lecture Notes in Computer Science (pp. 555-563). Heidelberg: Springer-Verlag.

von Neumann, J. (1966). *Theory of Self-Reproducing Automata (edited and completed by AW Burks)*. Urbana, IL: University of Illinois Press.

Wolfram, S. (2002). *A New Kind of Science*. Champaign, IL: Wolfram Media.

Chapter 13
A Zoo of Self-Replicators

INTRODUCTION

Our basic metaphor: in this chapter, we present a taxonomy of self-replicators - as if they were animals in a zoo. In the zoo, we play the role of an external observer (a zoologist) whose role is to describe the animals (artificial organisms) and their behavior. Different species reproduce in different ways - some sexually, some asexually. We observe differences in their developmental dynamics and differences in the way they adapt to their environment. In each case, what we see are life-like self-replicators, each adapted to a specific habitat.

In Chapters 7 and 8, we have seen how evolutionary techniques can create a broad variety of self-replicators. The complex ways in which these self-replicators grow, reproduce and become extinct bring to mind the behavior of biological systems. In Chapter 11, we examined their genetics. Here we examine their behavior at the individual and the species level, applying a range of observational and experimental methods and using the results as the basis for a taxonomy.

Like eukaryotic cells, artificial organisms can divide, creating offspring similar to themselves and transmitting their genetic heritage to the next generation. In our artificial world, we can identify several different models of mitosis. As in biology, it is these models that underlie the growth of individual organisms and of popula-

DOI: 10.4018/978-1-61520-787-9.ch013

tions of genetically similar organisms. We also observe meiosis: the basis for sexual reproduction in diploid cells. Meiosis allows recombination of the different elements in a population's gene pool as well as mutations giving rise to novel behaviors and patterns. In the natural world, the combination of mitosis and meiosis allows species to adapt to unfavorable environmental conditions. At times, different species, that normally reproduce asexually, hybridize, in a simulation of sexual reproduction. Artificial organisms and species behave in the same way. We can thus consider our simulation environments as artificial eco-systems with an equilibrium that depends on the sizes of populations belonging to different species. In some species, populations are relatively constant; in other they undergo short or medium-term fluctuations that may include exponential growth and collapse. Sometimes, these fluctuations are due to internal factors (the traits characterizing the species), sometimes to external causes (expanding populations of competing species, resource availability, disease etc.). When we observe fluctuating populations of artificial organisms, we are observing a behavior also found in plants, micro-organisms, funguses and the majority of animals. These species, which typically have very high rates of mortality, use the energy available to them to produce the highest possible number of offspring. Both in artificial organisms, and in animal populations, this reproductive strategy typically produces temporary spurts and collapses in population. In the opposite strategy, similar to the reproductive strategies of birds and mammals, energy is invested in the development of individual organisms and birth and death rates are relatively low. In these conditions, the artificial eco-system is relatively stable.

As zoologists studying artificial worlds, we use two distinct methods.

a. We can observe artificial ecosystems and classify the species they contain;
b. We can engineer new species, with their own ad hoc developmental dynamics. This way of proceeding allows us introduce novel processes of "sexual" reproduction leading to changes in the characteristics of the species. This is one of the tasks of the artificial laboratory described in Chapter 11.

Applying this method we have detected forms of asexual reproduction similar to those we observe in plants, micro-organisms, funguses, corals etc. When organisms reproduce asexually there is no exchange of genes and no genetic recombination. Offspring are clones, identical to the parent organism, which can thus be considered as the founder of a colony. Like their biological counterparts, artificial organisms display different kinds of asexual reproduction:

a. Fragmentation: A part of the artificial organisms breaks off to form a new individual, or a group of highly active cells that give rise to a new individual.

b. Gemmation: the organism develops "buds" that break away from the parent organism. When several buds break away at the same time, they give rise to a colony; when they break away individually they grow independently until they form new buds and begin the cycle again.
c. Polyembryony: the organism produces multiple nuclei of dividing cells. These give rise to organisms with identical genomes (as in the case of identical twins).
d. Propagules: the organism produces special organs (equivalent to, seeds, fruit etc.) that break away from the original organism, giving rise to new organisms.
e. Sporification: the organism creates special reproductive cells with the ability to generate artificial individuals. One example of sporification is the production of glider-guns.

Our second method, based on genetic engineering, allows us to observe the workings of evolution. In particular, it has allowed us to "hybridize" organisms with different genomes, and to study which of the resulting hybrids was best adapted to the environment. In biology, sexual reproduction uses specialized germ cells, each containing just half of the organism's genetic heritage. In our own work, we de-activated some of the genes in individual self-replicators and investigated the possibility of recombining the resulting genetic material. This procedure made it possible to produce offspring that were different from each other and from parent organisms.

BACKGROUND

In what follows, we present a brief pictorial history of research on self-replicators. The first CA-based self-replicator was Langton's hand-built system, published in 1984 (Langton,1984). Figure 1 shows the loop and the colony to which it gives rise together with other famous loops from Byl (1989), Chou and Reggia (1997; 1998) and Sayama (1998; 2000; 2004).

Unlike the systems proposed by these authors, the k-totalistic CAs, described in Chapters 7 - 8 - 10 and 12 are not hand-designed, but the result of evolutionary processes giving rise to artificial microworlds, similar to Conway's game of life. Within these micro-worlds, we observe a huge range of self-replicators and an equally large range of models driving the self-replication process. It is these self-replicators and models that we describe in this chapter. Their taxonomy is still incomplete. However, it is far enough developed to show that simulated evolution can produce artificial structures that behave like biological systems. On the basis of these observations, we believe that this kind of artificial laboratory can make a major contribution to biological research.

Figure 1. This sequence of images presents a brief pictorial history of research on self-replicators. a) Langton's loop, b) Byl's loop, c d) Chou and Reggia's loops, e and f) Sayama's loops.

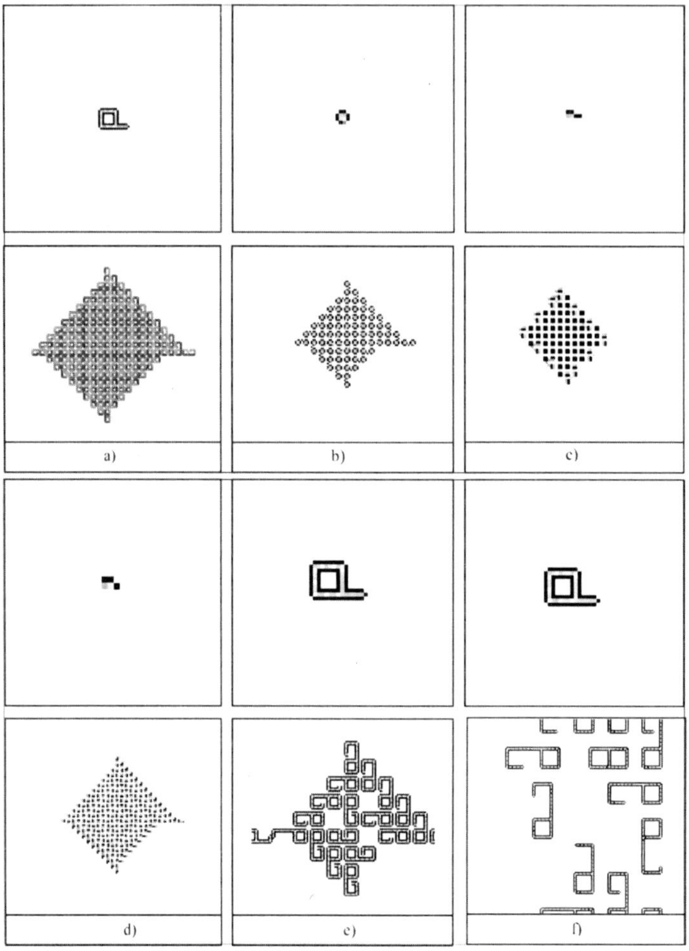

MAIN FOCUS

In this chapter, we present a gallery of images (originally published in Bilotta & Pantano, 2006), depicting a wide range of evolutionary patterns and self-replication models. The images were produced by applying the methods described in the previous paragraphs. Our goal in presenting them here is to demonstrate that there are many different kinds of self-replication. It is likely that each of these models has an equivalent in nature. A second goal is to show the generative and compositional richness of our own self-replicators, which give rise to structures that are more diverse and complex than any previously reported in the literature. In our work, we have used two methods: on the one hand, the method of the zoologist who observes and classifies different species and studies their population dynamics; on the other that of the genetic engineer who actively manipulates organisms' genomes to produce changes, affecting their developmental dynamics. When we hybridize different species, population sizes sometimes remain constant; in other cases they change producing glider dynamics, in others they lead to extinction.

Microworlds with Artificial Species and Population Dynamics

Figure 2 depicts a series of Artificial Microworlds (AMW) each identified by a code (AMW + a sequence number) and each with its own ecological characteristics. These microworlds contain life-like self-replicators, belonging to different species, each distinguished by its own characteristic developmental dynamics. Below we briefly describe some of the key traits of these systems.

AMW 1 (C13.1) is a system with k=10 and r=1 that displays a broad range of self-organizing features. In what follows, we denote the species present in the system by an S + a sequence number + the value of the state of the system. Thus S1=3 indicates that species 1 has a state of 3. This particular self-replicator begins from a single seed yet gives rise to very beautiful patterns, displaying a high degree of geometrical regularity. Applying our biological metaphor, we can compare S1 to a spore that generates a large number of new organisms that invade the environment. This picture is complicated by the presence of a second species S2 (S2=333), with linear replication processes that merge into the processes set off by S1.

AMW 2 (C13.2) is a system with k=4 and r=1 that contains two species S1 (S1=22) and S2 (S2=222) that are genetically similar and could be considered as variants of the same system. Both give rise to complex forms of self-organization and replication; both have the ability to spread rapidly through the environment.

AMW 3 (C13.3) also has k=4 and r=1. The system contains a single species S1 (S1=1) that begins with a single seed and invades the ecosystem. Other residual dynamics are generated by aggregation of small scale structures that do not belong to a true species but to what we call a Precursor Species (PS=2). S1 appears to use the precursor species as a nutrient during its development. As a result, the AMW displays the typical dynamics of a predator-prey relationship.

AMW 4 (C13.4) is an ecosystem which contains mobile structures which we refer to as glider-like species (from now on Glider Species becomes GS). When glider-like species collide they conserve their dynamics but change shape. Some of these gliders replicate by producing new gliders which produce still more, forming colonies

AMW 5 (C13.5) contains two genetically similar, highly invasive species (S1=333 and S2=333,333)

Figure 2. Examples of Artificial Microworlds, each identified by a code (AMW + a sequence number). Each AMW contains different species of lifelike self-replicator displaying different population dynamics. The rules of the AMW are listed in the Appendix of this Chapter, with codes C13.1-C13.6.

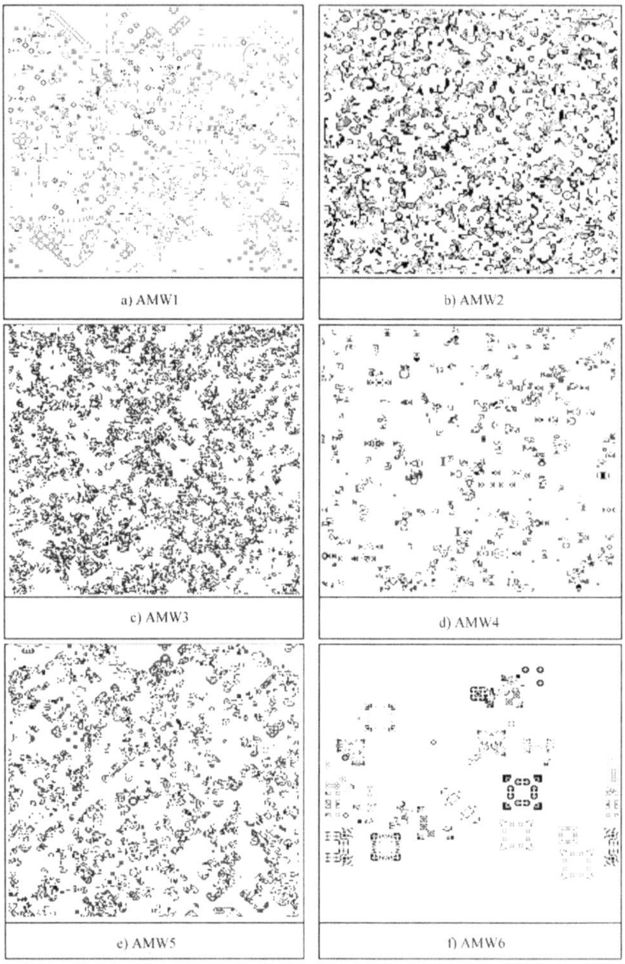

AMW 6 (C13.6) only contains species that go extinct. S1=0 goes extinct after just 72 steps (a fraction of the total number of steps in the evolution of the artificial system). Before it disappears it goes through a series of complex self-replication processes. Other, genetically similar species go extinct even more quickly.

As an example of our methods, consider AMW 7, (a k-totalistic CA with k=5 and r=, labeled with the code C13.7). The first 8 images in Figure 3 depict the first 8 steps in the evolution of the system. The next 4 show the state of the system after 30, 60, 90 and 120 steps respectively. In the first few steps, we see the global dynamics governing the evolution of the system. In this early period, the ecosystem is densely populated by structures struggling to replicate and survive. These give rise to the 8 species of self-replicators shown in Figure 4. As previously, each of these systems is identified by a code (S which means *species* + sequence number). The sequence numbers indicate the order in which the species emerged during the evolutionary process. Thus, high numbers refer to species that emerged only after many steps.

In what follows, we describe the developmental dynamics for each of these species. The codes of these structures are listed in the Appendix of this Chapter.

$$S1 = \begin{pmatrix} 222222 \\ 222222 \\ 222222 \\ 00333 \end{pmatrix}$$

, displays population dynamics that are shared by the other species present in the AMW, as if the system exploits redundancy in self-replicator behavior to ensure their successful reproduction. We observe a structure that spreads

Figure 3. AMW7. The artificial environment and the struggle for survival among emerging species. In the early stages, the environment is densely populated. In the later stages, we observe the emergence of species and of self-replication.

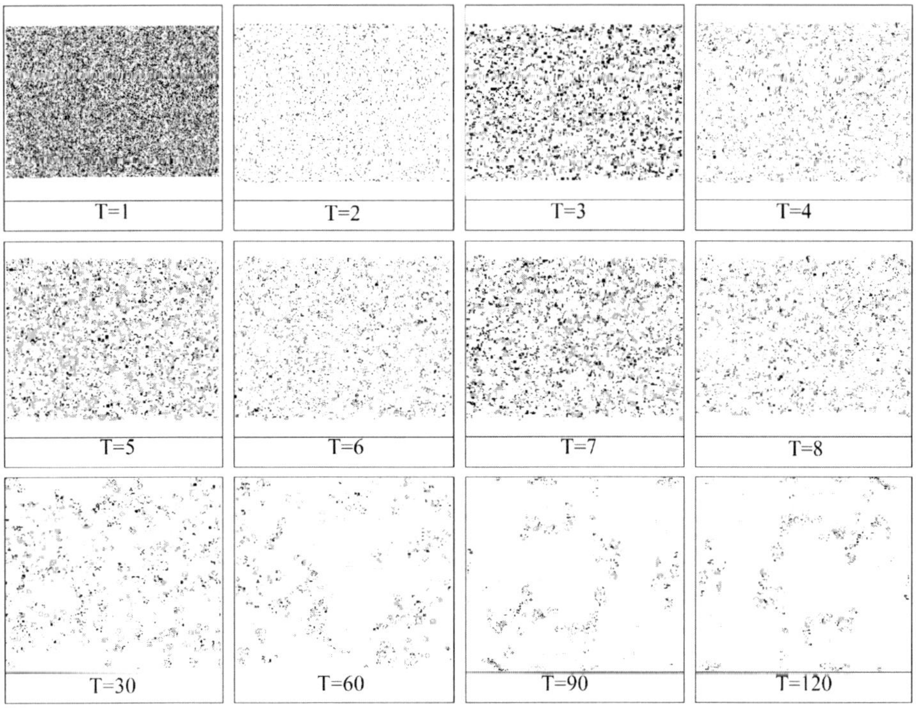

Figure 4. Self-replicator species in AMW7. Some species (S3, S6) are variants of a single system (S5). As a result, several replication processes are similar, though not identical. In most cases, the differences are minor. However, in the case of S6 they lead to the extinction of the species.

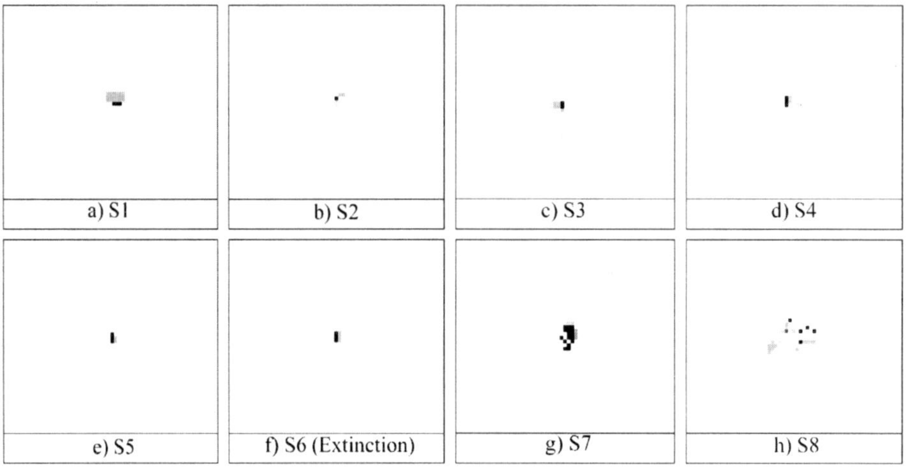

into the surrounding space (Figures 5 and 6), forming buds (rectangular structures in state 2). These structures split off from the parent organism to form a colony which expands leftwards. Changes in the orientation of CA S1 produce corresponding changes in the pattern of colony formation.

S1 gradually fills the available space, alternatively swelling and contracting. Intermittent gliders replicate and disappear. Other gliders seem to be generated directly from the initial system.

Figure 8 provides an overview of these dynamics, together with entropy curves for S1, S2 and S8.

Method 2.

To illustrate our second method, we will describe genetic experiments, again using species generated by AMW7. Figure 9 shows an experiment in which we "mate" S5 with randomly chosen elements from non-species, simulating a kind of sexual reproduction. The dynamics of the resulting system are similar to those of the systems we have just described. After a period of initial stagnation, the entropy curve climbs gradually.

By contrast, Figures 10 and 11 show two examples of hybridization. In the first (hybridization between S5 and S1), the result is a residual structure with changes in

Figure 5. Evolution of S1 from step 1 to step 16. S1 moves in space, first becoming smaller and then expanding again. The system does not display a characteristic geometry.

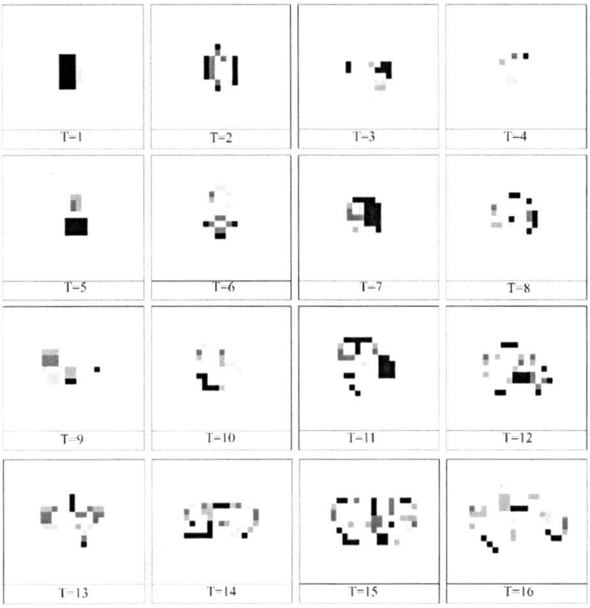

Figure 6. Evolution of S1 from step 17 to step 28. The structure invades an ever larger proportion of the grid (in biological terms, the environment). Replication is based on structures similar to those observed in the early stages of the evolutionary process (rectangular structures in state 2).

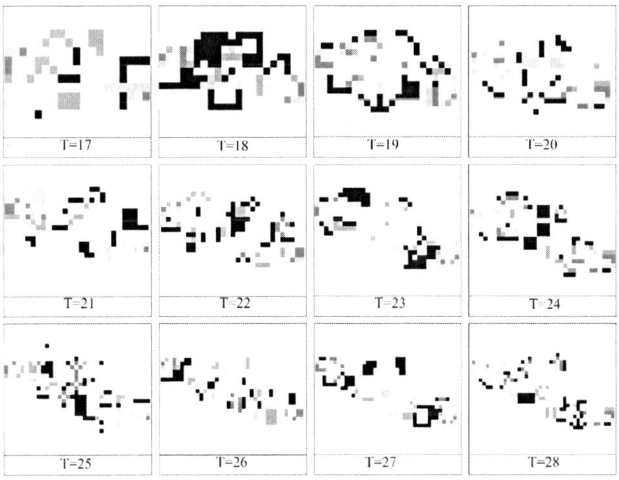

Figure 7. S1 after 40 and 110 steps respectively. As the simulation proceeds and the environment fills up, the replication process appears to change direction.

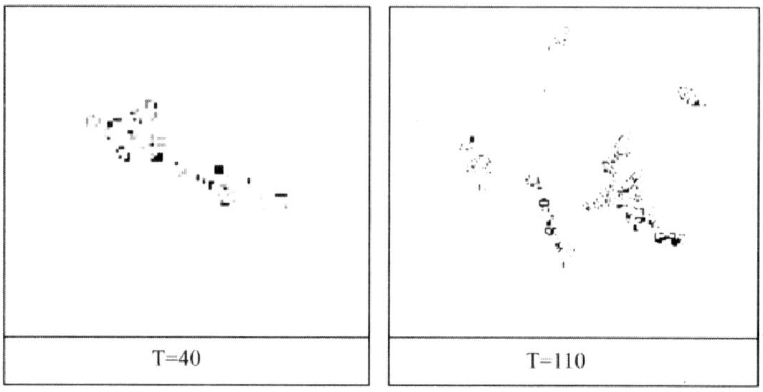

Figure 8. Overview of the species generated by AMW7 (S8, S1 and S2), population dynamics after 150 steps, Input Entropy (IE) for the same period. IE for S2 and S1 increases in parallel with the expansion of the self-replicating structure. The curve for S8 increases more slowly.

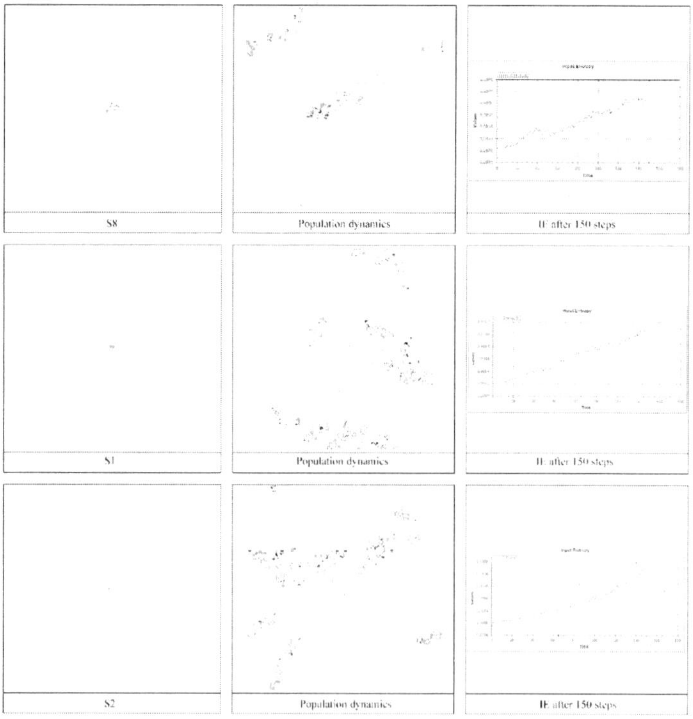

Figure 9. Merging of S5 with randomly chosen elements from non-species. The population dynamics do not appear to change. IE increases in parallel with the expansion of self-replicating structures.

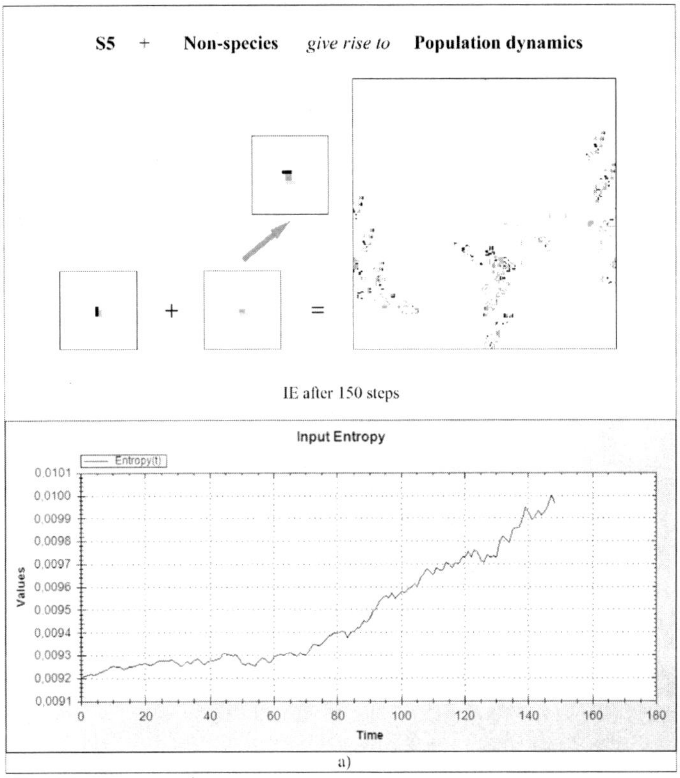

the structure of S1. In the second, (hybridization between S2 and S1) both genomes change, leading to modifications in the resulting structures. In neither hybridization experiment, do we observe changes in the basic replication dynamics of. In both cases, the way in which the system fills space continues to follow the original population dynamics.

Figures 12 through 16 depicts images of colony formation in AMW7. Colonies are distinct from self-replicators. We refer to them as growth systems (GS), identifying specific systems with a code in the form "GS" + sequence number. The codes of these rules are listed in the Appendix of the Chapter. There are an enormous number of these systems. The figure depicts those we detected in our own work. In some cases, we provide the number of rules they use (their genetic heritage). In all cases we show the initial conditions, and a 3-D image of the developmental dynamics for 80 simulation steps.

Figure 10. Hybridization between S5 and S1, leading to changes in the genetic structure of S1. The population dynamics remain unchanged.

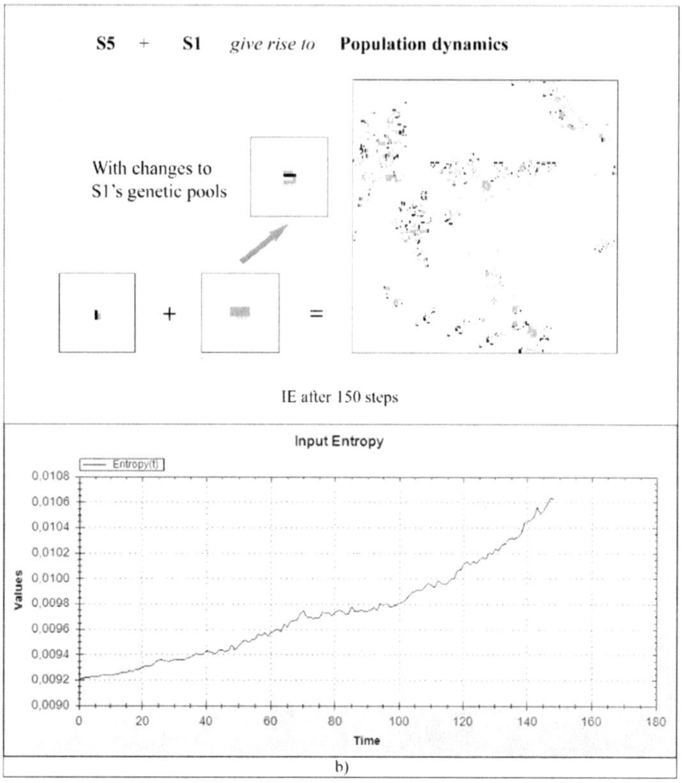

A Taxonomy of Self-Replicators

Classification is a scientific approach to the grouping of biological or artificial systems. In previous chapters, we have seen that while there exist classifications for 1-D CAs (Wolfram, 1984; Li & Packard, 1990; Gutowitz & Langton, 1995; Oliveira et al., 2001), there is no equivalent system for 2-D CAs. Given that we obtained our k-totalistic CAs using evolutionary methods, we can base our classification on their respective evolutionary trees. From a methodological point of view, a classification is a way of describing a system and of determining its role in a hierarchy or taxonomy. To achieve this goal, we begin with a binomial nomenclature similar to the system Linnaeus (*System Naturae*, 1735) used to indicate the genus and species of biological organisms. In our case, we use the symbol S to indicate the species and a sequence number to indicate the order in which species appear during the evolutionary process. We then enrich our scheme with a higher level taxonomical unit - the

Figure 11. Another example of hybridization between S2 and S1, giving rise to a changed structure. The dynamics are basically unchanged. However, the process appears slightly more active than before the change.

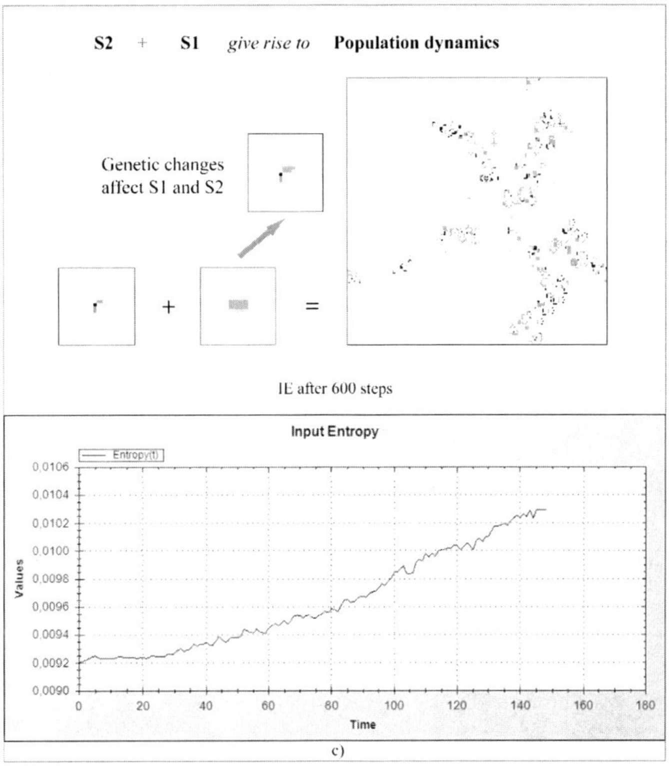

class. Species belonging to the same class share the same model of replication. For the purposes of our taxonomy, we distinguish 4 classes:

a. Species with simple replication,
b. Species with complex replication,
c. Species with chaotic replication,
d. Species that form colonies (hyper-complex replication).

The definition of these classes is based on extensive experimental observation of populations of artificial species in AMWs. These observations show that species belonging to the same class not only have common physical traits and common behaviors; they also share similar genes and similar number of genes. In more general terms, we have based our taxonomy on criteria similar to those used by biologists (Campbell & Reece, 2002; Cummings et al., 2002):

Figure 12. Growth systems for AMW 7. The initial data for each system, the code of the system, and 3-D images of the developmental dynamics for 80 simulation steps

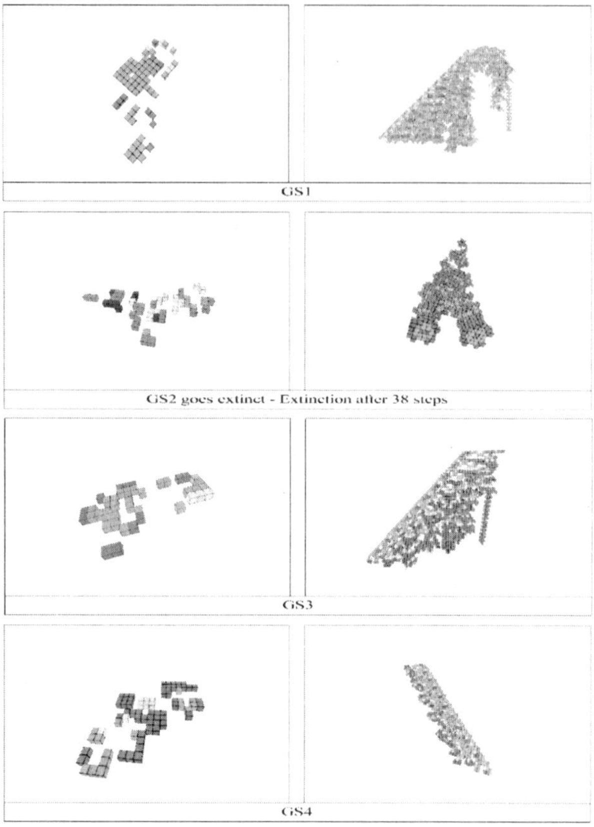

a. Organisms belonging to the same taxonomical unit have common ancestors that display similar behaviors. These behavioral similarities make it possible to trace their evolutionary tree.
b. Homologies between species of artificial organism suggest common ancestry. Important information on common ancestors can be derived from careful analysis of a species' physical structure (the developmental and replication model underlying its behavior). Examples include gliders that develop into self-replicators, or particles that are not originally self-replicators but acquire self-replication capabilities during the evolutionary process.
c. Cladistics: cladistics is based on the idea that members of a group have a shared evolutionary history and have closer ties to other members of the group than to members of similar groups. Groups of artificial systems defined in this way have common traits that are not present in distant ancestors. A cladogram is

Figure 13. Growth systems for AMW 7. The initial data for each system, the code of the system, and 3-D images of the developmental dynamics for 80 simulation steps

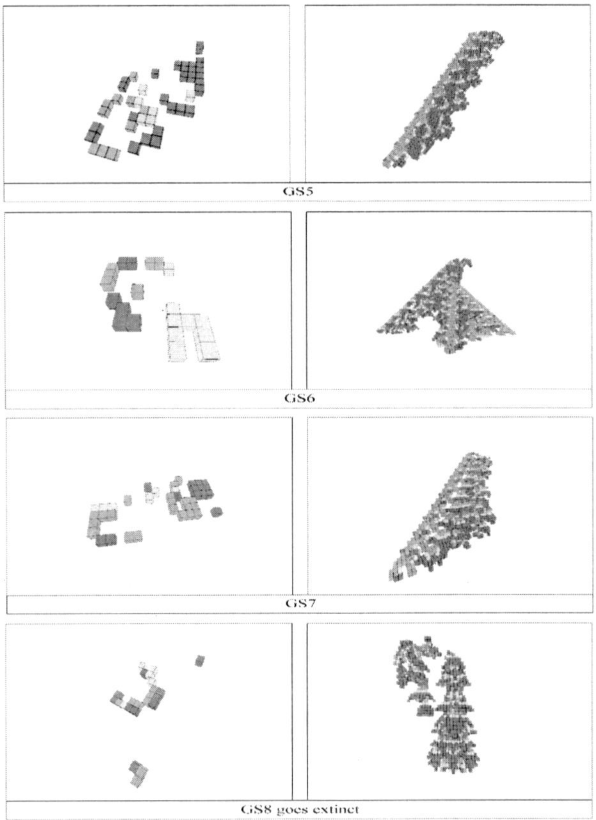

a branching diagram depicting hypothetical relationships based on the shared and derived characteristics of groups of artificial organisms.

d. Comparative analysis of artificial DNA sequences. Knowledge of evolutionary relationships and shared ancestors makes it possible to construct evolutionary trees. In this way the classification reflects the Darwinian principle of common descent and the evolutionary history of the organism (the systematic approach).

To classify the behavior of artificial organisms, we use two high level constructs, self-replication and growth, which we discuss in detail elsewhere in this book. These concepts allow us to identify, name, compare and describe the species we detect in our AMWs. The way these behaviors are sampled depends on the methodologies with which we distinguish different classes of behavior and on the way we study organisms' relationship to the environment and to other organisms. We present the results using the standardized form below.

Figure 14. Growth systems for AMW 7. The initial data for each system, the code of the system, and 3-D images of the developmental dynamics for 80 simulation steps

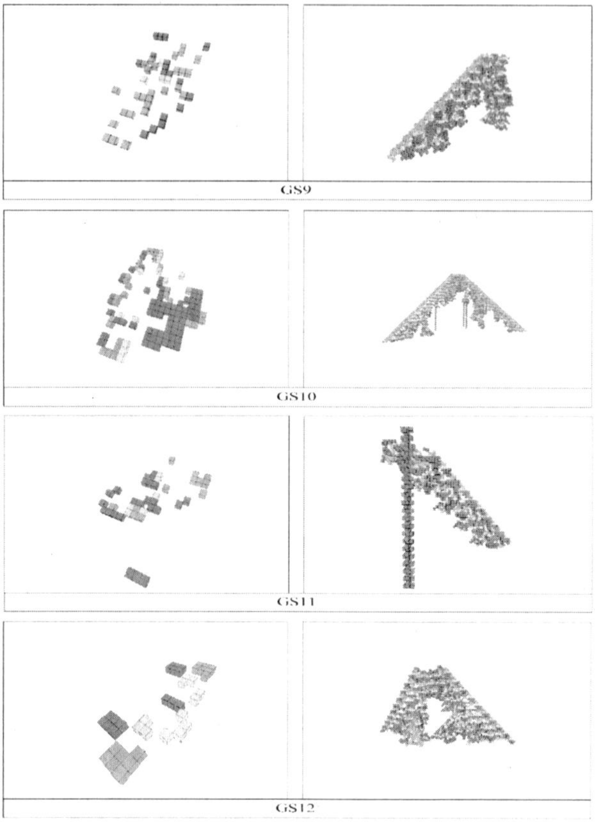

Standardized form for the classification of an artificial organism inhabiting an AMW

AMW N°, Simulation No, Generation N°

Species (S) (SC), Number.

1. Brief description of the species.
2. Anatomy (shape of the 1 life-like self-replicator), including an image of the organism and a description of its component parts.
3. Development of the organism (a representation of the principal steps); evolution of the organism (evolutionary tree).
4. Self-replication and growth: identification of the growth model for the organism and (where possible) the geometric structure of the model (see models described earlier).

Figure 15. Growth systems for AMW 7. The initial data for each system, the code of the system, and 3-D images of the developmental dynamics for 80 simulation steps

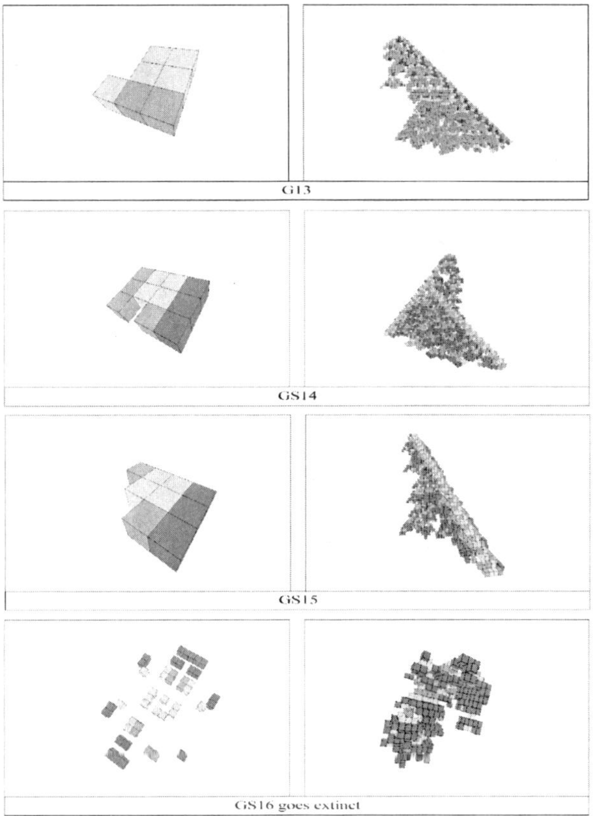

5. Relationship to the AMW.
6. Relationships with other species present in the environment.

In what follows, we describe examples of very interesting self-replicators that have not been previously reported in the literature on 2-D CAs.

Double Stranded Self-Replicators

This class of self-replicator develops in a linear way while moving in a specific direction. The resulting spatial-temporal patterns lie on the face of a pyramid. A two-dimensional view shows that the self-replication process is double stranded. Figures 12 through 16 depict the first strand in state 3 and the second strand in state 4. The two strands synchronize to create a Sierpinski Triangle configuration.

Figure 16. Growth systems for AMW 7. The initial data for each system, the code of the system, and 3-D images of the developmental dynamics for 80 simulation steps

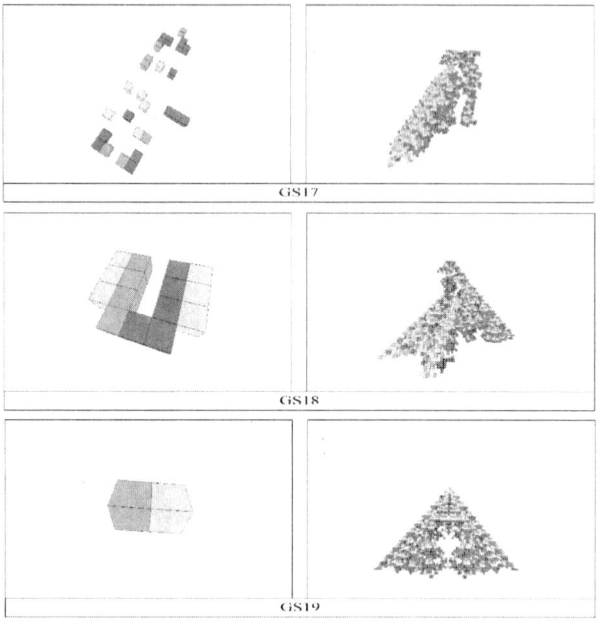

Each of these self-reproducing systems gives rise to endless sets of other self-reproducing automata. Automata belonging to the same set share the same "constructor" automaton. The process is made possible by a set of hidden rules that impose constraints on the way structures are put together. The final pattern is repeated many times at many different levels. At each level, we observe the same sub-structures we observe in the structure as a whole. This is the self-similarity we see in the Sierpinski Triangle. The implication is that all automata in the same set share the same formal "genetic language", driving their open-ended evolution.

In our work, we have created an artificial environment in which we can expose self-replicators to mutations (Bilotta & Pantano, 2005; 2006). Each self-replicator is identified by a set of conserved genetic rules that conserve its shape, by conventionally defined colors and by its replication behavior in the CA. To detect these rules (the self-replicator's "template") we have developed a computational tool which enables us to see which rules the system uses and how often. These studies show that different self-replicators are connected through a network of potential mutations.

Figure 17 depicts a k-totalistic CA (C13.8) with $D = 2 and k = 6$. The seed for the self-replicator consists of 3 cells: $A = (335)$. The figure shows the state of the automaton after 1, 2, 6, 14, 16 and 28 steps. The last image shows its development in time and space.

In the first step, the self-replicator rotates by 90°. It then returns to its original orientation and doubles. As discussed in detail in Chapter 10, this behavior is typical of the Sierpinski Triangle. In this case, however, the position of the self-replicator shifts over time. As a result, the system grows in a slanted direction. This behavior is different from the other behaviors we have studied up to now.

Linear Growth with Variants

Now consider rule C13.9. This describes a CA with $k = 5$ that generates a linearly growing self-replicator with $A = \left(22\right)$. Figure 18a depicts its development over time. After a series of intermediate phases, the self-replicator doubles after 6 steps. In the Sierpinski logic, this implies that after 12 steps, the two self-replicators will be further apart. This is what we observe. However, we also observe a new structure $B = \left(2222\right)$ occupying a central position in the gap between the two. This structure does not disappear but gives rise to a new, apparently chaotic replication process in which new self-replicators appear continuously but are then reabsorbed by the structure. The resulting process is a combination between complex behaviors

Figure 17. Self-replicator C13.8. Replication follows the Sierpinski model. However, the colony's position shifts in a direction perpendicular to the direction in which it is growing.

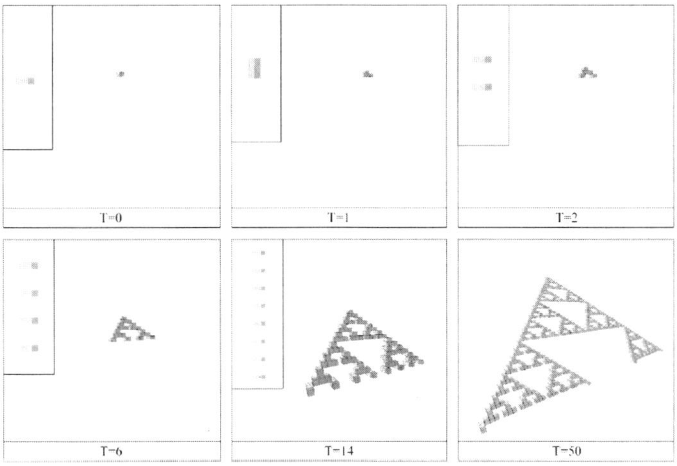

typical of linear self-replicators and chaotic behavior. We call this, *quasi-chaotic* behavior (Bilotta & Pantano 2006).

This is not an isolated case. In other cases too, we observe highly diverse behavior in the gap between the two attractors. Consider for example the self-replicator

$$A = \begin{pmatrix} 333 \\ 333 \end{pmatrix}$$

with rule C13.10 and $k = 5$.

The self-replicator develops linearly, following the Sierpinski logic (Bilotta & Pantano, 2005). However, if we change the initial data and consider:

Figure 18. This figure depicts the development of 2 quasi-chaotic systems. In the first steps of the simulation, the systems seem to reproduce regularly. Soon, however, we observe a shift towards chaotic organization.

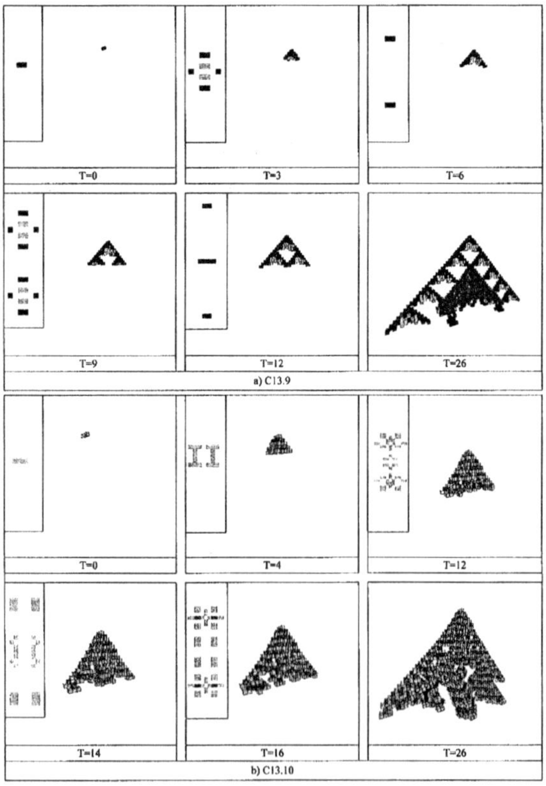

$A^{'} = (333)$

the development of the self-replicator becomes far more complex (see Figure 18b). In this case, we observe a series of type A self-replicators. These systems develop linearly but are subject to perturbations that break their internal symmetries. After 14 steps, we observe the emergence of 4 linear self-replicators belonging to type A^T. As in the previous case, the gaps between the self-replicators are filled by a central structure displaying chaotic behavior. In this case too, interactions with other structures lead to the continuous creation and destruction of self-replicators

Cross-Shaped Development Patterns

CA C13.11 with $k = 4$ displays another interesting symmetry. Consider the structure.

$$A = \begin{pmatrix} 22 \\ 22 \end{pmatrix}$$

After 8 steps it gives rise to 4 self-replicators with structure:

$$A^{'} = \begin{pmatrix} 22 \\ 22 \\ 22 \end{pmatrix}$$

The self-replicators develop linearly in four independent directions, following the Sierpinski logic. The result is a perfectly symmetrical cross-shaped pattern (Figure 19a). In this case, unlike the previous ones, the central structure does not give rise to chaotic processes. After a certain number of steps, it disappears.

C13.12 with $k = 4$ displays similar behavior. From Figure 19b, it is evident, that with the same initial conditions as the previous case (the same values of A), the CA produces the same cross-shaped pattern of development. Here too we observe four self-replicators that develop in four directions along two perpendiculars. This time, however, there are no new structures in the gaps between the self-replicators. Given that the self-replicators are still governed by the Sierpinski logic, the internal space within the structures fills with triangles. This behavior is particularly clear in step 12, where it determines a significant change in the shape of the self-replicators.

Figure 19. System with a cross-shaped pattern of development, based on a hybrid replication model

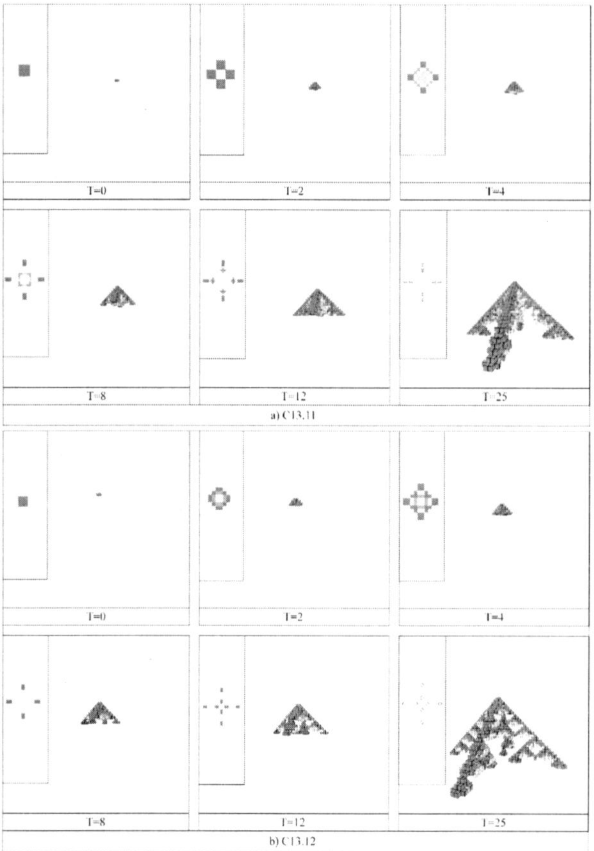

In reality, there are many CAs that produce this kind of perfect cross-shaped development, driven by Sierpinski logic. For example, consider C13.13 with k=5. Figure 20a shows the development of the CA starting with $A = 4$. The CA is governed by Sierpinski logic. After 4 steps we obtain 4 self-replicators with $A = (444)$

By step 9, the CA has generated new structures, each out of phase with respect to the others. As in the previous example, the CA develops in a perfect cross-shaped pattern. The 3-D image clearly shows the presence of new structures, protruding from the horizontal branches of the triangles.

Other rules produce interesting modifications to these patterns. In some cases, we observe the formation of new structures with their own symmetries. For instance, in a CA based on C13.14 with k=5, the initial structure $A = 4$ develops as shown in Figure 20b. Compared to the previous example, there are no significant changes

Figure 20. This collection of images depicts the growth of two self-replicators with a cross-shaped pattern of development, complicated by chaotic internal processes

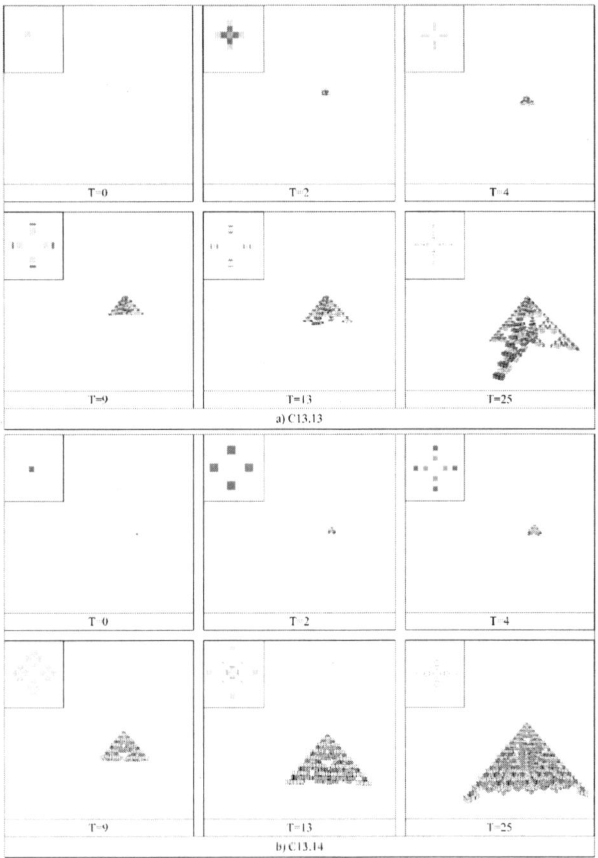

in the external symmetry of the system. However, step 13 is marked by the appearance of a period 2 pulsar. C13.13 and C13.14 are both products of the same run of our Genetic Algorithm. If we observe other rules from the same run, we discover several different patterns of development:

a. Cross-shaped development in which an internal pulsar and self-replicator guns give rise to complex dynamics;
b. Emergent chaotic behavior with continual production and destruction of self-replicators inside a cross-shaped colony;
c. Simultaneous production of gliders, self-replicators and pulsars inside a cross-shaped colony.

Multifractals

Multifractal self-replicators are systems that create complex structures occupying planes with different orientations in multidimensional space, organized around a single axis of rotation. As an example, consider a CA based on 13.15 with $k = 4, r = 1$

Figure 21a shows the development of the CA beginning with initial structure $A = (22)$.

After 8 steps, we observe the emergence of 2 structures that develop in opposite directions, producing a symmetrical figure. The gap between them is occupied by new structures. After a certain time (on step 23), these structures give raise to 4 new structures that develop along 2 parallel lines, perpendicular to the direction of the

Figure 21. Development of a multifractal system. This kind of self-replicator involves a combination of multiple self-replication processes, and chaotic perturbation.

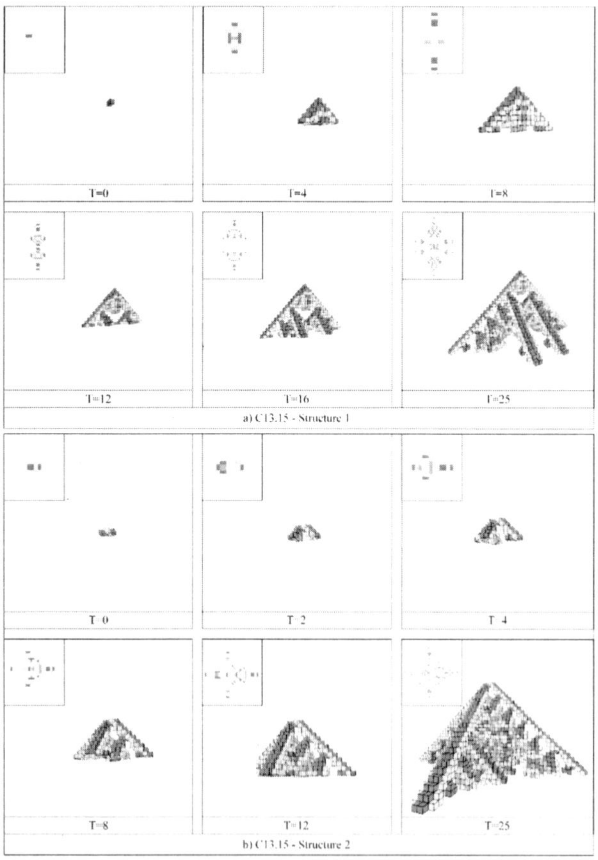

originals. Figure 21b shows the development of another self-replication with the same characteristics. The code for the new structure is:

$$B = \begin{pmatrix} 0030 \\ 2202 \\ 2202 \\ 0030 \end{pmatrix}$$

The way in which it develops is far more complex than that of the other structures we have examined so far. As is clear from Figure 23, the new structure begins to move in a well-defined direction, leaving behind a new emergent structure. This structure begins to change shape. At step 15, it produces two new structures identical to B - the original self-replicator. These two structures move in opposite directions along a line perpendicular to the direction in which the original self-replicator is growing. Step 15 is also marked by the emergence of a new structure which gradually decays, returning to the shape of the original self-replicator. The new structure moves along the same axis as the original self-replicator but in the opposite direction. As a result of this complex process, the initial structure B gives rise to 4 new structures of the same type. The new structures are out of phase and move in opposite directions, following lines perpendicular to the line of development of the original structure. This process continues for the rest of the simulation. We observe no chaotic behavior inside the structure. If we consider other rules produced by the same evolutionary run, we discover a wide range of interesting behaviors and symmetries.

Consider, for example, C13.15, with $k = 4, r = 1$. With initial condition:

$$C = \begin{pmatrix} 22 \\ 22 \end{pmatrix}$$

we obtain the behavior shown in Figure 22a.

The system develops in two perpendicular directions, giving rise to 4 structures that move in opposite directions. At step 14, they take on the same form as self-replicator B whose behavior we described above. The structures are in phase with each other and produce a perfectly symmetrical spatio-temporal pattern.

Now let us consider C13.16 with $k = 4, r = 1$ and initial condition $A = (22)$.

The development of this system is shown in Figure 22b. At step 8, the system produces 2 structures which develop into 2 self-replicators and 4 gliders at step 14.

Figure 22. Development of multifractal systems with chaotic internal processes

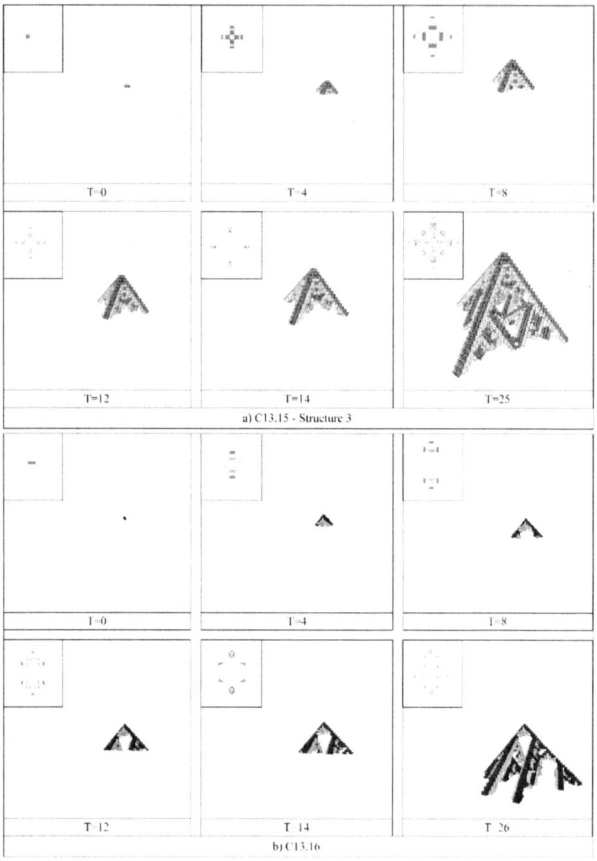

The self-replicators all move in the same direction. After a certain number of steps they double. The new structures produce 4 gliders that travel in opposite directions along two parallel lines, perpendicular to the direction in which the gliders are developing.

Self-Replicators with Built-In Mechanisms of Development

The evolution of these systems is very complex. They are characterized by two phases of non-linear growth. In the first, they grow until they produce copies of themselves, which grow and produce their own offspring. This is similar to the behavior we observe in biological organisms. However, in our case, the parent continues to grow, going through a sequence of morphological changes and generating new structures. The development of the offspring follows the same pattern (though

Figure 23. Rows 1, 2, 3: The initial development of the self-replicator. Rows 4 and 5. The second phase in the development and replication of the parent structure. Row 6: Replication of the parent structure. The parent structure gives rise to a new child structure once every 12 steps

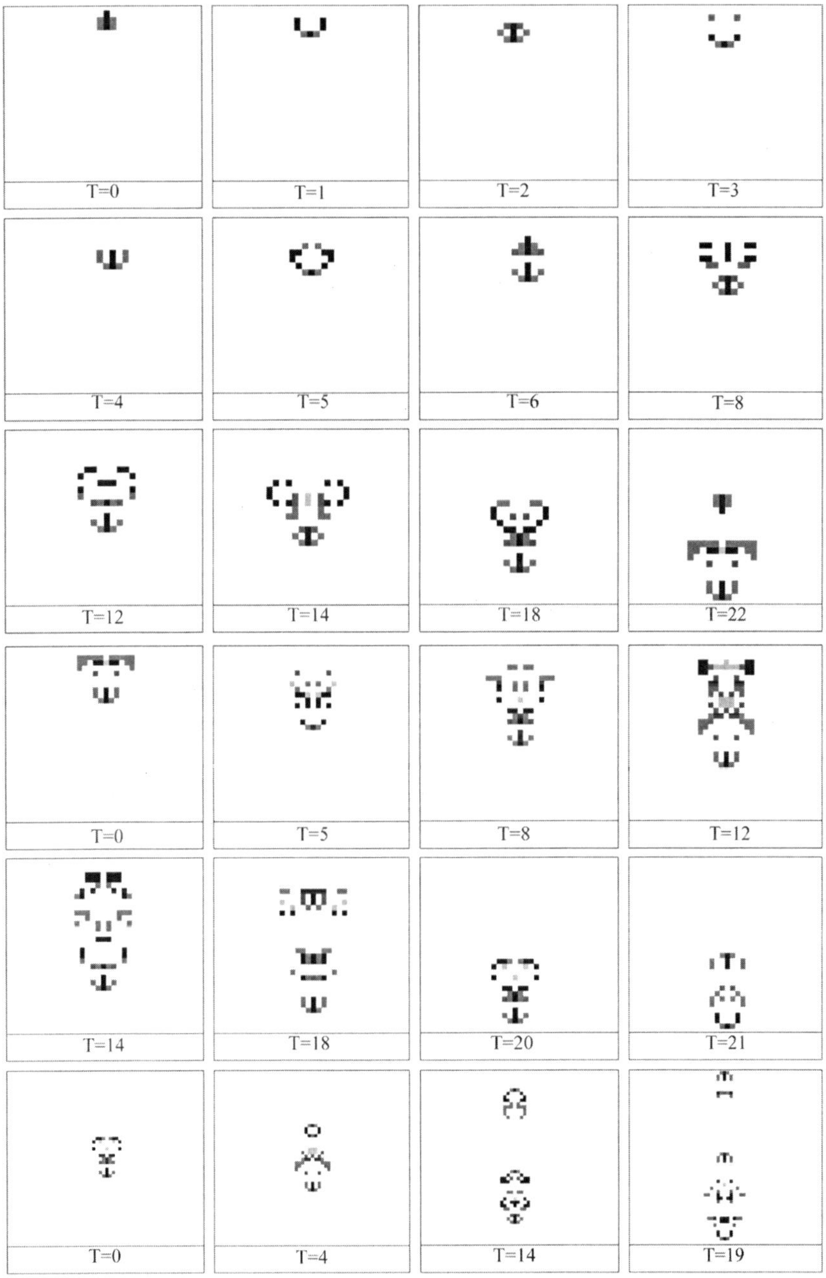

the changes are not synchronized with those of parent organisms). As a result, the development of these systems involves a range of nested developmental processes. The main process is set in motion by the self-replicators, while offspring follow their own asynchronous processes. Both are driven by mechanisms intrinsic to the self-replicator. The dynamics of the system are an emergent phenomenon that does not depend on any external procedure (Nehaniv, 2002).

As an example of this kind of complex behavior, consider C13.17 with $k = 5$.

The development of this CA can be divided into five stages (see Figure 23).

The initial structure is quite simple:

$$A = \begin{pmatrix} 0 & 2 & 0 \\ 4 & 2 & 4 \\ 4 & 2 & 4 \end{pmatrix}$$

After 21 steps, the system replicates (see rows 1, 2, 3 of Figure 23). But before this can happen, the system has to grow. This is a completely non-linear process. On the one hand, the whole structure expands; on the other, we observe the growth of an appendage which first undergoes a series of transformations and then expels a new structure identical to the initial one. In other words, the system has matured and is ready to give birth to a new structure. Like its parent, the child structure will begin its existence in an infantile state and mature for 21 steps before producing its own offspring. Child structures move and replicate in the opposite direction to the parent structure. Meanwhile the parent structure continues to grow and to develop, undergoing a series of transformations before giving rise to new replication processes (see rows 4 and 5 in Figure 23).

This second phase in the development and replication of the parent structure is even more complex than the first. After a series of transformations, we observe the emergence of a large stable structure, a child structure that travels in the opposite direction to its parent. The shape of the child structure is identical to that of the original structure at step 4 (see row 6, Figure 23). Figure 24 shows the complete spatio-temporal pattern for the CA. In the image, we can clearly perceive the development and replication of the parent structure.

It is important to observe that this asynchronous self-replication process is completely different from the process described in (Lee, et. al.,2007). What this work shows is that self-replicators can exist even when the CA is designed to evolve asynchronously. In our case, rather than being pre-defined, asynchronous evolution is an emergent feature of the self-replication process .

Figure 24. Replication of the parent structure. The structure moves leftwards, and produces three child structures. The first, which has reached maturity moves rightwards, producing a child structure that also moves in a rightwards direction. In the image, it is about to collide with the second child structure. The collision will annihilate both structures.

Loop Self Replicators

Loop self-replicators, such as the famous Langton loop, have highly complex patterns of replication. In our studies, we have actually found four different classes of self-replicating loop. In most cases, replication proceeds in four distinct directions, with each process generating many copies of the original structure as well as large numbers of other self-replicating entities. The same processes also produce many gliders with large-scale structures in the system acting as glider-guns.

As an example, consider C13.18 (Figure 25) with $k = 5, r = 1$, and initial structure:

$$A = \begin{pmatrix} 0424244 \\ 4204244 \\ 2200400 \\ 4203000 \\ 0410000 \\ \\ \\ \end{pmatrix}$$

Figure 25. Development of a colony displaying intense self-replication. As we will see, there is a clear distinction between colony formation through self-replication and through growth. In this latter case, self-replication processes are restricted to the outer edge of the colony and are nearly invisible.

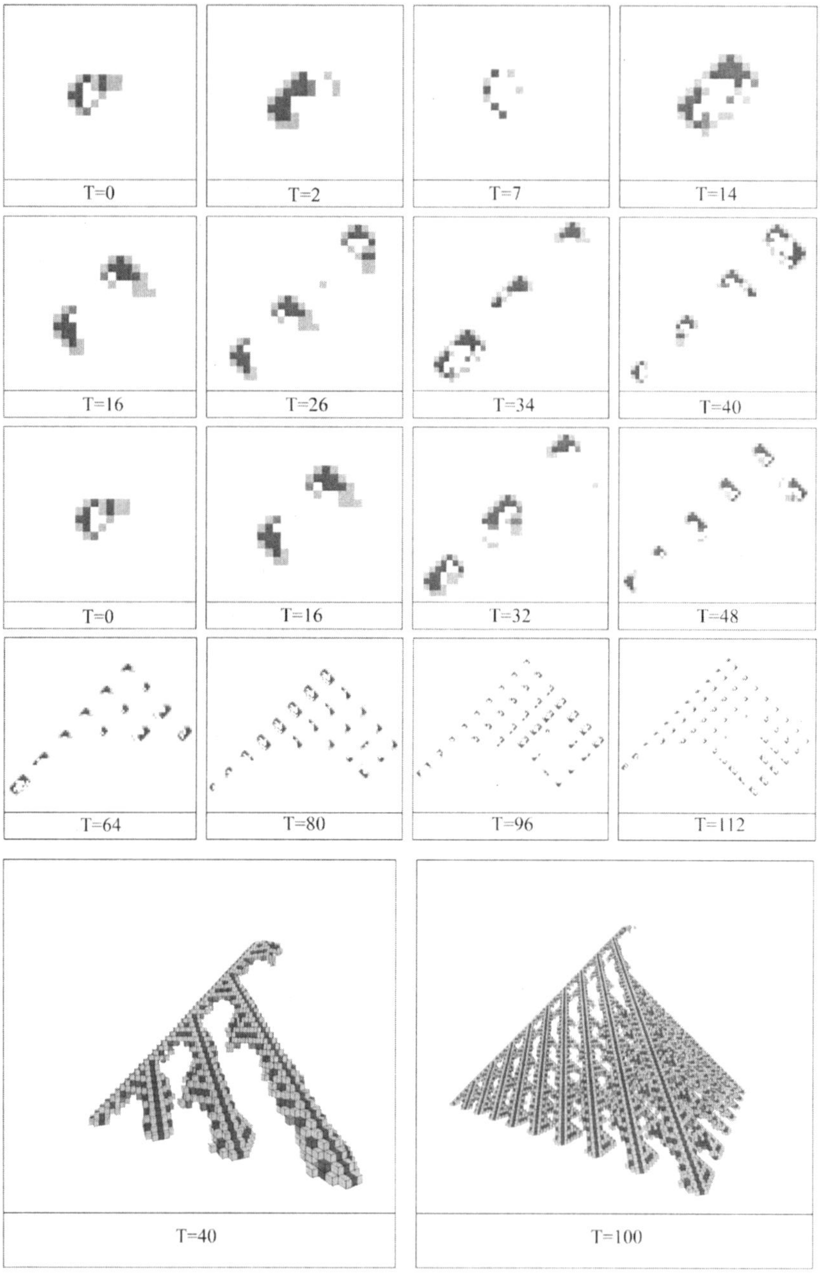

After a series of transformations, the structure doubles (Figure 25, step 16). By step 26, there are 3 structures each of which goes through its own transformations before beginning replication. The spatio-temporal pattern for the first 40 steps clearly shows symmetries in the replication process (Figure 25, step 40). In the image, we can perceive 3 independent structures. Given the lack of available space, the first structure replicates in a direction perpendicular to the original direction of development. This is apparent in the image for step 48. 5 structures continue to develop in the original direction, 1 develops along the perpendicular. The image for step 64 shows that the internal structures of the main self-replicator are developing in a direction perpendicular to its main direction of development. This process continues for several steps, producing interesting compositions of spiral symmetries (see the images for step 100).

FUTURE TRENDS

In this chapter we have seen that self-replication can take many diverse and complex forms. We need to classify them in some way. In previous chapters we focused on highly symmetric processes. In this case we analyzed:

a. New symmetries involving translations of the self-replicator in time and space;
b. Cross-shaped development;
c. The emergence of chaotic phenomena inside linear development processes;
d. Composition of emergent symmetries derived from initial structures that develop again into self-replicators;
e. Growth and development of complex organisms prior to the start of replication;
f. Expulsion of particles and gliders during the replication process.

Combining observation with the methods usually used to classify biological organisms, we have produced a more diverse and complex taxonomy than those we originally presented in (Bilotta & Pantano 2006; 2008). In the future, we intend to translate our methodology into software, going beyond the prototypes we used in the work presented here.

CONCLUSION

Our many experiments with Genetic Algorithms have allowed us to generate a wide range of 2-D CAs. These CAs have made it possible to create Artificial Microworlds (AMWs). In our microworlds, we have discovered populations of self-replicating, growing organisms with a range of behaviors similar to those we observe in nature. In our work, we have applied a mix of observational and engineering methods, Our observational methods have made it possible to classify the organisms we discovered; engineering methods have allowed us to manipulate their genotypes and phenotypes and thus to discover universal models of behavior, applicable both to biological and to artificial organisms. We have thus been able create a taxonomy of artificial systems, based on their behavior: the way they react to changes in the environment and their own internal transformations. The taxonomy we have created uses techniques from cladistics and systematics to represent similarities and relations among organisms In general, we can say that the more similar the behavior of two species of artificial organism, the closer they will be genetically and the more similar will be their patterns, structures and dynamics.

This chapter presents a number of important results. First it describes a number of self-replicating and growing systems, not previously reported in the literature for 2-D CAs. Second it classifies these systems by increasing order of complexity. Third it proposes a classification methodology for artificial systems that replicates the historical trajectory followed by biology, genetics and zoology. The application of this methodology leads to a number of conclusions.

a. The general behavior of artificial systems, at the individual and the population level, together with their complex patterns of development, demonstrates a close similarity with biological systems.
b. Artificial systems reflect some of the basic characteristics of life. This implies we can study them using the same methods we use to study natural systems.
c. AMWs contain two kinds of species, both described in this chapter. On the one hand, we have self-replicators - based on several different models of self-replication, on the other, systems that grow via cellular division. Much work remains to be done. This includes studies comparing the dynamics of artificial systems to biological processes such as cancer, the growth of stem cells, and the development of embryos and bacterial colonies.
d. Perhaps in the not too distant future, it will be possible to use these systems as the basis for new bio-engineering technology.
e. AMWs make it possible to study the complex evolutionary processes, typical of natural systems, in simplified environments. This makes them a useful tool for theoretical biology.

REFERENCES

Bilotta, E., & Pantano, P. (2005). Emergent Patterning phenomena in 2D Cellular automata. *Artificial Life Journal*, *11*(3), 339–362. doi:10.1162/1064546054407167

Bilotta, E., & Pantano, P. (2006). Structural and Functional Growth in Self-Reproducers. *Complexity*, *11*(6), 12–29. doi:10.1002/cplx.20131

Bilotta, E., & Pantano, P. (2008). Biological traits in artificial self-reproducing systems. In F. Orsucci, & N. Sala (Eds.), Complexity Science in Information Communication Technologies: Interaction Between Living Systems and Information Technologies (pp. 109-123). Hershley, NY: IGI Global, ISBN 978-1-59904-627-3.

Byl, J. (1989). Self-Reproduction in small cellular automata. *Physica D. Nonlinear Phenomena*, *34*, 295–299. doi:10.1016/0167-2789(89)90242-X

Campbell, N. E., & Reece, J. B. (2002). *Biology* (6th ed.). San Francisco, CA: Benjamin Cummings.

Chou, H. H., & Reggia, J. A. (1997). Emergence of self-replicating structures in a cellular automata space. *Physica D. Nonlinear Phenomena*, *110*(3-4), 252–276. doi:10.1016/S0167-2789(97)00132-2

Chou, H. H., & Reggia, J. A. (1998). Problem solving during artificial selection of self-replicating loops. *Physica D. Nonlinear Phenomena*, *115*(3-4), 293–312. doi:10.1016/S0167-2789(97)00237-6

Gutowitz, H., & Langton, C. G. (1995). Mean Field Theory of the Edge of Chaos. *Paper presented at ECAL*. Berlin: Springer-Verlag.

Langton, C. G. (1984). Self-Reproduction in Cellular Automata . *Physica*, *10D*, 135–144.

Lee, J., Adachi, S., & Peper, F. (2007). Reliable Self-Replicating Machines in Asynchronous Cellular Automata. *Artificial Life*, *13*(4), 397–413. doi:10.1162/artl.2007.13.4.397

Li, W., & Packard, N. (1990). The structure of the elementary cellular automata rule space. *Complex Systems*, *4*, 281–297.

Nehaniv, C. L. (2002). Evolution in Asynchronous Cellular Automata . In Standish, R. K., Bedau, M. A., & Abbass, H. A. (Eds.), *Artificial Life VIII*. Cambridge, MA: The MIT Press.

Oliveira, G. M. B., De Oliveira, P. P. B., & Omar, N. (2001). Definition and Applications of a Five-Parameter Characterization of One-Dimensional Cellular Automata Rule Space. *Artificial Life Journal*, *7*(3), 277–301. doi:10.1162/106454601753238645

Salzberg, C., & Sayama, H. (2004). Complex genetic evolution of artificial self-replicators in cellular automata. *Complexity*, *10*(2), 33–39. doi:10.1002/cplx.20060

Sayama, H. (1998). Introduction of Structural Dissolution into Langton's Self-reproducing Loop. In C. Adami, R. K. Belew, H. Kitano, & C.E. Taylor (Eds.), *Artificial Life VI: Proceedings of the Sixth International Conference on Artificial Life* (pp. 114-122). Cambridge, MA: MIT Press.

Sayama, H. (2000). Self-Replicating Worms that Increase Structural Complexity through Gene Transmission. In M. A. Bedau, J. S. McCaskill, N. H. Packard, & S. Rasmussen (Eds.), *Artificial Life VII: Proceedings of the Seventh International Conference* (pp. 467-476). Cambridge, MA: The MIT Press.

Stevens, P. F., Donoghue, M. J., Kellogg, E. A., & Campbell, C. S. (2002). *Plant Systematics: A Phylogenetic Approach* (2nd ed.). Sunderland, MA: Sinauer Associates.

Wolfram, S. (1984). Universality and Complexity in Cellular Automata. *Physica*, *10D*, 1–35.

APPENDIX

Initial Data for Species and Glider Like Species

S1

222222
222222
222222
00333

S2

022
1
4
4

S3

3441
3441
3004

S4

14
14022
10022
0
022

S5

1
14
14

S6

14
14
14

S7

0322

0111
01114
00114
10114
0101
001
011

S8
0000001
000002
00000200000104
00002302010001
00004
0004
02040000012222
222
220000002
2

S5+ NonSpecies
111
044
044
0333

S5+S1
444
1111
0002
2222
333

S2+S1
00330
02222
02220
1
4
4

GS1

00000000000000000000220
00000000000000000000221
00000000000000000000220
00000000000000000000203
00000000000003002100000
00000000000003002000000
00000000000003002000300
00000000000000000033000
00000000022222200000000
00000000022222200000000
00000000022022222200000
00000000022222200000000
00000000022222000000000
00000000000222200000000
00000000000222102000000
00000000001032304000000
00000000004100004000000
00000000000000002000000
00000000003000000000000
00000000000410100000000

GS2

0000000000000000000220000000000000
0000000000000000000400000000000000
0000000000000000020000000000000000
0000000000000000020000000000000000
0000000000000000020031200000000000
0000000000000100020332000000000000
0000000000111000220000030020000000
0000022200022003000200000040200000
0000000000022001000030003400000000
0000000000122001000003030400420000
0000000000000000000000002000000000
0000000000000000000000002000000000
0000000000000000000000000000300000
0000000000000000000000000003000000

GS3

00000022000000000000000

00000000000000000000000
00000000420000000000000
00000044002000010000000
00000044004000000333000
00024440040000000444000
00024004040000000004000
00000044040400000001000
00000200444000440220000
00000000222000000000000

GLS4

00001440000000000000000
00020040000000000000000
00020040222000000000000
00000000002000000000000
00041000002000000000000
00043000042000000000000
00000000000220000000000
00000040000222000000000
00000023022222000000000
00000002220222000200000
00000000000000000400000
00000222033300024042000
00000222033300024042000
00000222000000000000000
00000000000002200000000
00000000000002001000000
00000000000002222222000
00000000000002222200000

GS5

00144000000000000
02000020000000000
02000002200000000
00000100200000000
04000233000000000
04000013000000000
00404300022200000
00000000200000000
00200300000300000

00200000000020000
00010000000020000
00000020002220000
00000000002220000
00000000022220000
00000000000020000
00000000000000000
00000000000000000
00000000000020000
00000000000020000

GS6

0000000000000303000
0000000000000303000
0000000002200333000
0000000002000300000
0000000020000000000
0000000000100000000
0000000400000000000
0000000400000000000
0000000100000300000
0000000022044000000

GS7

000022044000000000000000
000100000000000000000000
000400000000001440020000
000444000000040440000000
000333000000040110000000
000000010030042002220000
000000000003003002220000
000000000020001002220000
000000000000000100000000

GLS8

000000000004
000000000024
000000000000
000000000000
000000000000

000000000000
000000000000
002132200000
002004400000
000000042000
000000042000
000000042000
000000300000
000023000000
000303000000
000000000000
000000000000
000000000000
000000000002

GS9

000041010000000000000000
000000000000000000000000
000010000200000000000000
000000000020000000000000
000004100002000000000000
000000001002000000000000
000000001000000000000000
001000111000020000000000
000000001000000000000000
000000000040042000000000
000000300000400000000000
000000000330200000000000
000000400220000000000000
000024040201110000000000
000000403200000000000000
000000200000000000000000
000000000010001000000000
000000000001000000000000
000000000001000000000000
000000000000000200000000
000002220000000000000000

GS10

00000000014430000000000

00000000200430000000000000
00000000200430000000000000
00000000000000022200000000
00000000400000022222200000
00000000122020022222200000
00000002030030000022200000
00000000000000000022200000
00000024000000222022211000
00000024000000222222222100
00000000400000222222222100
00000000000000020102022000
00000000400020021000222000
00000024000000021000022000
00000024000000200000122000
00000000000000230000000000
00000002030000000000000000
00000000120003000030000000
00000000400001000000000000
00000000000000000000000000
00000000200430000000000000
00000000200430000000000000
00000000014430000000000000

GS11

000000000000222000000
000000000000000000000
000000000000000000000
000000000000000000000
000000000000000000000
000000000000000000000
000022220000000000000
000204003110000000000
000000003130000000000
000200000220220000000
000020000220000000000
000000000003000200000
000000000000000020000
000000002220400000000
000000002230000000100
000000000230000000000

00000000000000011100
00000000000000010400
00000000000000000000
00000000001000000000

GS12

0000000441000000
0000000441000000
0002220004000000
0002220000000000
0000000330000000
0000000300300000
0000000000030000
0000000000030000
0000000000300000
0000000222000000
0000000000000000
0000000000300000
0000000000300000
0000000300000000
0000000330000000
0002220000000000
0002220004000000
0000000441000000
0000000441000000

GS13

0411
0044
0044

GS14

1012
1442
1442

GS15

0012
0442
1442

GS16

000021012000000000000
000020002000000000000
000000000000000000000
000300031000000000000
000002000000000000000
000000000000000000000
000000300000000000000
000303000222000000000
000000000000000000000
000000000000000000000
000000000000000000000
000100000000000000000
000103020000000000000
000000000000000000000
000000000000000000000
000100000000000000000
000200000000000000000
000220220000000000000
000000200000000000000
000000000000000000000
000001000000000000000

GS18

01220
41023
41023
41023

GS19

14

Rule Codes

C13.1: $D = 2, k = 10, r = 1$. This code is very long. If some reader wants it, send an e-mail to the authors, who will provide the system.

C13.2: $D = 2, k = 4, r = 1$

0001200000000000000000200000200200000200000300000000000101000020

0000000103100000000100000000000000003000000000200033000000000000000
00000000010033011100020000010000000000013000001000000003030000302
00000000200000000020000000

C13.3: $D = 2, k = 4, r = 1$

000100000000000000000000000030010022200000030000010000033202000
0000001200003001000000000300000000000100003000002000000000001000000
000000000000000000000001030000020200002020310202000110200001000100
0001020020020202000003010

C13.4: $D = 2, k = 4, r = 1$

230123222202022222322230220202223222222222222222222222222222323
220222212202212322223331220220223223321213201212120222302230121322
2223123110222102232212232222102022223232122222202122222221122212202
2221212221232202301222232

C13.5: $D = 2, k = 4, r = 1$

200122021210320001032000010100001100000330013030010020300300222
021302110223030000003002210101001031032303103201000000032002100003
000110202000020003301333102000120312000110330310002031100000010110
0323030202020003200000000

C13.6 $D = 2, k = 4, r = 1$

333303132333331233333333302320303333333332033302330312223313233
303233222303312123313112230320313312013313331131330223003323132332
3313313323333313333233300303333333333320333031331333133331002303333
123333303330333333133323

C13.7 $D = 2, k = 5, r = 1$

000000000000000000000042100400000440004300003000000000000000000000
0000000000000000000000000000000000000000000000000000000000000032
0000130000000000100004000011120004000003000000000002212004000000000
0000003000000000000000000200230000000000430300000000044201240000000
0000000000040000000000300000004000020000000000000000000000000130200
0000000000000000000000000000000000000000000040400303000000003000003
00000000000001003200030000000400000000000000000000000000013003000000
020000000000000030000030000040000000042000000002002312010000003000
00002000000040000000000004000340020000400000000002000000000402000000

00000000000000000000000000000401000030000000000000000300000000000000
00000000000030300100000000000000032000032100000000000020104003120000

C13.8 $D = 2, k = 6, r = 1$

500042410333030000000000551125400005004030010401040405013300000000
004103410340442000000000300020000301001100210000000000300003000001100
300523000101001003000000415504000003201300002030000020000000000005500
101103020000050000000000050100051000004220004011501504333340000000000
034000500233302142002504000001000001300400030211211453351100200000
050000000000300002000001500000252200011000000000000000403004040030200
330001110000000000000400525500500043000001000510000040054001002140 2
410000000400000000000000000000400004200000000000000000000041411000000004
030404501000000000000004500425431151000500202000050302000020002 2025
002300104420000000000000023000040000000000044002100000120000000200330030
004015010000000003000000503000300010202220102341013350000000000050000
000554000000000530200000154244554500300342000004000000000000000000000
000010450404010000100000200041435342350033010013540305240000020001
000020400000500000004000000000012100300401100000502403000005000000
000002020500540000000505050000100000000000000000100100000000000000000
0503031020010000543005000000000055005004001503030020450000001400543
52000500000000000000000000300000000000122425010310000000000000000110
0000004000000020000025000000000000000000000000001000455420033010000 0
000002210000000000000000050000100120051000000000140203110330005000054
050000032101201000000001200302302300010000000003021245454240330031 00
001001000200000000000100300000040000033400100005000000000000000000005
000500100003113000000144143030001050025000040000000000000225030000000
000450400400000000300040002300000404000000030000000121000000000100000000
000000300100300000000015430040001000033001000005002001005521132000 0
000000000000050000000004000000002032500332220000000000000200000000000013
020000000000010040000000355302302235252114200000000200505003020050000
053005400002451131000000300000100003301020000000540000000000000200010
000000000000000300000000015553000420000204000000005400000000000000000
041134001200000000000000050001154000000000000000000000013002300000000000
000000001000301003535050300002041000000000004000000001000020000000344
5100032340000000000500000

C13.9 $D = 2, k = 5, r = 1$

0000000042000020201430004000001030021000200224000002000000000000
4000000010000313341000444000000001000430100000300400001210004314212

302034100000400000000231000010000000000000003000000003100000000300
000043030034400333000040000000000000000011000000200000000000000000
000000000000004300020043400044200300000000140040231114121000000
000002100000000000000021000200000422231001030203201001400000000011
004010300100000000000000000040000441100022300000330000304100042 04
020400100000103000300303342000020000000000004000000000000104400000
231040400120120000000010000000044133000100013000100000000100004000
000000002420004002000000001000340440400001000000404003004000000000
000440004204003013004000000000000000000000000000000000042004000000

C13.10 $D=2,k=5,r=1$

0133000042000020201430024400001030021010200220000002000000000003
400000001030031334100044400000041000433100020000400001210004314212
332234100020400000200231401010300000120000230000000131400020000 30
000004343403440033300004000000000000000031100000020000000000000000
000300000000000044300023043000044200000000000142040231114121000302
000003210000000002100021001200000422231000130203201001400000000001
100401030010000020000000400004000044110042230000033400030410004 20
402040010000010300030030034200002000000000000040000000000010440003
023104040112012000000101000330124413300010001300010000002010004 01
000000000042000400210001000100034044240000100000341400300400000400
400244000420400301300400000000000000000200000200000000110200430000 0

C13.11 $D=2,k=4,r=1$

1012000033020002203000000000020000000300001200002330012000023212 0
011002001321310012120332023231023300302000003111210022003200000000
01100010220001133100230300000210000030212132000020131000000000201301
000000010000000000320330000

C13.12 $D=2,k=4,r=1$

1001000000000000000000000000000000000000200000323221032100120000 0
000000000100003000000000001000302020200300330033000020002002000002 11
030200102000000000000000000102000030112310100000000000000000000003330
000000001000000000223000000

C13.13 $D=2,k=5,r=1$

00020010040000200000420000400012400004002310100000033124100034000 0
000033224342211422110420000300033300444132310400001001003230400030

1000000000000000200000000000440000000030000004000400010003000000000000
0014220023134204000300000401010000302012000140000000204300000113 1
0000034410302300024441134440000300000000300000000000130000002000 0
0000300400000000000000000200100000000000400000000000000000000000242
2431241240032010000013000101003340102104210013243122210302320003 3
2414221000430002330012333320042324312011212030042013203230401311 2
2003000200102130300440000030100100000000300032010004000100200020 0
0002000100230120441400103442433011013001310430431310001030000100 0
0000000200000000302000033000013000000000440030000000000000030420000

C13.14 $D = 2, k = 5, r = 1$

000200000000002030004000002040020000400040004000000002004000003 0
0120100400402400200040002010001000001100000000000300000000200000434
100010000000400400000000000002400100100000000000000000000240041000
000300000030000010034000000000040020000000000000000000000000000020
000000200200400030000000000000003000000203000200000000000000400000
000030030000000000023000010014340001432431000000000143114100003024 2
2431241240033010000013000101003240102104210013042122210302220003 3
4414024000030002330012334320042322312441012030044003234233431311 2
2000004412000340000240440000000212420232400020200442000323000303 1
0000000100330120444430100444432311013304000100103000120000003100 1
0040000000403043300020021000000010000004400300030001002003032000 0

C13.15 $D = 2, k = 4, r = 1$

10120000330200022030000000000000000000320103013002100033132000110
1323200200000231322100033220223000000122112300023000010220000110 01
0031110212000101333101133113331113001000001202001332200020330012 01
0330103010000000320300000

C13.16 $D = 2, k = 4, r = 1$

10120000130300022030200000000000000000000120000233002010002321 20
0020000000000000000000001323002330030200002321121002200320000000 0
0110011022000111200000011203000122003021213300000013100300000013 01
0330103010000000320000000

C13.17 $D = 2, k = 5, r = 1$

200100030000000020010000244003400000000120003000002033000000000
0000000000040000000000400000000200002000000000000000300000002241001 13

030020000000440002000000011013000304000000000330000000400400004000000
440010000000000400000000000000000000000000000000000000000000000100303
400000000000200000302021000000000000010000000004000000000000000000400
000002002423002000000300100300000000000000100000000000000000000000
000300003104100300000000000000030020200430034100000034000000000000
101000000000040001000000000000100000002000000000000000000330000000004
000040000102300120002102010000000200000100002020000240000200000420
212201004232100000000000000000220004030000000400000004300002103000
000000014003000000000300000000000032200000000420000022004000200000040

C13.18 $D = 2, k = 5, r = 1$

000000000000000100002230003030000044000000040430000000000000000000
032400200040020000004020400000004000020000020012200040002000024000
042000040000040000000000000200000000000002000034032030004001443031000
010010040000000000042420030400001120000000000004304000000000000040000
320002000000401000000000000004001000020022143210000000001000000000
200000000000000000000000000000000000000000000000004033000000000300
000102000000000000000032000000000000000000004000000000000000030000000
000000001200200000000000002004020000000004000000000000000000000013230
340020001000000030012000000000033002010004000000000400000200000010
000000000000000000200000000300001000400000000000000004421300012000000
000000001010000000130400000003000030213200000400000000004330000000000

Chapter 14
Rhythms of Life

INTRODUCTION

It is widely recognized that the birth of modern science dates to the moment when Galileo first timed physical processes taking place in space. In biology, it is only recently that scientists have felt the need for experimental and mathematical methods describing the development of living organisms in terms of dynamic processes. When we analyze the development of self-replicators, we see that they develop their characteristic patterns and self-similarity through processes that resemble a set of oscillators, operating on different time scales. The periodic behavior of these large scale processes and the relations between them both depend on information processing and on local activity. The end result is the steadily increasing complexity we observe in all biological development processes. Physiological rhythms are essential to the life of the organism. Some are maintained for its whole life and even a brief interruption signifies death. Others only operate for short periods of time - some under the control of the organism, some not. Rhythms interact with each other and with the external environment. In most cases, abnormal and new rhythms are associated with serious pathology. To understand the typical rhythms of living organisms, we have to integrate mathematics and physiology. One important approach is non-linear dynamics. Non-linear dynamics allow us to study biological observables as a func-

DOI: 10.4018/978-1-61520-787-9.ch014

tion of time. Time series for physiological variables show four distinct classes of dynamic behavior: steady state, oscillations, chaos and noise. These are the same behaviors we observe in 2-D CAs, as described in previous chapters. “Steady state” is a way of describing homeostasis - the tendency of biological organisms to maintain the stability of their internal environment (for example, in terms of the presence of specific molecules). It also refers to a solution for an equation that is a constant. More technically, homeostasis is the ability of a system or living organism to adjust its internal environment to maintain a stable equilibrium, thanks to regulatory processes that come into operation when external conditions change. The investigation of the processes that maintain physiological variables within a narrow range of values is a lively area of physiological research. Many physiological phenomena - such as the beating and electrical activity of the heart - are approximately periodic. Other cyclical behaviors - such as breathing, sleep, the menstrual cycle - are equally familiar. Less well-known oscillatory behaviors include insulin release, release of male and female sex hormones, peristaltic waves in the intestine, and the electrical activity of the cortex and the autonomous nervous system. Each of these physiological oscillations is associated with a periodic solution to a mathematical equation. Of course, the phenomena we observe in physiology are highly variable; we never observe a perfect steady state or a perfectly regular oscillation. Even in systems we class as stationary or periodic, there are always fluctuations around the stable value or the regular cycle. There are also systems whose behavior is so irregular as to escape classification. For instance, blood pressure tends to be maintained at a constant value, but is nonetheless subject to fluctuations, reflecting the activity and emotional state of the organism. What is more, some physiological rhythms influence (and are influenced by) other rhythms. In arrhythmia, for instance, heart beat accelerates during inhalation. This is a relatively simple phenomenon. When we measure the electrical activity of the brain with an electroencephalogram, the situation becomes more complex. The instrument measures mean electrical activity in localized areas of the cortex and shows how this activity fluctuates over time. In many cases the fluctuations are fairly irregular. Understanding this irregularity is a difficult task. Mathematics has two distinct methods for studying irregularities in physiological rhythms. The first is to consider them as noise - the result of random processes such as the opening and closing of ion channels in neurons or in heart cells. The second is to apply the concept of *chaos* - an explanation for apparently random fluctuations in deterministic systems. We can observe chaos even in the absence of noise from the environment. Associated concepts include sensitivity to initial conditions and perturbations, strange attractors, multiple routes to chaos and the fractal boundaries of basins of attraction (for a review see Bilotta et al., 2007). Attempts to describe physiological rhythms in mathematical terms have led to the development of several models, including those incorporated in modern cardiac pacemakers. These models take account of a wide set of dynamic processes, including the processes that regulate ion exchange in neurons, and many others. These models are highly sensitive to perturbations and can generate chaotic behavior. In 1914, Brown (1914) demonstrated that cats have a Central Pattern Generator (CPG) that continues to generate rhythmic motor activity, even after afferent nerves have been cut. Even today we do not have a clear understanding of the mechanisms driving the CPG. It has been suggested that the observed behavior is the result of a cycle of mutual inhibition. It is known, furthermore, that negative and mixed feedback processes in systems, subject to interruption, can give to both chaotic and oscillatory behavior. In this chapter, we will propose an alternative interpretation based on our own observations of 1-D and 2-D CAs. The interpretation we propose is based on insights into the general patterns of development and the rhythms that these systems display. In chapter 15, we will see how we can translate these patterns into other languages - such as sound and music. In what follows, therefore, we will analyze the rhythms associated with the development of individual CAs and with the longer timescales on which CAs evolve. In both cases, what we will be talking about are "rhythms of growth."

BACKGROUND

Studies by naturalists have shown that rhythm is a ubiquitous feature of life, from single celled organisms to higher animals and plants. In many cases, observed rhythms have been attributed to internal oscillators or biological clocks. In mammals, experiments with rats and hamsters have identified an oscillator in the suprachiasmatic nucleus (SCN) in the hypothalamus, which appears to be responsible for circadian rhythms. (Rietveld and Groos, 1980; Meijer and Rietveld, 1989). Working with autoradiography images from rats, Moore (1973) discovered the retinal hypothalamic projection, a direct connection between the SCN and the retina. The results of bilateral ablation or surgical isolation of the SCN in rats supports the idea that it is a pacemaker. It has also been observed, again in rats, that electrical stimulation of the SCN leads to phase changes in circadian rhythms. In reality, every organism displays many different rhythms in its behavior. This has led to the suggestion that there may be many structures like the SCN, each with the ability to automatically generate and regulate a specific kind of oscillatory behavior. It seems likely that these structures exist not only in the brain but also at lower levels of biological organization, for instance at the tissue, cellular and sub-cellular levels. In each case, they act as pacemakers, helping to filter the influence of rhythmic elements in the external environment. Halberg (1977) calls them "synchronizers"; Pittendrigh and Daan (1977) refer to them as "entraining agents", Aschoff (1978) uses the term

"Zeitgeber" (time giver). In human beings, they contribute to our ability to perceive time and movement in dynamic stimuli from the environment. For example, studies of people listening to music (Jones and Boltz, 1989; Drake et al., 2000; Large, 2002; Khalifa and Foster, 2006) show that attention can be modeled as a process which gradually synchronizes a set of internal biological oscillators. It is this process that underlies our ability to process acoustic information, allowing us to predict the dynamics of the music and to focus our attention on peaks in the oscillation, perceived as musical accents. This is how we perceive the regularities of rhythmic structure in music. From the basic beat, we construct new rhythms with frequencies that are multiples or sub-multiples of the beat. These self-generated oscillators contribute to the perception of metrical structure. Every time the music confirms our expectations, we increase the attention we dedicate to the main beat. Recent research has confirmed the existence of these attentional processes, suggesting the existence of many biological clocks, in different regions of the brain, all running at different frequencies. If this suggestion is correct, our sensation of the flow of time does not depend on any individual brain region but on mechanisms distributed across the cortex. In other words, there is no master clock; rather there are many different clocks, each with their own "time." It is these clocks that allow us to stretch or compress our sense of time, to meet our needs or to match external stimuli. The rhythms we observe are the fruit of the interaction between non-linear stochastic mechanisms and highly variable stimuli from the environment. Nonetheless many questions are still open. How do these physiological rhythms originate? How do they interact with each other and still manage to stay in sync with the external environment? In sum, rhythms exist at different levels of biological organization. They are endogenous. Given that each species has mechanisms of its own, they are probably inherited.

We can group biological rhythms into three main categories

a. Ubiquitous rhythms: endogenous rhythms associated with the circadian cycle, operating independently of external stimuli ;
b. Pacemakers, including master and slave rhythms;
c. Synchronizers (Zeitgeber), that exploit external stimuli detected via appropriate sensorial systems. Zeitgeber rhythms are genetically programmed and are regulated by exogenous synchronizers known as Zeitgeber (biological clocks or metronomes). Synchronizers act through the central nervous and endocrine systems, modulating the parameters for the endogenous rhythm, and contributing in this way to organisms' adaptation to their environment.

In this chapter, we will demonstrate the rhythm of life in artificial systems. To this end, we will consider rhythm as an emergent dynamic structure or a set of emergent dynamic elements that gives rise to a certain pattern. In the former case, we associate rhythm with a single agent; in the latter it depends on the coordinated action of multiple agents. In reality, both have a plausible role to play and may actually alternate during developmental and evolutionary processes. A rhythm maintains a process in a particular configuration, it gives it regularity. If this is so, rhythm is closely related to the "flow of things." In fact it is rhythm that regulates this flow, giving it an appropriate form. It is rhythm that maintains the dynamic relationships among the components of a system, even as it moves. It is an attempt to impose regularity on change. Rhythms are not constant - they can change over time. But what do we mean by change? Suppose we begin with a particular pattern generated by an emergent process. If we introduce changes in the process, the change will influence the rhythm and the structure of the pattern it produces (Bilotta & Pantano, 2006). Initially the change may be slow, producing only minor degradation in the pattern. If we are talking about musical patterns, it will take a trained ear to detect subtle degeneration in the melody and the rhythm. This kind of variation, on a small time-scale, does not usually lead to any substantial transformation of the piece. Gradually, however, we begin to see larger scale variations, changes in the pattern so profound, as to make it unrecognizable. Now the initial shape is lost and replaced by new, inter-related offspring shapes. Even when all these shapes are the product of the same evolutionary process, they can sometimes be so different from the original, that the only way to make the link is through an artificial genetic analysis. Variations can have many different effects on patterns. Sometimes they work from the inside outwards, sometimes from the outside inwards. The effects can spread in just one direction, or from the center towards the edge, or in spots. Some of the patterns are orderly, some complex or chaotic, some random. Sometimes the same basic dynamics can produce different combinations of different kinds of variation at different times.

MAIN FOCUS

In this chapter, we will examine the rhythms we see in the development and evolution of 1-D and 2-D CAs. In our treatment, we use two fundamental concepts. The first is the concept of the spatio-temporal pattern that a CA produces. The second is the idea that the development of this pattern has a rhythm. If we keep to our metaphor of self-replicators as artificial agents, we observe that the rhythm changes in line with the system's interactions with the external environment. We will begin by discussing these variations and the effects they produce on the structure of patterns. We will then examine the dynamics of these structures as they develop. This analysis will allow us to identify general models of CA growth that apply, not only to the development of individual CAs, but to their evolution over multiple generations.

Let us take an initial pattern and introduce variations, as shown in Figure 1. We consider this pattern as a system, which has many dimensions. In this simple example, we define all the possible positions which each circle assumes in the configuration, during the process of mutations, as dimensions of the system. In this case, the variations affect just one dimension of the system. However, we should be aware that we can potentially vary any dimension. It is multiple variations at different levels, sometimes acting synchronously, sometimes asynchronously - that determine the final shape of the system. The range of possible variations is enormous and impossible to predict (Figure 2).

What is the function of variation? From an evolutionary viewpoint, variation provides a basis for adaptation: in a dynamically changing environment it is the fittest variants that survive. We find the same process in the dynamics of the brain. Learning leads to apparently adaptive variations in the structure of the brain. By analogy, patterns too have to change (or be changed) in response to external forces. It appears, furthermore, that learning is one of the most important factors capable of producing variations in cognitive processes. It is this variation that allows us to handle novelty when we meet it, and to make the changes necessary to integrate our new and pre-existing knowledge. In this way, variation allows to handle enormous quantities of information. Maybe, variation in biological, physical and economic systems has the same function. Where it is present, it allows the system to manage new information. We can view variation, just like rhythm, as a machine that generates new schemes of organization. Some are meaningful, distinctive, and salient. Others are not. Sometimes variation acts linearly; sometimes its effects are complex

Figure 1. Variations affecting a single dimension of pattern, considered as a system with many dimensions

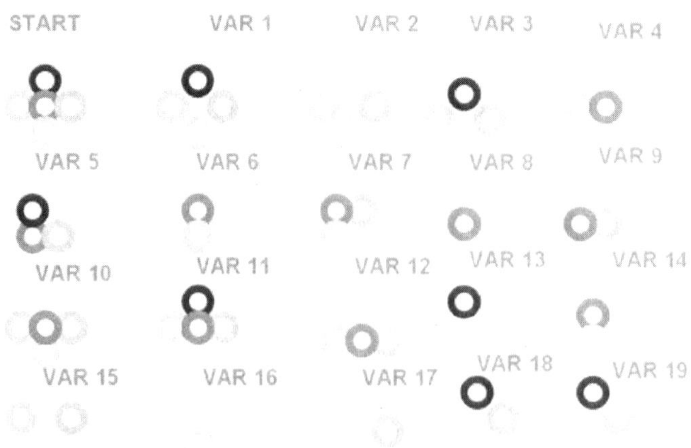

Figure 2. Dimensions can affect different levels of the system, and work in opposite directions

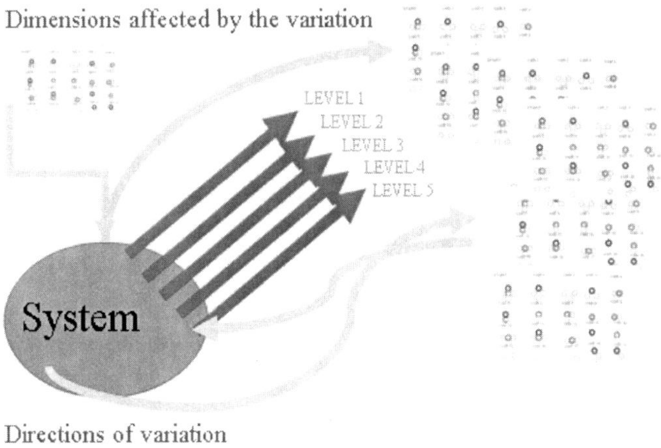

or chaotic. Chaos actually moulds variation. Does variation have a goal? Is there a method underlying the organizational schemes it produces or are they generated randomly? How much time does it take to produce them? Can we consider time and the rhythm of variation as in some way synonymous? Is variation an intrinsic feature of the phenomena we are observing, or does it only appear in the presence of an observer? How many levels of variation can we identify? One for each parameter controlling the system undergoing the variation? In this sense, rhythm and variation overlap. In other cases, variation can actually destroy rhythm, in a process of creation and destruction that simultaneous transforms the patterns present in the environment and our own ability to understand and perceive them. Genetic variations can give rise to new quantitative and qualitative traits. The former refer to characteristics that can be measured (e.g. the size of a structure) or counted (e.g. the number of elements it contains). The latter are associated with presence or absence of characteristics that are discontinuous in nature. The former are multi-factorial - the fruit of the action of multiple genes. The latter usually depend on variants of an individual gene. In other words, qualitative variation is generally "mono-factorial." It is mono-factorial traits that give rise to polymorphisms - discontinuous variations - in a population. Polymorphisms can affect the structure, physiology and behavior of a specific trait in an organism.

In the remaining part of this chapter, we will report a series of experiments, investigating different models of growth in terms of their periodic rhythms, and the complex, chaotic and noisy rhythms underlying the coordination between different components of an organisms and different organisms in a population. We will also present experiments on mono-factorial variations in k-totalistic self-replicators, showing how genetic changes lead to changes in the shape and behavior of the system. Here too we will examine how the variations we introduce affect growth rhythms. (Bilotta e Pantano, 2006).

Experiments

In what follows, we will examine the periodic rhythms produced by 1-D CAs and by 2-D self replicators and the complex, noisy, rhythms driving coordination among different components of an organism and different organisms in a population.

Periodic Rhythms

As we saw in Chapter 2, emergent structures in CAs are arranged in clearly defined spatial patterns, standing out against the background. 1-D CAs generate a pattern on a 1-dimensional grid, 2-D CAs on a 2-dimensional grid. Over time, the patterns change, giving rise to emergent dynamics, as previously described. In the case of gliders, we observe a return to the initial pattern - often shifted by a certain number of positions on the grid. This shift makes it possible to measure the speed of the glider, which varies periodically over an interval we can measure.

Consider rule C4.1 (already discussed in Chapter 4) - well-known as rule 20 for Boolean CAs. C2.1 refers to a 1-D CA with $D = 1, k = 2 and r = 2$. Figure 3 depicts some of the gliders produced by the system.

To detect these rhythms, we have built a computational tool that allows us to measure the number of structures of a given class - in this case gliders. In a second stage, we added a clock model. The model visualizes population sizes actually present in the sequence and incorporates them in a rhythmic structure. In this way, the number of marks on the clock reflects the rhythmic structure of the phenomenon under investigation.

To gain insights into this process, we can examine the behavior of the gliders described below.

1st Glider - 10010111

The glider oscillates between two states. If we count the number of cells in state 1 for each step in the development process, we obtain the sequence:

5 5 5 5 5 5 5 5 5 5 5 5 5 5 5 5 5 5 5 5 5 5 5 5 5 5 5 5 5 5 5 5 5 5 5 5 5 5 5 5 ...

The equivalent sequence for cells with state 0 is:

Figure 3. Row 1 (a, b, c, d, e) shows the gliders generated by rule 20, together with their associated Input Entropy Curves (Row 2, Figures 3f, g, h, i, j). Row 3 (Figures 3k, l, m, n, o) depicts the internal clock for each system. The functioning of these clocks is described in the main text.

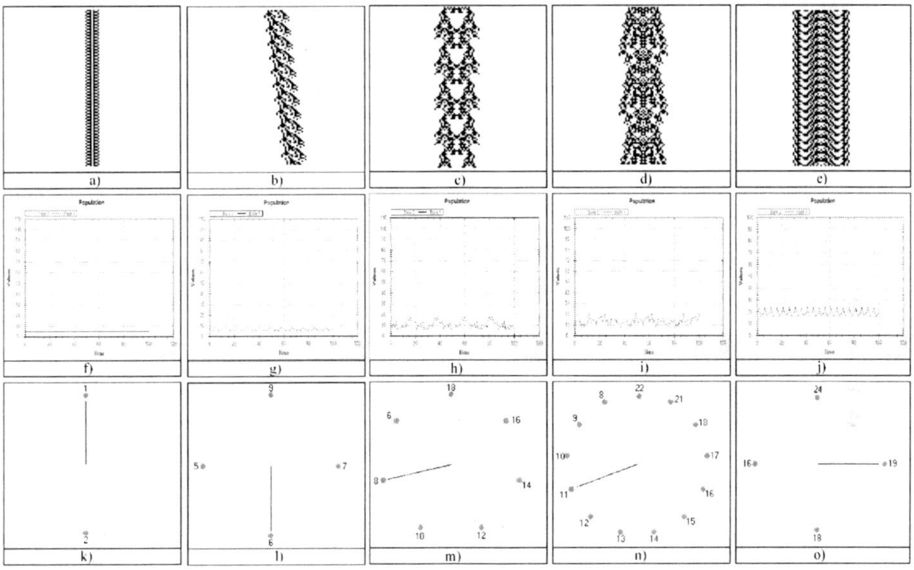

95 95 95 95 95 95 95 95 95 95 95 95 95 95 95 95 95 95 95 95 95 95 95 ...

The same number repeats on every step. We can thus represent the dynamic behavior of the glider with a clock, with a single hand fixed in one position (Figure 3k).

2nd Glider – 10111011

This glider has a more interesting behavior. In this case, we observe 9 different configurations, as we can see from the sequence below, showing the number of cells in state 1.

6 6 5 7 9 7 6 5 6 6 6 5 7 9 7 6 5 6 6 6 5 7 9 7 6 5 6 6 6 5 7 9 7 6 5 6 6 6 5 7 9 7 6 5 6 6 ....

Below, we see the sequence for cells in state 0.

91 93 94 95 94 94 95 93 91 93 94 95 94 94 94 95 93 91 93 94 95 94 94 ....

The glider's behavior is clearly periodic - it returns to its initial configuration once every 9 steps. To represent this behavior, we can use a clock face with 9 marks. On each step, the clock hand moves clockwise by 40°. After 9 steps it returns to its starting position.

If we analyze the sequence in greater detail, we discover it contains just 4 unique values:

5 6 7 9

The best way of representing it is thus to use a clock face with just 4 marks (Figure 3i). In this case, the hand no longer moves smoothly. Rather it jumps from one position to another, following a well-defined *rhythm*. It moves in characteristic intervals:

1:1 6:5 5:7 7:9 7:6

The inverse intervals are also present. In brief, the behavior of the system is characterized by:

a. A period
b. A minimum value for the number of cells in a given state
c. Rhythmic population dynamics
d. Characteristic intervals between population sizes in subsequent periods.

In what follows, we will see the crucial importance of these observations.

If we analyze the population dynamics, we observe symmetries in the dynamics of populations 1 and 0 in Figure 4. As the data for the two populations are different, the symmetries in population 0 are a mirror image of those in population 1.

Figure 4. Populations 1 (a) and 0 (b) for the glider generated by sequence 10111011

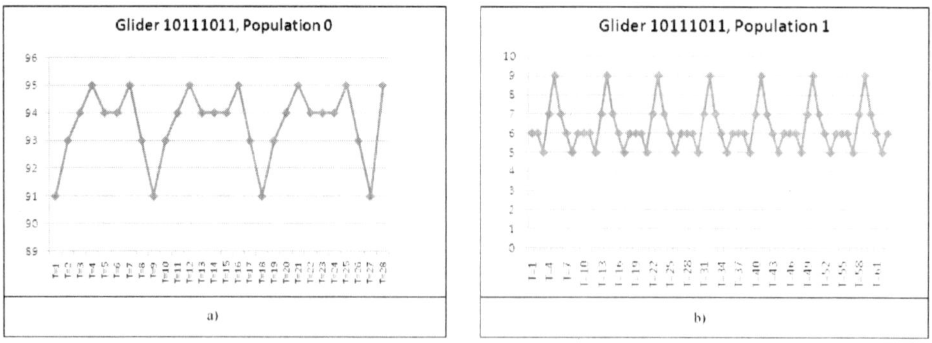

Consider population 1, with the mapping on the musical notes:

5→A, 6→B, 7→C, 9→D.

We thus have:

BBB ACDCA BBB ACDCA BBB.

This is a rhythmic sequence in which groups of three values alternate with groups of 5.

We now repeat this procedure with population 2, using the mapping

91→A, 93→B, 94→C, 95→D.

This yields the sequence:

ABCD CC BDA BCD CC BDA BCD CC.

If we exclude the first block, which represents a settling down period, we find a rhythm with alternating groups of 2 and 3 values.

3rd Glider – 11111000011111

This system has a period of 22. The sequence is:

10 8 12 8 14 6 6 8 6 10 8 10 8 16 18 14 18 14 12 12 14 6 10 8 12 8 14 6 6 8 6 10 8 10 8 16 18 14 18 14 12 12 14 6 10 8 12 8 ....

From this sequence, we can extract 7 values (Figure 3m):

6 8 10 12 14 16 18

Below we show the intervals detected in the sequence:

5:4, 2:3, 4:7, 7:3, 3:4, 1:2, 8:9, 9:7, 7:6, 7:3.

It is interesting to note that some of these intervals - the octave (1:2) and the fifth (2:3) - correspond to the musical intervals studied by Pythagoras. In this case too, the system displays rhythmic behavior between 7 values (Figure 3m) with a period of 22 steps.

4th Glider – 11110100100101111

This glider is very large and has a very long period (38 steps), as we can see from its sequence:

11 14 13 12 15 8 11 8 15 14 11 22 15 12 15 14 15 10 17 16 17 18 13 16 21 12 15 8 9 16 13 14 13 16 11 10 11 8 11 14 13 12 15 8 11 8 15 14 11 22 15 12 15 14 15 10 17 16 17 18 13 16 21 12 15 8 9 16 13 14 13 16.......

The sequence contains 13 distinct values

8 9 10 11 12 13 14 15 16 17 18 21 22

We have used these values to construct a strange clock face with 13 marks (Figure 3n). Pythagoras' classic intervals 1:2 3:2 are still present. However, the situation is now far more complex and we observe many other intervals as well.

5th Glider – 101001111010101001001010101111100101

This glider too is very large. However, its period (5) is very small, as we see in the sequence below:

19 24 19 16 18 19 24 19 16 18 19 24 19 16 18 19 24 19 16 18 19 24 19 16 18 19 ...

The sequence contains just 4 values.

16 18 19 24

The clock face has just 4 marks (Figure 3o).

None of the classical Pythagorean intervals are present.

Rule 20 can generate many of these systems. For instance, we can start with strings of a given length, all in the rest state 0, and observe if they degenerate into periodic structures. This method is extremely fruitful.

Rule 20 is not the only rule capable of generating multiple periodic structures. In reality, this ability is common to the majority of complex rules and is a fundamental characteristic of many CAs.

To conclude, a single organism, regulated by a local rule, can contain multiple internal structures, each with its own periodic behavior and characteristic rhythms. One way of visualizing these behaviors is to display them on a "clock face" with one mark for each value present in the sequence of population values. We also observe the emergence of what we have called Pythagorean intervals - special ratios between small population values. In other cases, the intervals involve larger numbers.

The structures we have just observed are very similar to biological organisms. In what follows we will see how they can enter into more complex symbiotic relationships.

Complex Rhythms Associated with Coordination among the Components of an Organism

If we re-examine the sequences described in the previous section, we observe that the overall rhythm is often the result of coordination among simpler rhythms tied together by connector sequences. Consider the sequence for the second glider:

6 6 5 7 9 7 6 5 6 6 6 5 7 9 7 6 5 6 6 6 5 7 9 7 6 5 6 6 6 5 7 9 7 6 5 6 6 6 5 7 9 7 6
5 6 6 ....

We observe the presence of a simple sequence (6 5) that appears twice in the main sequence. The first time it is connected to the next sequence with a (7 9 7), the next time with a 6 6.

However there is also an alternative interpretation.

6 6 5 7 9 7 6 5 6 6 6 5 7 9 7 6 5 6 6 6 5 7 9 7 6 5 6 6 6 5 7 9 7 6 5 6 6 ...

This kind of complex coordination is even more evident in the 3rd glider:

10 8 12 8 14 6 6 8 6 10 8 10 8 16 18 14 18 14 12 12 14 6 10 8 12 8 14 6 6 8 6 10
8 10 8 16 18 14 18 14 12 12 14 6 10 8 12 8....

In this case, the main sequence contains 3 repetitions of (10 8), one at the beginning of the sequence, and two in the central segment. (18 14). If we extend our search to intervals, we find that interval 1:1 also repeats twice.

Coordination among Multiple Organisms

In Chapter 9, we saw that emergent particles like gliders or regular domains often combine to form more complex organisms. Let us examine the implications in terms of emergent rhythm.

Rule C14.1 with k=3 and r=1 and initial data 100001 generates the glider shown in Figure.5a. The glider has a period of 48.

The sequence for the glider is given below:

16 18 16 10 14 10 8 12 16 8 16 2 8 8 6 6 2 4 2 0 4 0 8 4 6 2 6 0 6 0 4 0 6 0 16 10 10 6 8 10 6 8 4 8 14 22 20 20 16 18 16 10 14 10 8 12 16 8 16 2 8 8 6 6 2 4 2 0 4 0 8 4 6 2 6 0 6 0 4 0 6 0 16 10 10 6 8 10 6 8 4 8 14 22 20 20 16 18 16 10 14 10 8 12 16 8 16 2 8 8 6 6 2 4 2 0 4 0 8 4 6 2 6 0 6 0....

The clock face for this glider has 12 marks (Figure 5d):

0 2 4 6 8 10 12 14 16 18 20 22

The sequence for population 2 is:

4 2 2 6 0 8 0 4 0 8 0 10 0 14 0 12 0 14 0 6 2 8 6 4 6 2 8 6 6 0 2 0 4 2 2 6 2 4 6 6 0 12 0 6 4 4 2 0 4 2 2 6 0 8 0 4 0 8 0 10 0 14 0 12 0 14 0 6 2 8 6 4 6 2 8 6 6 0 2 0 4 2 2 6 2 4 6 6 0 12 0 6 4 4 2 0 4 2 2 6 0 8 0 4 0 8 0 10 0 14 0 12 0 14 0 6 2 8 6 4 6 2 8 6 6 0 2 0 4 2 2 6 2 4 6 6 0 12 0 6 4 4 2 0 4 2 2 6 0 8 0 4 0 8 0 10 0 14 0 12 0 14 0 6 2 8 6 4 6 2 ...

The clock face has 8 marks:

0 2 4 6 8 10 12 14

These are a subset of the values for population 1.

If we merge two of these gliders (see Figure 5b), beginning with the new sequence 10010000000000000000000000001001, we obtain a new glider again with a period of 48. This time the clock face has 13 instead of 12 marks (Figure 5e):

0 4 8 12 14 16 20 24 28 32 36 40 44

All except one corresponds to a value which is exactly double the value in the previous case. 14 (2*7) is a completely new element.

In some cases, gliders combine in ways that are even more complex, as when we begin with the generative code 10012200222222. Figure 5c shows the result. This structure too has a period of 48. The internal clock has 11 marks (Figure 5e).

2 3 4 5 6 7 8 11 12 13 17

Unlike the previous case, there is no clear emergent synchronism with the original rhythm.

Figure 5. A glider generated by rule C14.1, with initial data 100001 (Figure 5a). The clock face has 12 marks. Population dynamics for population 1 are shown on a clock face with 12 marks. The equivalent clock face for population 1 has 8 marks. The second glider with code 100001000000000000000000000100001 can be obtained from the first by repeating the initial data twice (Figure 5b). This process has already been explained in Chapter 9. This time the clock face has 13 marks. A third glider with initial data 10012200222222 (Figure 5c) has a clock face with 11 marks.

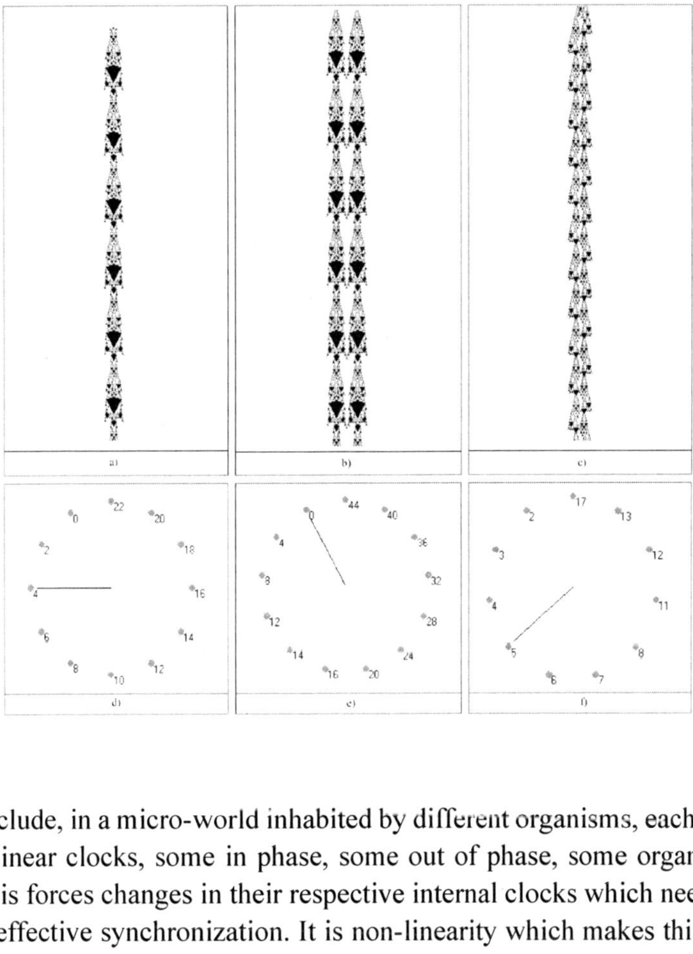

To conclude, in a micro-world inhabited by different organisms, each with their own non-linear clocks, some in phase, some out of phase, some organisms may "fuse." This forces changes in their respective internal clocks which need to adapt to ensure effective synchronization. It is non-linearity which makes this synchronization possible.

An examination of the images in Figure 6 confirms this picture.

Consider rule C2.8 with $D = 1, k = 4 \, and \, r = 2$ and initial conditions 1010001.

This rule gives rise to extremely complex structures with a period of 106 (see Figure 6a). The clock face has 9 marks:

0 1 2 3 4 5 6 7 8

Figure 6. Glider generated by rule C2.8 with initial state 1010001 (Figure 6a). Glider with initial state 101000100000000000000000000001310303 (Figure 6b). Glider generated by rule C14.2 with $D = 1, k = 4 \, and \, r = 2$, *and with initial data 2020210333333333333332.*

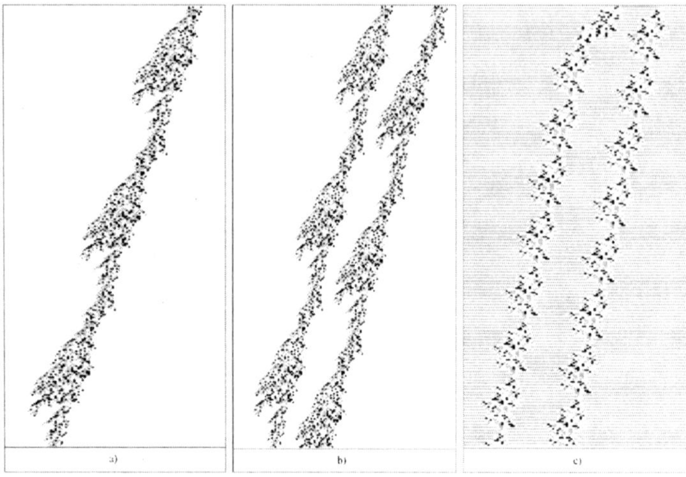

Population 3 also has a clock face with 9 marks.

0 1 2 3 4 5 6 7 10

Apart from the last, all the values are the same.

Now consider 2 out of phase gliders, with generative code 101000100000000 000000000000000001310303. We now obtain the image shown in Figure.6b. The clock face for population 1 has 11 marks:

1 2 3 4 5 6 7 8 9 10 11

The clock for population 3 now has13 marks:

0 1 2 3 4 5 6 7 8 9 10 11 13

In other words, the fusion alters the internal clocks allowing effective synchronization between the two structures. This implies a significant change in their relationship.

The situation with rule C14.2 ($D = 1, k = 4 \, and \, r = 2$) is even more complex. This rule has a periodic domain that oscillates between state 0 and state 3. If, by chance, two of these domains are present at the same time, the frontier region between the two is occupied by a new glider of great complexity. It is this glider which allows the two to co-exist. As we see in Figure 6c, the presence of another glider allows the presence of structures inhabiting a 0-3 periodic domain. This time the generative code is 20202103333333333333332. The two gliders begin in state 0 and periodically swap to state 3. Apparently they are out of phase. By contrast, the behavior of the regular domain displays its own periodic behavior.

Growth Rhythms and the Development of Organisms

Let us now turn to growth processes and see if we can find rhythm there. We will begin with the sequence generated by rule 8 introduced in Chapter 12. As we saw, this rule produces a sequence similar to the Sierpinski Triangle generated by ECA90:

1 2 2 4 2 4 4 8 2 4 4 8 4 8 82 4 4 8 4 8 8 16 4 8 8 16 8 16 16 32 2 4 4 8 4 8 8 16 4 8 8 16 16

This sequence contains values 1, 2, 4, 8, 16, 32….. As we see below, these are all powers of 2.

$$1 = 2^0 \;\; 2 = 2^1 \;\; 4 - 2^2 \;\; 8 = 2^3 \;\; 16 = 2^4 \;\; 32 = 2^5$$

This is not an unexpected result. The Sierpinski Triangle is a self-similar fractal structure expressed by a power law (Schroeder, 1991). As in the sequences generated by our 1-D CAs, this sequence too has a rhythm. But now what we are interested in is the way the system grows - and the effects of self-similarity. On the one hand, we find repetitive behavior, on the other growth. Our modeling should reflect both sides of the story.

In the previous sections, we saw that we can represent periodic rhythms on a clock face with an hour hand and a certain number of marks. If we want to represent growth we have to add a minute hand, showing the value of the power we are considering. Figure 7 illustrates this arrangement.

With square-shaped development process generated by additive rule 1, described in Chapter 13, we obtain the following sequence.

$$\left[1441641616644161664166464256416166416646425616646425664256256\ldots\right] \quad (14.1)$$

Figure 7. A schematic representation of time and growth rhythms for the self-replicator described above

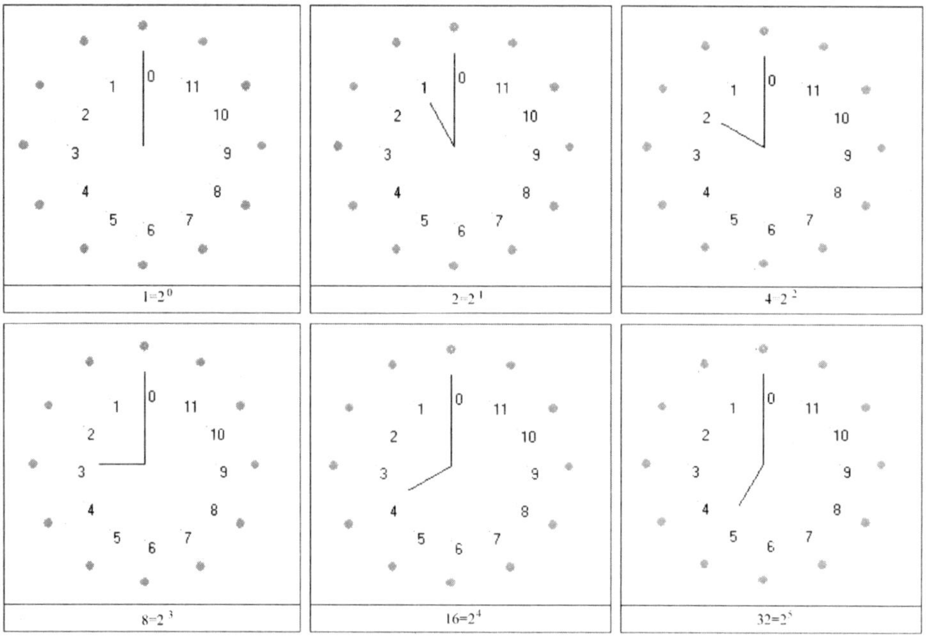

The values in this sequence are:

$$1 = 4^0,\ 4 = 4^1,\ 16 = 4^2,\ 1 = 64^3,\ 256 = 4^4 \ .....$$

In other words we still have a power law. This time, however, the base is 4.

If we consider rule C2.5, discussed in Chapter 11, the situation becomes more complicated. This time, the sequence is:

$$\left[1943643616144436161441614464576436161441614464576161446457664576256....\right] \quad (14.2)$$

We can use this sequence to create two new sequences, the first containing values in an even position in the sequence, the second values that are in an odd position. The first sequence is:

1 4 4 16 4 16 16 64 4 16 16 64 16 64 64 256.

The second is:

9 36 36 144 36 144 144 576 36 144 144 576 144 576 576.

The first is identical to the Sierpinski sequence for the self-replicator with the square-shaped pattern of development generated by additive rule 1. If we divide the values of the second sequence by 9, we obtain the same values as those in the first:

1 4 4 16 4 16 164 16 16 64 16 64 64 64

In other words, we have another Sierpinski sequence. This should not surprise us. In Chapter 11, we have already seen that the module for this rule has a temporal dimension of 2, and dynamics identical to those of the Sierpinski Arrowhead.

Now consider the logarithmic sequence for the main series:

0 1.58 1 2.58 1 2.58 2 3.58 1 2.58 2 3.58 2 3.58 3 4.58 1 2.58 2 3.58 2 3.58 3 4.58 2 3.58 3 4.58 3 4.58 ....

What we see here is a primary scale that grows in a sequence 1 2 3 … and a secondary scale consisting of just two values: 0 and 0.58. Returning to our clock metaphor, we can represent the main sequence with the hour hand and the secondary scale with the minute hand, which oscillates between just two marks on the clock face (see Figure 8).

In other words, we can represent the growth process on a primary scale characterized by ratios of 1:2 and a secondary, periodic scale with ratios of 4:9 and 5:9. To be strictly rigorous, we should position the minute hand a little more to the left on the circle (see Figure 8).

At this point, we have all the tools we need to describe power law-based growth processes.

Consider self-replicator $A = (2332)$ generated by a k-totalistic rule C14.3 with k=4 and r=1. Figure 9a shows the self-replicator at the beginning of the self-replication process and after 30 steps.

If we consider population 2, at the various steps in the growth process, we obtain:

$$\left[26412412824412824824164841282482416488241648164832\ldots\right] \quad (14.3)$$

If we divided these values by 2 we obtain a new sequence:

1 3 2 6 2 6 4 12 2 6 4 12 4 12 8 24 2 6 4 12 4 12 8 24 4 12 8 24 8 24 16…

Figure 8. A schematic representation of the time and growth rhythm for the self-replicating system described by sequence (14.2)

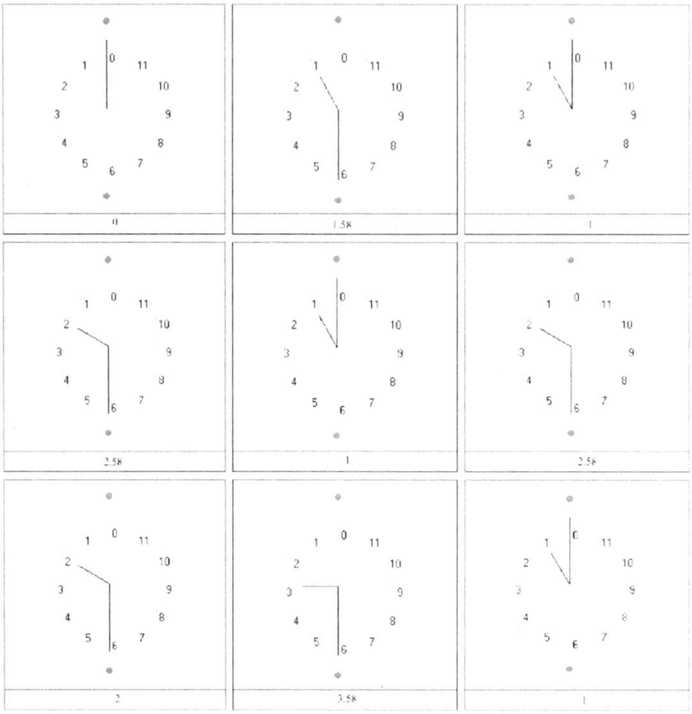

The working of the base 2 power law is clearly visible. As in the previous case, we can extract two sequences, one corresponding to values with an even and the other to values with an odd position in the sequence. As in the previous cases, the sequences correspond perfectly to the Sierpinski model.

Calculating logarithms, we obtain:

0 1,58 1 2,58 1 2,58 2 3,58 1 2,58 2 3,58 2 3,58 3 4,58 1 2,58 2 3,58 2 3,58 3 4,58 2 3,58 3 4,58 3 4,58

This is the same result we saw in the previous series and the clock for these dynamics is identical to the clock already shown in Figure 8. The only differences are that now we have a ratio of 2:3, the same as in Pythagoras' perfect fifth and the base of the logarithm is now 2 instead of 4. Finally we have to remember that:

$$\frac{2}{3} = \sqrt{\frac{4}{9}}.$$

Figure 9. A schematic representation of time and growth rhythms for self-replicator C14.3. The group of Figures labeled with a) shows the evolution of the self-replicator. The Figures labeled with b) show different growth clocks.

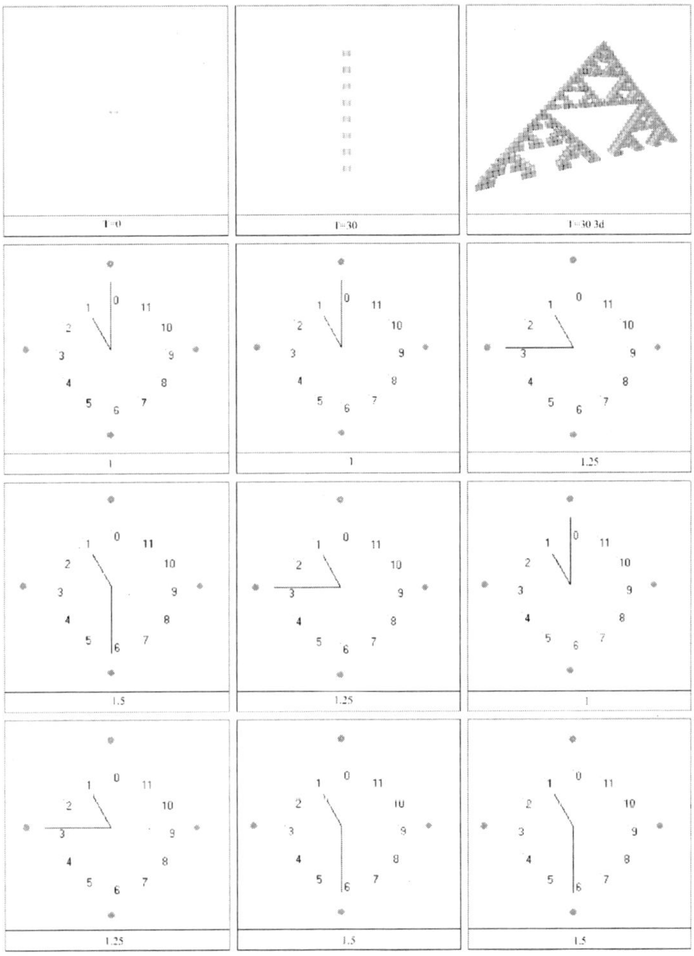

To see how these rhythms are expressed in a different situation, consider the universal constructor C10.6. This rule refers to a k-totalistic CA with $D = 2, k = 8 \, and \, r = 1$. The time series for population 1 is:

4 0 8 0 16 0 16 0 16 0 32 0 64 0 32 0 16 0 32 0 64 0 64 0 64 0 128 0 256 0 64 0 16 0 32 0 64 0 64 0 64 0 128 0 256 0 128 0 64 0 128 0 256 0 256 0 256 0 512….. (14.4)

Here we detect a power series with base 16, 7 distinct values and a period of 16. This is the time scale for the module of this particular self-replicator. The behavior of the system is based on the Sierpinski Arrowhead.

If we take the 7 elements in the series, and extract those in an odd position we obtain:

16 16 32 64 32 16 32 64 64 64 128 256 64 16 32 64 64 64 128 256 128 64 128 256 256 256 512…

Taking base 16 logarithms, we obtain:

1 1 1.25 1.5 1.25 1 1.25 1.5 1.5 1.5 1.75 2 1.5 1 1.25 1.5 1.5 1.5 1.75 2 1.75 1.5 1.75 2 2 2 2.25….

As we can see, this sequence yields yet another rhythm. This time the "perfect fourth" 1:4 (or 3:4) emerges spontaneously.

Variations

In Chapter 11, we saw that self-replicators only use a limited number of elementary rules during their development. When we vary these rules, we obtain mutations, which may, on occasion, give rise to new self-replicators. In what follows, we will examine the effect on the rhythm of the system.

Consider rule C14.4 referring to a k-totalistic rule with $D = 2, k = 5 and r = $.

With $A = \left(3\right)$, this rule behaves like a Sierpinski Arrowhead (see Figure 10a).

The sequence for population 3 is:

$$\left[944361643616161446443616161446416144646457625643616161446416\ldots\right] \quad (14.5)$$

The sequence clearly displays a power law with base 4 and a period of 3. As usual, we consider the logarithms:

1.58 1 1 2.58 2 1 2.58 2 2 3.58 3 1 2.58 2 2 3.58 3 2 3.58 3 3 4.58 4 1 2.58 2 2 3.58…

Considering the intervals in the sequence we find not only an octave (1:2), but also the ratio 4:9, equivalent to a fifth in base 2.

Figure 10. Self-replicator C14.4: a) variant R10_0 (equivalent to C11.2); b) variant R702_1; c) variant R711_3; d)

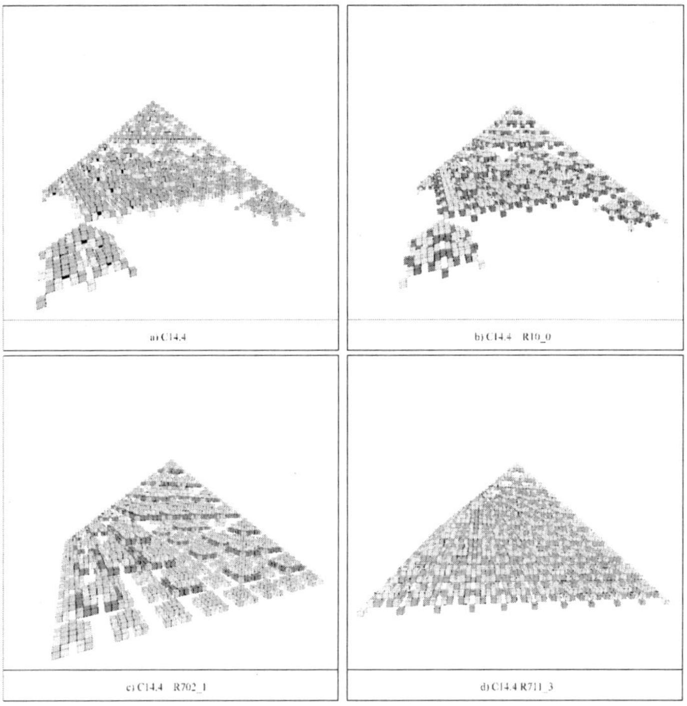

This system has 15 elementary rules, meaning we can perform 75 mutations (Bilotta & Pantano, 2006). Mutation R10_0 produces a new self-replicator with code C11.2. This system has already been discussed in Chapter 11. Even though the system now uses 13 rules (as we saw in Chapter 11) the sequence for population 3 is still identical to the sequence in the previous case we examined. The reason is that the two patterns are nearly identical and differences between populations 0 and 1 are very small.

However, if we consider mutation R702_1, the situation changes. Although the system continues to behave as a Sierpinski Arrowhead, it now uses 24 rules. The new sequence for population 3 is:

$$\left[94443616164361616161446464436161616144646416144646464576256\ldots\right] \tag{14.6}$$

In many ways, the sequence is similar to the previous one. However, we now observe the presence of new elements and the base sequence now contains 4 instead of 3 elements. This is very visible in the 3-dimensional pattern shown in Figure 10c.

Considering the logarithms we obtain:

1.58 1 1 1 2.58 2 2 1 2.58 2 2 2 3.58 3 3 1 2.58 2 2 2 3.58 3 3 2 3.58 3 3 3 ....

As we could have expected, the intervals in the sequence remain the same but the rhythm changes.

Mutation R711_3 produces a more complex situation (Figure 10d). This system uses 13 rules, producing the sequence shown below:

9 4 4 36 16 4 81 36 4 144 64 4 225 100 4 324 144 4 441 196 4 576 256 4 729 324 4 900 400 4...

Here no power law is visible.

Finally consider R702_2. This mutation creates a self-replicator which uses more rules (34) than any other system we have studied. Once again the pattern is reminiscent of a Sierpinski arrowhead. The sequence for population 3 is:

9 4 4 8 64 12 12 8 64 24 24 32 256 68 28 8 64 24 24 32 256 96 56 32 256 96 96 128 1024 324... (14.7)

If we start from the fourth value in the series and divide all the values by 2, we find a base 4 power law. Side by side with the octave and the fifth we find new ratios such as 3:8.

To sum up, mutations sometimes lead to changes in the rhythm of development, sometimes they leave it unchanged.

From the Rhythm of Development to Musical Scales

Symmetry is important in all its shapes and forms. It also plays a key role in determining the tones we include in our musical scales.

In the history of music and musical instruments, what are the criteria which have guided the choice of tones and, more importantly, the intervals between them? The study of musical instruments from the ancient past suggests that notes were placed at intervals dividing the scale into equal intervals. However, this method did not last. It did not take long to realize that what matters from a musical point of view is not that distances should be equal but that they should produce consonance. Two notes are said to be consonant if they are pleasant to the ear and restful when played together. Dissonant notes, by contrast, produce a sense of instability and tension. Mechanisms for the perception of sound are the same for all human beings from any part of the world. All human beings, anywhere in the world can perceive the consonance produced by the octave, the interval between two notes where one note has double the frequency of the other.

The octave is not the only universal consonant. If we are looking for a central note in the middle of the interval our choice nearly always falls on the note a fifth above the lower note in the octave and a fourth below the higher one. According to Helmholtz's theory, two notes are perceived as consonant if they share many of their harmonics. The interval with the greatest number of harmonics in common is the octave followed by the fifth. This physical observation provides an objective foundation for our immediate perception. We can go back to Pythagoras' research on vibrations in strings. If we take a string that produces a certain tone, all we need to do to obtain a tone that is one octave higher, is press on the midpoint of the string. Let A be the length of the string that produces the first tone, and B the length of the segment that produces the second.

Thus A: B = 2:1.

To raise the tone by a fifth, we have to press two thirds of the way along the string. If C is the length of the segment that produces the new tone, we obtain A: C = 3:2. Finally the intervals between the tones produced by C and B, is a fifth. This corresponds to a ratio of 4:3. We thus have three main consonant intervals corresponding to the ratios 2:1; 3:2 and 4:3.

If we begin with an initial frequency ν, we use the recursive formula:

$$\nu_{i+1} = \frac{3}{2}\nu_i$$

to generate new frequencies. If we set $\nu_0 = 1$ and use a base 2 logarithmic scale, we obtain a quasi linear progression with the frequency values almost perfectly positioned along a straight line. If we begin with a base frequency of 32.703 Hz. we obtain a series corresponding to the frequencies in the Pythagorean scale. Figure 11a, depicts the values on a logarithmic scale. The values 1, 2, 3, to the right of the horizontal lines show the various octaves.

Let us consider the various intervals (the decimal part of the values). From

$$I_i = log_2(\nu_i) - [log_2(\nu_i)]_{whole\ part}$$

Figure 11. Frequencies on a logarithmic scale. Values 1, 2, etc. to the right of the horizontal lines show different octaves. In Figure 11b, the decimal part of the values is divided into 12 intervals. Figure 11c shows the logarithmic sequence for the self-replicator described earlier. The sequence consists of a single note, C. Figure.11d shows the intervals obtained from sequence m2, which contains 2 notes.

we obtain 12 intervals (Figure 11b). As the great musicians who preceded Bach discovered, the intervals making up the octave are not perfectly divided into 12 parts. However, if we divide the interval in twelve using the formula $\sqrt[12]{2}$ we obtain Bach's classical well-tempered scale, which finally resolved the ancient problem of transposing from one key to another.

If we begin with C1 (which has a frequency of about 32.7032Hz) and multiply by $\sqrt[12]{2}$, we obtain the next note in the scale. Continuing in this way, we can obtain the frequencies (including sharps and flats) for all the notes in every octave.

In what follows, we will see how we can use this approach to produce natural scales for self-replicators and their corresponding sequences. But first, let us return to the system we were considering earlier: a self-replicator based on the Sierpinski Arrowhead (sequence (14.1)). Figure 11c shows the base 4 logarithmic sequence for the system. We can consider it as a melody made up of a sequence of 'C's played in different octaves. By contrast, sequence (14.2) (Figure 11d) contains 2 intervals (Figure 11e). The base 2 logarithms for sequence (14.3), give the musical sequence shown in Figure 11f. Here again, we have two intervals: an octave and a fifth. The result is the melody shown below.

C1 G2 C2 G3 C1 G3 C2 G3 C2 G3 C3 G4 C1 G2

Sequence (14.4) is more complex. Now all the intervals are contained in just two octaves (Figure 11g). As well as the octave, we have four other intervals (Figure 11h).

FUTURE TRENDS

The approach described in this chapter shows that the development of self-replicators has a rhythm that we can express in melodies using various intervals in an emergent musical scale. Variations in the pattern produce variations both in the rhythm and the melody. Interpreting self-replicators in this way provides us with a new way of investigating the complex dynamics of our micro-worlds. As we will see in the next chapter, we can use these rhythms to produce melodies that are even more complex than those presented here, indeed we can even bring together different populations to produce harmony.

In this setting, it is important to note the fundamental role of power laws, a role that brings to mind classical theories of unlimited population growth. These theories suggest new ways of describing the development of self-replicators and self-replicator colonies. This is an issue that deserves further research.

CONCLUSION

Everything that in nature - including our own human activity - is movement; everything is subject to the laws of rhythm. Rhythm can be defined as order in movement, organized dynamics. Rhythm implies periodic repetition of simple or complex structures in a coordinated sequence. Repetition of rhythmic sequences plays a fundamental role in human perception, and produces motor induction (Parncutt, 1987; 1994; Kelso, 2008; Treffner & Turvey, 1993). It is rhythmic regularities that allow us to move in time with the signal we perceive. If we consider 1-D and 2-D CAs, we can consider rhythm in the same way.

The rules for considering a CA sequence as rhythmic are the same as those we apply to natural phenomena.

a. It must be organized in a sequence (each element has to have a clear position in the overall structure).
b. It has to be structured: each element in the sequenced has to be in a well-defined relationship with the other elements.
c. It has to be periodic (structures have to repeat in time)

Rhythm

a. Orders the structure of self-replicators and gliders, driving the creation of patterns and three dimensional structures;
b. Provides insights into the structural relationships between the parts making up the system, helping to clarify their geometry
c. Shows the presence of periodical patterns in self-replicator growth dynamics.

In cellular automata as in nature, there is always a quantitative side to rhythm - the number of phenomena with a given duration in time. But there is also a qualitative aspect - rhythm allows us to perceive phenomena, not as chaos, but as organization and structure.

We can sum up our basic concepts of rhythm as follows:

a. Rhythm is present in all dynamics aspects of biological organisms and artificial systems;
b. Rhythm lies at the base of organisms' behavioral dynamics over short periods of time - their development. However, we can also find it on other, longer time scales.
c. Any rhythm we detect in 1-D or 2-D CAs is structurally significant - in the sense that it is rhythm that underlies the organization of identifiable structures.
d. Variations can degrade rhythm, generating chaotic rhythms or situations in which there is no rhythm at all (random sounds or noise).

The life-like rhythms we have discovered in 1-D and 2-D cellular automata demonstrate that their elementary components are very similar to those underlying human cognition and that they perform very similar computations. Here we should emphasize two important aspects of our work, one concerning measurement, the other process. In our research, we have discovered that to produce rhythm there is one basic criterion that has to be met. We must have a way of measuring the gap between events, in time and space. Once this basic modular structure exists, the only way it can structure the patterns it produces is to replicate and repeat itself on the basis of precise mathematical relationships. When we study the long term development of a CA, it is as if we are bringing the pattern into being, finally allowing it to prevail over its components. Pattern becomes process, the organized development of elements, tied together in ways that depend on the measurement itself. In this way, the process is the result of an operation, which is in some way organized, at least at a superficial level. The emerging internal law we found in the Pythagorean intervals lies at its very essence. What we see is the revelation of an enigma.

REFERENCES

Ashoff, J. (1978). Circadian rhythms within and outside their rangers of entrainment. In I. Assenmacher, & D.S. Farner (Eds.), Environmental Endocrinology (172-181). Proceedings, Berlin, Heidelberg, New York: Springer Verlag.

Bilotta, E., & Pantano, P. (2006). Structural and Functional Growth in Self-Reproducers. *Complexity, 11*(6), 12–29. doi:10.1002/cplx.20131

Bilotta, E., Pantano, P., & Stranges, F. (2007). A Gallery of Chua Attractors. Part I. *International Journal of Bifurcation and Chaos in Applied Sciences and Engineering, 17*(1), 1–60. doi:10.1142/S0218127407017161

Daan, S., & Pittendrigh, C. S. (1976). A functional analysis of circadian pacemakers in nocturnal rodents. V- Pacemaker structure: a clock for all seasons. *Journal of Comparative Physiology, 106*, 333–355.

Drake, C., Jones, M. R., & Baruch, C. (2000). The development of rhythmic attending in auditory sequences: attunement, referent period, focal attending. *Cognition, 77*, 251–288. doi:10.1016/S0010-0277(00)00106-2

Graham-Brown, T. (1911). The intrinsic factors in the act of progression in the mammal. *Presented at Royal Society of London, Series B, 84*(572), 308–319. doi:10.1098/rspb.1911.0077

Halberg, F., Carandente, F., Cornélissen, G., & Katinas, G. S. (1977). Glossary of chronobiology. *Chronobiologia, 4*(Supplement 1), 1–189.

Ibuka, N., Inouye, S. T., & Kawamura, H. (1977). Analysis of sleep-wakefulness rhythms in male rats after suprachiasmatic nucleus lesions and ocular enucleation. *Brain Research, 122*(1), 33–47. doi:10.1016/0006-8993(77)90660-6

Jones, M. R., & Boltz, M. (1989). Dynamic attending and responses to time. *Psychological Review, 96*(3), 459–491. doi:10.1037/0033-295X.96.3.459

Kelso, J. A. S. (2008). Synergies: Atoms of brain and behavior . In Sternad, D. (Ed.), *A multidisciplinary approach to motor control*. Heidelberg: Springer.

Khalifa, Y., & Foster, R. (2006). A Two-Stage Autonomous Evolutionary Music Composer. In Rothlauf, F. (Eds.), *EvoWorkshops*. Berlin, Heidelberg: Springer-Verlag.

Large, E. W., & Palmer, C. (2002). Perceiving temporal regularity in music. *Cognitive Science, 26*(1), 1–37. doi:10.1207/s15516709cog2601_1

Meijer, J. H., & Rietveld, W. J. (1989). Neurophysiology of the suprachiasmatic circadian pacemaker in rodents. *Physiological Reviews, 69*(3), 671–707.

Moore, R. Y. (1973). Retinohypothalamic projection in mammals: a comparative study. *Brain Research, 49*, 403–409. doi:10.1016/0006-8993(73)90431-9

Parncutt, R. (1987). The perception of pulse in musical rhythm . In Gabrielsson, A. (Ed.), *Action and Perception in Rhythm and Music* (pp. 127–138). Stockholm: Royal Swedish Academy of Music.

Parncutt, R. (1994). A perceptual model of pulse salience and metrical accent in musical rhythms. *Music Perception, 11*, 409–464.

Rietveld, W. J., & Groos, G. A. (1980). The central neural regulation of circadian rhythms . In Scheving, L., & Halberg, F. (Eds.), *Principles and application of chronobiology to shifts in schedules with emphasis on man. Alphen an den Rijn: Sijthoff.*

Rusak, B., & Groos, G. (1982). Suprachiasmatic stimulation phase shifts rodent circadian rhythms. *Science, 215*, 1407–1409. doi:10.1126/science.7063851

Rusak, B., & Zucker, I. (1979). Neural regulation of circadian rhythms. *Physiological Reviews, 59*(3), 449–526.

Schawartz, W. J., Davidsen, L. C., & Smith, C. B. (1980). In vivo metabolic activity of a putative circadian oscillator, the rat suprachiasmatic nucleus. *The Journal of Comparative Neurology, 189*(1), 157–167. doi:10.1002/cne.901890109

Schroeder, M. (1991). *Fractals, Chaos, Power Laws*. New York: W. H. Freeman and Company.

Treffner, P. J., & Turvey, M. T. (1993). Resonance constraints on rhythmic movement. *Journal of Experimental Psychology. Human Perception and Performance, 19*, 1221–1237. doi:10.1037/0096-1523.19.6.1221

APPENDIX

Rule Codes

C14.1: $D = 1, k = 3, r = 1$

0111111101111110102110102020

C14.2: $D = 1, k = 4, r = 2$

0000032002310021000001002203100000000033000000001000023130001020
002013300000310000010200120000001010010020000003200230201000310000
030002303330303031300000003020100200301001300030002000200000203000
002031001320031020302000310000122020100020103000321013000000300200
00300000020010201101112030100000222001110301020003000230000130300
100031001023002030001000200030003010003310003001300001310000000020
003201120303003000200200012023300323001030100121102002200000302000
21200000330012301000000003020300100231000120010000000020001023000111
002030110310003000203000002200212110101201100023011032003030001121
302322003032320000001000301210031330100010000033000110302010301020
020332100030000033200030230001100320021102113110203011033011200000
0030300100100000210033030030300030202000000000000000000003300300203
032100000000000000300000202010001033001003001000213313302010011000
012010000022200020000212100033002200000230000020200313002021000010
031020003013202003102011001210300220002020302021002210200000031000
1000100022000012000110003000213033003

C14. 3: $D = 2, k = 4, r = 1$

00000000000030000200201010020000100030000000100000000022222020030
00000020000000000000000000000000003001000000100000000000000000000
2020030210023001200212300000000000021000003000003100200000002000000
0033003003002000201102000

C14.4:: $D = 2, k = 5, r = 1$

01000103010000000000000220000030000000000200011330001000000000400
00040000000000000000000400041223000000000000000000000330003100000002
401012002002200000000000100000000030003010202000200000001000212043000
00000000300133340004000000203000000001000000000000000000003401011000
4000000000000000000040000000000000100030000000000010000000110203003000
0000002100003000000300000040000000002000001000000000034000400001000

003000000100000020000000000100001200001000000100040200000300000000
000034400000300000000000000100100000323010000000000000340000200001
300000000020000000040044001020002332020024001030301004004000042000
002000010004020400100003000000000000002201000010000000030000000004
0002000000030000040400000000000000000000003001000000004400001003020

Chapter 15
From Rhythm to Sound and Music

"Music is the arithmetic of the soul, which counts without being aware of it" (Leibniz)

INTRODUCTION

In early 1950s, Iannis Xenakis became the first composer to use stochastic processes to generate pieces of music, working by hand. The first entirely computer-generated composition was *Illiac Suite for String Quartett*, realized by Lejaren Hiller in 1956 (Hiller, 1970, 1981). In Hiller's approach, all kinds of musical processes were coded and implemented by computer. In the '80s, David Cope created a computer-aided system allowing anyone to create new compositions in the styles of past composers. The system worked even for users with no skills in programming or composition. Cope thought that computers could be a good tool for studying musical style, encoding it into programs to increase musical creativity. So he began to study the essential elements that define musical style, creating an explicit definition he could encode in a computer program. What he was searching for were "patterns.." To investigate them he analyzed samples of a composer's works, puts then on a database, and

DOI: 10.4018/978-1-61520-787-9.ch015

used computational functions to analyze the composer's style. Xenakis, Hiller and Cope are just three examples of the tight relationship between technology, scientific models and music (Cope, 1991; Schwanauer and Levitt, 1993; Roads, 1996). It was their work that laid the foundations for *algorithmic music composition,* a style in which algorithmic encoding and decoding schemes are used to represent scientific theories in computer generated music. Theories that have been treated in this way include fractals (Mandelbrot, 1975), dynamical systems (Kramer, 1994; Witten, 1996; Bilotta et al., 2005; Cupellini et al., 2008), and Cellular Automata (Bilotta & Pantano 2001; Bilotta & Pantano 2002a; Miranda 2001). Since this initial experimentation, a growing number of composers have shown interest in using this approach in their own music (Bolognesi, 1983; Dodge, 1988; Leach and Fitch, 1995; Manaris et al. 2003) or in generating sounds (Voss and Clarke, 1975; 1978). In this way, computers have turned into a virtual environment in which researchers can recreate and experiment with biological processes. It is inside computers that we can produce "*Artificial life.*" And it is here that we can simulate natural and biological phenomena, studying them in ways that would never be possible in the natural world. In this spirit, the authors have used computers to develop theoretical hypotheses, concerning the construction of musical languages, the sound and musical quality of the resulting music, and its acceptability to musicians and non-musicians (Bilotta et al., 2000a; 2000b). Computers have also enormously broadened the scope of musical creativity. One particularly interesting area of activity is the creation of new instruments (Bilotta et al., 2002b, 2002c). Prior to the introduction of the computer, any musical instrument had to be *physically reproducible*, a challenging requirement. Today, by contrast, we can use digital synthesizers to create a wide range of instruments covering a broad and diverse spectrum of sound. One interesting area of work is chaos and complexity theory, which can be used to produce sounds with very special characteristics. This work is also very interesting for the development of music technology (Bilotta, et al., 2008; NIME 2008).

In previous chapters, we have described the use of mathematical models to simulate the key traits of biological processes. In this chapter, we show how we can map these models to musical structures. In other words, we return to Pythagoras' idea of a tight relationship between music and mathematics. It was Pythagoras who first discovered the relationships allowing different notes to merge into harmony pleasant to the human ear (Frova, 1999). In his view, it was these numerical relationships that govern the universe. All that remains for human beings is to learn to read the music.

As we discovered in our investigation of CA rules, self-replicators display fractal-like behavior. Using statistical methods, we have used two and three dimensional shapes with different fractal dimensions to produce different kinds of sound and music. We presume that self-replication and self-similarity are consistent characteristics not only of natural and artificial systems but also of music. In this chapter, therefore, we map the self-replication and self-self similarity we find in our CAs into music, investigate whether we can use this method to generate creative musical compositions, and attempt to interpret the complexity we find at different levels of organization.

BACKGROUND

Pythagoras' Concept of Number

The first philosopher to interpret nature in terms of numbers and measurements was Pythagoras, in the VI century B.C.E. Pythagoras was influenced by the arithmetical discoveries of the Babylonians, some 1500 years earlier and by Egyptian geometry which was even more ancient. Born in Samos, he emigrated to southern Italy, where he attracted a group of disciples, who dedicated their lives to the mysteries of philosophy and mathematics. The group, with its mystical, esoteric rites, was absolutely closed to the non-initiated so it was more than a century after the death of their master that some of them began to break the oaths of secrecy to which they had been bound Only then did some of Pythagoras' oral teachings found their way into written texts. Fragments of these texts, with commentary by Aristotle and other philosophers, allow us to reconstruct his teachings. First, however, we should ask why these teachings are so important and what they have to do with music.

Pythagoras was interested in the properties of the integers and discovered several theorems of great importance for arithmetic and number theory. The method he used was to represent sequences of numbers as geometric shapes in the plane, in which each unit was represented by a stone in a specific position. This allowed him to arrange the integers symmetrically as rows of stones, one above the other. The result is a triangle, with a single stone at the vertex. In the second row, we have two stones, representing the number 2. The third row has 3 stones representing the number 3, and so on (Figure 1). We observe a surprising similarity to some of the spatio-temporal patterns produced by cellular automata

Figure 1. Triangular diagrams

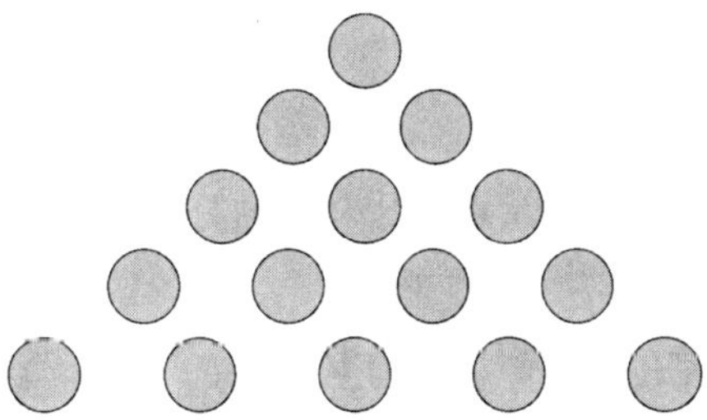

If we sum the stones in the rows, we obtain the triangular numbers 1, 3, 6, 10, 15, etc. Each triangular number is the sum of all the integers from 1 up to the position of the number in the series. For example, 6, the third number, is the sum of 1, 2 and 3. We discover that numbers and number series are intrinsic to physical bodies, just as the stones in Pythagoras triangle are an intrinsic part of the triangle. Units - or individual stones - are bodies in their own right. However they can also be arranged in lines, or used to create shapes in the plane. We can then use these shapes to describe bodies in space. In other words, we pass from individual units to lines to 2-dimensional shapes (such as triangles and squares) to pyramids, cubes and other solid bodies. Given that the Greeks did not have the concept of zero, units had a double function. On the one hand, they represented a 1-dimensional building block; on the other they acted as a dimensionless point of contact between two sections. For Pythagoras, these perfect bodies were nothing more than an idealization of physical models actually existing in the cosmos. It follows that numbers and their properties are intrinsic to all physical objects and that number is the basic principle underlying everything that exists. Pythagoras translated these concepts into a sort of religious faith that was to have a profound influence on Plato. Aristotle summarized the main elements of this belief as follows: (Aristotle, *Metaphysica*, 986a)

"Evidently, then, these thinkers also consider that number is the principle both as matter for things and as forming both their modifications and their permanent states, and hold that the elements of number are the even and the odd, and that of these the latter is limited, and the former unlimited; and that the One proceeds from both of these (for it is both even and odd), and number from the One; and that the whole heaven, as has been said, is numbers." (Ross, 2007).

From Pythagoras Concept of Number to Modern Scientific Thought

For Pythagoras, the universe is made up of matter and whatever the essence of this matter, it can always be represented in terms of geometry or arithmetic, that is by numbers. Any substance we can perceive with our senses belongs to a numerical series. Whatever the state of a physical system, however it changes, it can always be represented by numbers or a mathematical formula. It is this idea of Pythagoras' that underlies our modern mathematization of physical phenomena. And it is this idea that underlies our modern mathematics-based science.

A key element in Pythagoras' philosophy is the role he assigned to odd and even numbers. To understand it, we have to analyze the way we can arrange physical points into rows. It is here that we can find the origins of the concept of number.

Each unit is represented by a particle. Each particle is separated from other particles by a gap. Regardless of whether the number consists of a single unit or multiple units, it is this gap that lies at the heart of the concept of number. The followers of Pythagoras postulated the existence of empty space. This was necessary to isolate the number (or the physical body representing the number) from other numbers and bodies. Here we find the core of Greek atomic theory, according to which the cosmos is a set of solid physical bodies separated by empty space. Empty space isolates individual numbers from every other number. According to Pythagoras, numbers are an expression of two fundamental physical phenomena: extension and form. The matter that fills space with its extension is unlimited - and is represented by even numbers. The main property of even numbers is that they can always be divided into two equal parts. If we cut one of the rows in Pythagoras' triangle in half, we touch a single limit point. So long as the row contains an even number of points, the dividing line passes through the space between two points. On the contrary, if the row contains an odd number of points, the dividing line passes through one of the points. This creates a limit. As the originally unlimited line becomes a section with a fixed length, so it creates a shape. At a higher level, the line becomes a limit for a surface and the surface becomes a limit for a solid. Given that the set of alternating even and odd numbers contains all the integers, Pythagoras sees it as the origin of all that is infinite and all that is finite, of matter with its infinite extension and of form that limits this extension.

Harmony

Another key concept in Pythagoras' philosophy is harmony. Harmony is a fundamental property of numbers that enables them to bring together the opposite qualities of substance and form. According to Philolaus, a follower of Pythagoras, (Philolaus, *fragment* D44B6), *the substance making up the cosmos has different components, to know which we would need the abilities of the gods rather than those of mortal men.* What holds these different components together is the ordering process represented by harmony. If all the components of the universe were the same, there would be no need for harmony. It is only harmony that unites the cosmos.

Nothing would be understandable, neither objects in themselves nor relations among objects, without number and its substance. In the soul, it is number that harmonizes all these things with our perception, allowing us to know them and their relations (Philolaus, fragment D44B4).

In this fragment, Philolaus points to the relationship between number and the world of the senses. It is a relationship he discovers in Pythagoras' world and which could have laid the foundations for the mathematical interpretation of the physical world. It might even have helped to provide an explanation for our bodily senses and perceptions - in the sense we use these terms in modern theories of neuro-physiology and neuropsychology. However, Plato was to direct this line of thought in another direction, abandoning the opinion that the harmony of the universe is revealed through the senses, and putting forward an opposite view. For Plato, the senses are an obstacle to knowledge and the only way to understand the universe is through pure mathematics. And yet in the Timaeus he makes an effort to defend Pythagoras' teaching that it is the harmony of numbers that gives form to matter. When he discovers the Platonic solids, he applies Pythagoras' generating principle, showing how we can construct any solid belonging to a certain dimension from elements belonging to the next lower dimension. Pythagoras reveals how the harmony of numbers translates into mathematical relations that reveal themselves as ratios or as geometric relationships (for a modern discussion of these relations in music, art, sound and thought, see Hofstadter, 1999). Putting to one side the later development of Pythagoras' arithmetic and geometry (Geymonat, 1971), the important point for our discussion is that the Pythagoras' theory of proportions derived from his discoveries about musical harmony. It was the first ever application of mathematics to a basic physical phenomenon. Pythagoras' disciples inferred the laws of musical harmony from experiments - a major novelty in its own right, given the weak propensity of the Greeks to systemic experimentation. According to Pythagoras' disciples, their master took strings of different lengths and thicknesses, stretched them with screws, and studied the sound they made. In other experiments he studied the vibrations of columns of air of different lengths in wind instruments and in identically shaped recipients filled with different quantities of water. Some of his experiments were purely qualitative; others, however, produced genuine measurements. Pythagoras observed that the shortest string always produces the highest note and the longest the lowest. He went on to discover the three perfect chords: the octave (corresponding to a length ratio of 1:2), the fifth (2:3) and the fourth (3:4). His followers also had clear ideas on the relation of pitch and the frequency at which the string vibrated. A fragment from Theon of Smyrna, (Theone, Smyrn., 61, 11, D47A19a) tells us that: " Eudoxus and Archytas and their followers thought that the ratio of concords was in numbers, agreeing themselves too that the ratios are in movements and that the swift movement is high pitched, since it strikes the air continuously and pricks the air more swiftly, but the slow is low-pitched since it is more sluggish" (quoted in Archytas of Tarentum, Pythagorean, philosopher and Mathematician King, Carl A. Huffman, Cambridge University Press., 2005). This fragment tells us that the followers of Pythagoras associated the highest number with the highest pitch, and the shortest string, according to the law that frequency is inversely proportional to the length. These discoveries had a profound effect on mathematics. As shown earlier, the perfect chords are associated with ratios of 1: 2, 2: 3, and 3: 4. These ratios involve the numbers 1, 2, 3, and 4 which together form the Pythagorean quartet (or tetrachord): the first four rows of dots in Pythagoras' triangle. The sum of the numbers in the quartet is 10. By composing the numbers in the quartet, we obtain three proportions of fundamental importance for mathematics. The numbers 1,2,3 are in arithmetic proportion in the sense that the difference between the first and second term in the series is equal to the difference between the second and the third. 1, 2, 4 are in geometric proportion: the ratio of the second to the first is the same as the ratio of the third to the second. Finally, harmonic proportion involves the numbers 3, 4, 6. which form the basis for the octave (3: 6), the fifth (4: 6) and the fourth (3: 4). Three numbers are said to be in harmonic proportion when the ratio of the difference between the second and the first to the first, is equal to the ratio of the difference between the third and the second to the third. A fragment from Archytas provides a rigorous definition of these proportions, explicitly linking them to music and to the laws governing the universe. This is the origin of the "music of the spheres" - an idea that was later to inspire Kepler's conjectures on the relation between the orbits of the planets and the platonic solids.

The idea of a harmony of the spheres exerted a profound influence on the Greeks' conception of the cosmos. They had a deep faith in the essential unity between geometry, music and astronomy. They believed, furthermore, that it was this unity that made human beings a part of the cosmos. According to Pythagoras followers, there is a deep tie between the laws of musical harmony, based on simple arithmetic and geometric proportions, and human psychology and physiology. It was this sense of euphoria that justified Leibnitz's claim that "music is the hidden exercise of arithmetic by the soul, which counts without being aware of it."

We are still far from a true theoretical understanding or explanation of these observations. Helmholtz asked: what is the secret harmony of the octave which allows us to guess the identity of a note when we repeat it at a higher pitch? Why is this particularly perfect harmony based on a ratio of 1: 2 - the simplest of all possible ratios? If the visible spectrum covered more than an octave could we produce a similar sensation with light? The Greeks never raised these problems. And yet they were intuitively aware of the link between human consciousness and the external world, and of the role of human consciousness in organizing external reality. For them, biological metaphors - such as the "breathing of the cosmos" - were more than mere allegories. What they were trying to say was that the cosmos has its own rhythm, that at a fundamental level, the laws of the cosmos are the same as the laws governing life, and thus that in some way, the universe is conscious of the harmony of the spheres. The idea of the world as a living organism was present in every period of Greek science.

To summarize: the first mathematical-physical law the Greeks discovered was a law describing qualities with the same dimension: the ratio between the lengths of strings or musical frequencies. Three hundred years later they formulated a second law - Archimedes' law of the lever. Here too, the law was formulated in terms of pure arithmetical relationships, in this case between lengths and weights. By contrast, modern physics is based on dimensional magnitudes such as velocity, force, pressure, etc. which are derived from fundamental dimensions such as length, time and mass. The meaning of these dimensions does not coincide with that of the pure numbers. Nonetheless they are immensely valuable for understanding nature. It was Aristotle who first understood the importance of velocity and Archimedes who defined specific weight - the relationship between the weight and volume of a body. Nonetheless, these dimensional physical magnitudes provide the best possible illustration of Pythagoras' idea that physical attributes are based on number. Seen from the viewpoint of modern science, the scientific method used by Pythagoras and his school was entirely correct - so correct in fact that some of its implications have only become apparent in modern times. The followers of Pythagoras departed from experiments with various kinds of apparatus, expressed the results in general terms, and used them to propose universal mathematical laws. One wonders why they did not go further, why they limited themselves to astronomy and music, instead of applying their methods to the whole of nature and the rest of science.

Plato borrowed heavily from Pythagoras' work on the importance of number, arithmetic and measurement in the interpretation of nature. And yet he was also one of the first authors to put a damper on Pythagoras' approach to science. For Plato, truth is to be found exclusively in the world of "ideas." If number is a property of physical bodies, in the "world of shadows", the study of number is devoid of value. In Plato's system, there is a complete separation between "the world of ideas" and our own physical world; the former is associated with number as such; the latter with numbers as used by Pythagoras and his school. For Plato, there is no connection between the latter and the absolute precision of pure number. The numbers used by Pythagoras and his school are impure and physical measurements senseless

There can be little doubt that the popularity of these ideas delayed the merging of mathematics and the experimental method. These same ideas explain the tendency of the Greeks to consider induction as superfluous and to grossly over-estimate the power of deduction

Synchrony and Rhythm

It is not just through Pythagoras that harmony became part of modern science. It is well-known how non-linearity can give rise to synchrony. It is the non-linear interaction between multiple oscillators that causes clocks in the same room to strike in rhythm (Pikovsky et al., 2001, Strogatz, 2000). Thanks to the frequency of this kind of phenomenon, synchronization has become one of the key concepts of the non-linear sciences. One of the most important challenges for technologies trying to control chaos is exploiting synchronization among chaotic oscillators (Murali et al; 1995). Rhythm and synchrony plays an important role, even on the nano-scale where we observe the growth of crystals and the self-assembly of nano-particles (Robert et al., 2004).

Obviously synchrony is important not just in physical phenomena but also in biology, where we observe effects such as the simultaneous flashing of fireflies which transforms certain kinds of African tree into spectacular Christmas light shows (Strogatz, 2003; Stewart, 2007). There are many ways of describing biological synchrony in mathematical terms (Pikovsky et al., 2001; Winfree, 2006). From Circadian rhythms to the seasonal and monthly cycles of human life, every manifestation of life displays some kind of rhythm. And of course, rhythm is not limited to the time scales typical of biological organisms. Non-linear synchrony can also be found in geology and evolution, or in the metabolism and reproduction of biological cells.

MAIN FOCUS

In previous chapters, we have seen that many self-replicators display a fractal structure and that their growth has a characteristic rhythm, observing well-specified laws. In this last chapter, we show how we can use this rhythm to generate music resembling a "canon" in which the same themes repeat with different tempi and in different keys. The result provides a highly effective representation of the fractal nature of the self-reproducing systems.

Canons E Fugues

For the sake of simplicity, let us consider the self-replicator with $A = (111)$ and k-totalistic rule C8.1. This structure develops immediately into a second structure $A^{'} = (222)$ that replicates with a rhythm already described in Chapter 8. In Chapter 10, we saw that it follows a linear pattern of development, governed by the block logic for ECA 150. This development can be measured in terms of the number of copies of the self-replicator generated by the CA in different phases of its development. Figure 2 plots this data against time.

As is apparent in the graph the sequence below (obtained for population 1, taken every 2 steps, starting from the third step), the general pattern of replication is rhythmic. The number of replications changes on each cycle, with longer term changes

Figure 2. a) number of replications of self-replicator $A = (111)$ *at different stages in the development process. b) number of replications of self-replicator* $A' = (222)$.

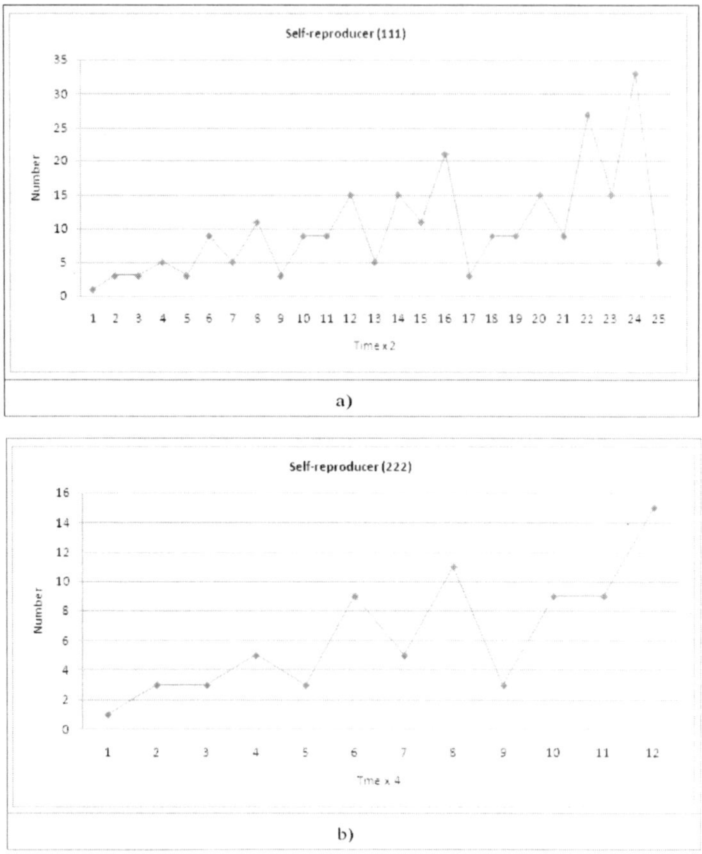

on a number of different time scales. The sequence below refers to population 1, sampled once every two steps, beginning with step 3.

9 99 27 15 33 **9** 27 27 45 15 45 33 63 **9** 27 27 45 27 81 45 99 **15** 15

This sequence contains three building blocks, each containing 7 elements that we will call **A**, **B** and **C**.

A= 9 9 15 9 27 15 33
B= 27 27 45 15 45 33 63
C= 27 27 45 27 81 45 99

A, **B** and **C** are all built from smaller sub-blocks. **A** contains 2. The first (**A1**), contains 3 elements,

A1= 9 9 15

The second (**A2)** has 4

A2= 9 27 15 33

B can be obtained by composing **B1** and **B2**:

B1=3x**A1**
B2= (9, **A2**+2***A1**)

C can be obtained directly from **A**

C=3***A**

Now consider 2 further blocks, **S** and **S**

S= 45 45 75 33 99 63 129 **9**27 45 27 81 45 99 **27** 27

Every time we reach the last element in one of the 7-element sequences, we return to **A**.

S1=5***A1**
S2, works differently:
S2= 33 99 63 129

S is obtained directly from A and is similar to **C**: **S**=3***A.**

We can say that this self-replicator consists of 3 basic themes and a variation.

As we mentioned earlier, self-replicator A gives rise to a second self-replicator A' **'.**Figure 2b shows the number of replications for each replication cycle:

5 9 11 915 21 9 15 27 33 15 25 33 43 9 15 27 33 27 45 45 63 15 15

This second self-replicator is made up of 3 blocks: **D, E, F.**

D= 5 9 11 9 15 15 21
E = 15 27 33 15 25 33 43

F= 15 27 33 27 45 45 63

As in the case of the previous self-replicator (11), each block is separated by a number, dividing it from the next block .

We can divide **D** into 2 sub-blocks **D1** and **D2.**

D1= 5 9 11
D2= 9 15 15 21

To obtain **D2**, we apply the rule:

D2= (9, 6+**A1**); **E1** as:**E1**=3***D1**; and **F**=3***D**,

E2 varies and it remains unchanged.

15 25 33 43

Self-replicator (222) consists of two themes and a variation (a break in the symmetry). If we also consider the presence of A1, generated by self-replicator (1 1 1), we could say that it has 3 themes. Taking the sequence of numbers for the development of the system and translating it into music, we obtain a result similar to a canon or a fugue.

According to the literature, a canon is a precisely defined musical form in which one voice (the *dux* or *leading voice*) is imitated by another voice (the *comes,* or *following voice*). Literally speaking, a canon defines the rule governing this process. Canons come in many kinds. In the *strict canon* the leading voice is given; the following voice repeats the line of the *dux* a number of beats later. Often the melody is pitched at a fixed interval (for example an octave, a fifth or a fourth) above or below the leading voice. In a *unison canon*, the following voice perfectly imitates the melody of the *dux.* In a *canon at the octave*, the theme is raised or lowered by an octave. At the end of the canon, the voices can end one after the other or together. In a *circular canon,* the voices, instead of finishing, start again from the beginning, giving the idea that the canon can be continued *ad libitum (canon pertetuus).* In the *twisted canon,* the last note of the dux is a tone higher than the first, causing the voices to rise by an octave on each repetition. In the *enigmatic canon,* there are no fixed entry points or intervals between the voices. The *mixed canon* consists of a strict canon with free variations.

The elements that characterize a canon are the number of voices, the number of beats separating the entries, the intervals between the voices, and the "motion" of the *comes* with respect to the *dux*. In a *direct canon*, the melody of the *comes* moves in the same direction as that of the *dux*. In an *inverted canon*, the *comes* advances in contrary motion.

Some of these canons can be produced from the numerical sequences we discussed earlier.

Let us define the convention that:

$$5 \leftrightarrow D \ , \ 9 \leftrightarrow C \quad , \ 11 \leftrightarrow F \ , \tag{15.1}$$

and change the notes within the chords intervals whenever we find multiples of the basic numbers. In this way, we obtain:

$$18 \leftrightarrow E, \quad 27 \leftrightarrow G, \quad 15 \leftrightarrow F, \quad 33 \leftrightarrow A, \tag{15.2}$$

This is the unison canon, shown in Figure 3a. Figure 3a shows a "canon at the octave" obtained from the same numerical sequence. Figure 3b show an *inverted canon*, from the same source.

We can use the same kind of scheme to produce a fugue. These are only examples of what it is possible to achieve. In reality, self-replicator behavior can be used to derive a huge range of modular interlocking themes and rules of composition that we can bring together in novel compositions. For instance, we can use the same technique we have described here to generate music from other self-replicators, such as those described in Chapter 14. In this case too, the music is a representation of self-replication and colony formation.

Coding Fractal Sounds

It has been suggested that it may be possible to exploit self-similarity in the structures of self-replicators and symmetries in their development to produce fractal sounds (Voss and Clarke, 1975; 1978). To achieve this goal, we can exploit the dynamics of the self-replicator, as displayed during the individual steps in the development process. As we saw, these partial dynamics create a graph. If we associate each point on the graph with a sine wave, and apply the inverse Fourier transform, we obtain fractal sounds such as those shown in Figure 4. To code this sound, we create a script in an audio programming language, like *cSound* (Roads, 1996), which can process the output from the transform. We then feed the resulting code to a synthesizer.

Let us examine this process in greater detail. As we saw in Chapter 2, k-totalistic CAs are governed by rules that we can represent in a string of length N_T. Consider a space of sine functions, with frequencies between 1 and N_T:

Figure 3. a) scheme for a unison canon showing the entry of the following voce an octave higher than the original theme. b) unison canon, c) scheme for an inverted canon, d) the inverted canon with three different time signatures.

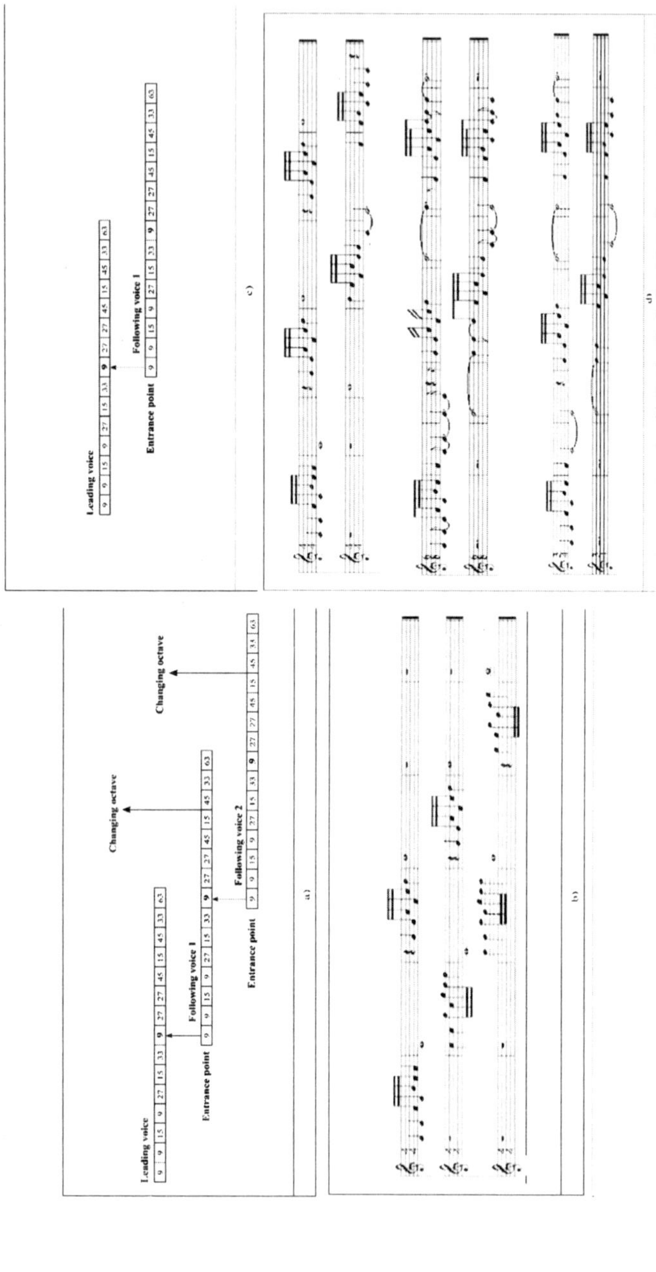

Figure 4. Sound visualization from a self-replicator for C2.2

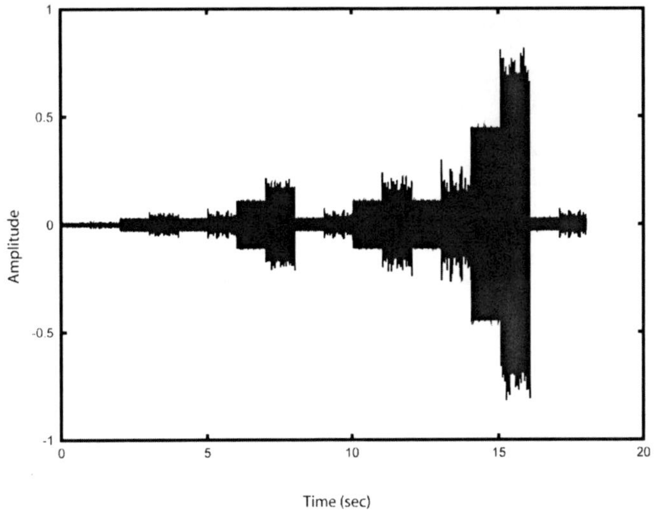

$$A_{\upsilon} Sin 2\pi \upsilon t \text{, with } 1 \leq \upsilon \leq N_T \qquad (15.3)$$

where A_{υ} is a parameter of undetermined value and t is time. Now consider a self-replicator that uses a consultation network, ρ comprising γ nodes, Rs_i, where s_i indicates the position of the ith node in l and $1 \leq i \leq \gamma$. Suppose that T is the period of the self-replicator. We thus have T dynamic networks $RD_1, RD_2, ..., RD_T$, all of length $\leq \gamma$. To simplify, we can include the complete set of nodes. We thus obtain T strings RDE_j all of length γ. If the node is used at time j, the value of the corresponding character in the string is set to 1; otherwise we set it to 0.

Using strings RDE_j we can construct T matrices SRD_j, each composed of γ columns, where we replace the 1s with a function in the form defined by (18.3). In this function A_{υ}=1. $\nu = s_i$ indicates the position of the ith character of the consultation network for the self-replicator:

$$SRD_j = (.., Sin 2\pi s_{i-1} t, Sin 2\pi s_i t,, Sin 2\pi s_{i+1} t, ...), i = 1, 2..\gamma \qquad (15.4)$$

Now consider column vectors VRD_j where each column represents the frequency with which the network uses node γ at time τ. We can now construct the sum of the sine functions as a function of time.

$$f_\tau(t) = RDE_{\tau \bmod T} * VRD_\tau \tag{15.5}$$

Given that each self-replicator is cyclical with a period of T, ($RDE_\tau = RDE_{\tau \bmod T}$), * is the usual row by column matrix product

Now consider C2.2: a k-totalistic CA with $D = 2, k = 4 and r = 1$. In Chapter 10, we saw that this CA generates a self-replicator composed of a single white cell that transforms into a 3 * 3 square of white cells. At the end of the replication process, it undergoes a further transformation into a shape with four white cells at the corners of a 5 by 5 square.

				0000000
		00000		0100010
000		01110		0000000
010	→	01110	→	0000000
000		01110		0000000
		00000		0100010
				0000000

As we saw in Chapter 12, the development of the self-replicator is similar to that of the Sierpinski Arrowhead. The pattern it produces is clearly fractal in nature. The consultation network for the self-replicator has 7 nodes:

R55, R164, R200, R210, R216, R2191, R220.

Thus:

$$s_1 = 55, s_2 = 164, s_3 = 200, s_4 = 210, s_5 = 216, s_6 = 219, s_7 = 220$$

Figure 5 shows the associated dynamic networks, which have a period of 2. The strings for the dynamic networks are:

Figure 5. Consultation networks for $A = (1)$

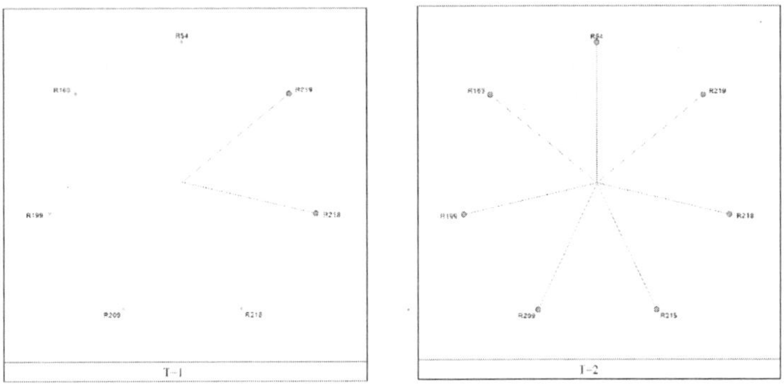

$$RDE_1 = 0\ 0\ 0\ 0\ 0\ 1\ 1$$
$$RDE_2 = 1\ 1\ 1\ 1\ 1\ 1\ 1$$

We thus obtain the row matrices:

$$SRD_1 = (0, 0, 0, 0, 0, Sin2\pi 219t, Sin2\pi 220t)$$

$$SRD_2 = (Sin2\pi 55t, Sin2\pi 164t, Sin2\pi 200t, Sin2\pi 210t, Sin2\pi 216t, Sin2\pi 219t, Sin2\pi 220t)$$

The column vectors below represent the frequency with which specific nodes are used:

$$VRD_1 = \left(0\ 0\ 0\ 0\ 0\ 9\ \mathrm{h}\right)^T$$

$$VRD_2 = \left(1\ 4\ 4\ 4\ 8\ 4\ \mathrm{h}\right)^T$$

$$VRD_3 = \left(0\ 0\ 0\ 0\ 0\ 36\ \mathrm{h}\right)^T$$

$$VRD_4 = \left(4\ 20\ 25\ 8\ 20\ 4\ \mathrm{h}\right)^T$$

$$VRD_5 = \left(0\ 0\ 0\ 0\ 0\ 36\ \mathrm{h}\right)^T$$

$$VRD_6 = \left(4\ 16\ 16\ 16\ 32\ 16\ \mathrm{h}\right)^T$$

where B^T indicates the transpose of matrix B.

The last term shows transitions to the rest state, normalized to h. The value of this term depends on the number of cells in the CA. Omitting it, we obtain a set of 1*6 row matrices and a second set of 6*1 column vectors. Applying (15.5), we construct functions $f_T(t)$. The result is a sequence of signals, each corresponding to a specific dynamic network, as shown in Figure 4.

As we saw in Chapter 12, the column vectors representing frequency of use of nodes present symmetries that we can describe in terms of relation (11.6). Let SK be the fundamental sequence of the population for a Sierpinski Triangle. Thus:

$$SK = \{1, 4, 4, 16, 4, 16, 16, 64, ...\}$$

Applying (12.6) we obtain:

$$\begin{aligned} VRD_{4n+1} &= \alpha(n+1)VRD_1 \\ VRD_{4n+2} &= \alpha(n+1)VRD_2 \\ VRD_{4n+3} &= \alpha(n+1)VRD_3 \\ VRD_{4n+4} &= \alpha(n+1)VRD_4 \end{aligned} \qquad (15.6)$$

where $n = 0, 1, 2......$ and $\alpha(i) \in SK$.

We can thus obtain relations similar to (15.4) for the $f(t)$ defined by (15.5)

$$\begin{aligned} f_{4n+1}(t) &= \alpha(n+1)f_1(t) \\ f_{4n+2}(t) &= \alpha(n+1)f_2(t) \\ f_{4n+3}(t) &= \alpha(n+1)f_3(t) \\ f_{4n+4}(t) &= \alpha(n+1)f_4(t) \end{aligned} \quad n = 0, 1, 2......\text{and} \qquad (15.7)$$

SK has a fractal structure. As a result, the symmetries of (15.6) and (15.7) are also fractal. As we saw in Chapter 11, the Sierpinski Triangle is an attractor for the fractal. It is important to note that 4 time steps are double the base period of 2. This is because SK intercepts the module of the self-replicator which itself has a period of 2.

Changing Fractal Sounds

Chapter 11 examined what happens to the self-replicator when we mutate string C2.2. Most mutations are neutral, in the sense that they have no effect on behavior.

However, mutations in the proximity of nodes in the consultation network produce several interesting effects. For instance:

a. The self-replicator changes behavior, while maintaining the same shape and self-replication capability. This is what we observe with mutations R55_1, R55_2 and R210_1
b. The self-replicator loses its self-replication capability and shifts towards ordered behavior
c. The self-replicator loses its self-replication capability and shifts towards chaotic behavior

For our purposes, the most interesting of these behaviors is behavior (a).

R551

After this mutation the consultation network has 9 nodes:

R55, R100, R164, R185, R200, R210, R216, R219, R220.

Figure 6 shows the dynamic networks for the self-replicator, which now has a period of 4.

The column vectors for frequency of use of specific nodes create a matrix X, as shown below.

$$X^T = \begin{pmatrix} 0\ 0\ 0\ 0\ 0\ 0\ 0\ 0\ 9 \\ 1\ 0\ 4\ 0\ 4\ 4\ 8\ 4 \\ 0\ 0\ 0\ 0\ 0\ 0\ 4\ 37 \\ 0\ 4\ 24\ 5\ 16\ 8\ 20\ 4 \\ 0\ 0\ 0\ 0\ 0\ 0\ 0\ 36 \\ 4\ 0\ 16\ 0\ 16\ 16\ 32\ 16 \\ 0\ 0\ 0\ 0\ 0\ 0\ 16\ 148 \\ 0\ 16\ 104\ 20\ 85\ 16\ 44\ 4 \\ 0\ 0\ 0\ 0\ 0\ 0\ 0\ 36 \\ 4\ 0\ 16\ 0\ 16\ 16\ 32\ 16 \\ 0\ 0\ 0\ 0\ 0\ 0\ 16\ 148 \\ \ldots\ldots\ldots\ldots\ldots\ldots \end{pmatrix} \tag{15.8}$$

Figure 6. Dynamic networks for mutation R55_1

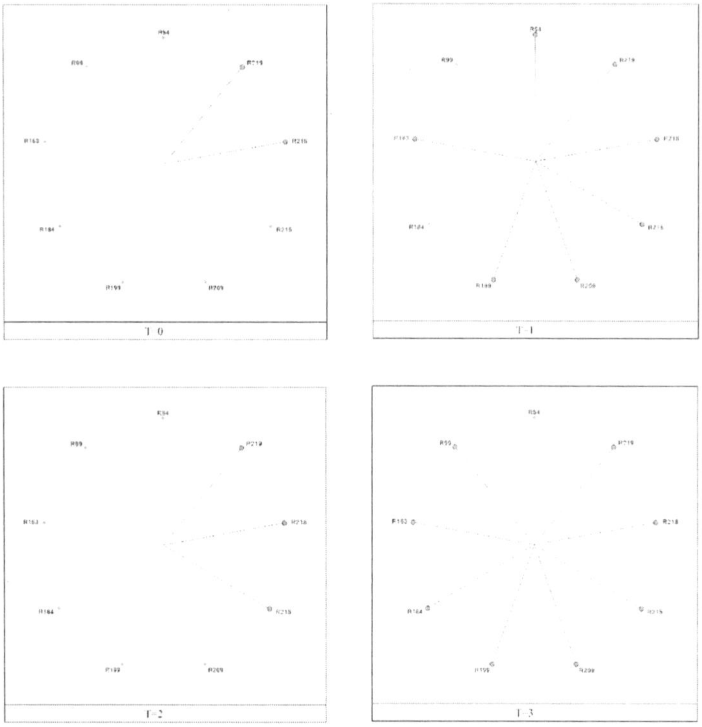

(15.4) and (15.5) continue to apply. Despite modifications, the fractal nature of the wave form is conserved (see Figure 7a) and the results described in the previous section continue to apply.

R55_2

The consultation network for this mutation has 11 nodes:

R55, R100, R164, R185, R200, R210, R215,R216, R218, R219, R220.

The self-replicator has a period of 4, as we can see from the dynamic networks shown in Figure 8.

The column vectors make up the matrix shown below:

Figure 7. a) Wave form for mutation R55_1. b) Wave form for mutation R55_2. The sound has been displayed ad a two-dimensional representation.

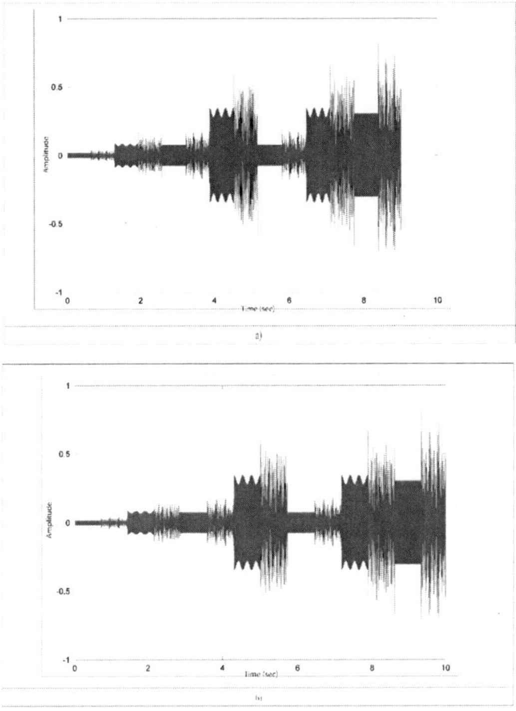

$$X^T = \begin{pmatrix} 0\ 0\ 0\ 0\ 0\ 0\ 0\ 0\ 0\ 9 \\ 1\ 0\ 4\ 0\ 4\ 4\ 0\ 8\ 0\ 4 \\ 0\ 0\ 0\ 0\ 0\ 0\ 4\ 0\ 5\ 32 \\ 0\ 4\ 8\ 8\ 20\ 12\ 0\ 24\ 0\ 4 \\ 0\ 0\ 0\ 0\ 0\ 0\ 0\ 0\ 0\ 36 \\ 4\ 0\ 16\ 0\ 16\ 16\ 0\ 32\ 0\ 16 \\ 0\ 0\ 0\ 0\ 0\ 0\ 16\ 0\ 20\ 128 \\ 0\ 16\ 40\ 32\ 101\ 32\ 0\ 60\ 0\ 4 \\ 0\ 0\ 0\ 0\ 0\ 0\ 0\ 0\ 0\ 36 \\ 4\ 0\ 16\ 0\ 16\ 16\ 0\ 32\ 0\ 16 \\ 0\ 0\ 0\ 0\ 0\ 0\ 16\ 0\ 20\ 128 \\ \ldots\ldots\ldots\ldots\ldots \end{pmatrix} \tag{15.9}$$

Figure 8. Dynamic networks for mutation R55_2

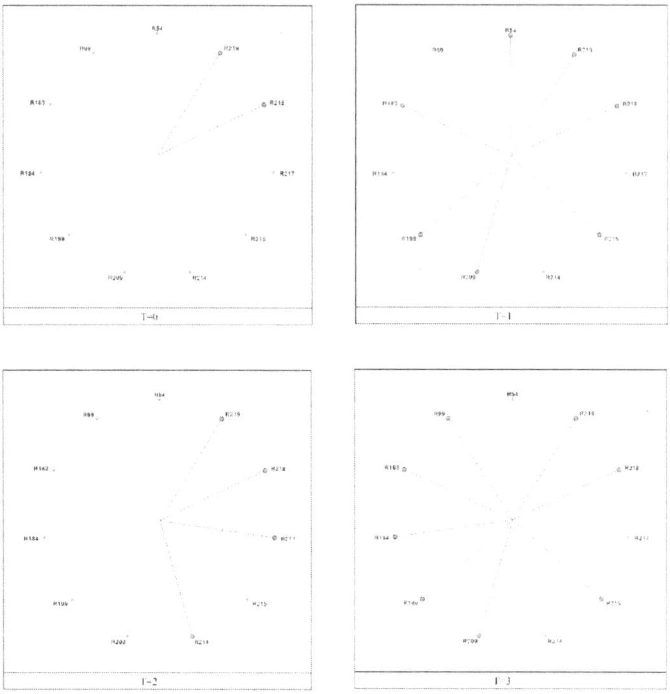

Again relations (15.4) and (15.5) are maintained and again, there is no damage to the fractal nature of the wave form, even if it is has undergone a number of modifications.

The results can be seen in the wave form shown in Figure 7b.

Obviously we can conduct the same kind of analysis for other mutations.

In the evolutionary experiments described in Chapter 11, we modify the basic self-replicator by replacing the original rules with new rules that conserve the CA's ability to self-replicate while generating a new consultation network with as many nodes as possible. For example, the network for the double mutation R55_2_210_2 has 16 nodes.

This procedure allows us to observe the modifications in fractal sounds and sound patterns associated with the mutation.

FUTURE TRENDS

We are currently working to create computational programs for the automatic generation of canons and fugues and to exploit self-similarity and scaling invariance in CAs.

Our system is extremely efficient in generating multiple patterns from a few basic elements. What counts is the way these elements are grouped together: a musical phrase which contains 8 sounds is fundamentally different from a single chord which merges these sounds. If we continue the process, producing compositions that last minutes instead of seconds, the result will be different again.

If we transpose the scaling invariance of CAs into sound and music, we can increase the complexity of our compositions, generating shifts in rhythm and key. In other words we can proceed in the same way as Bach, Webern or Stockhausen.

CONCLUSION

Over the last three decades, the use of computers and mathematical models to compose music ("algorithmic music") has become increasingly popular. In this chapter, we have shown how we can translate the implicit organization of the 3D patterns from CAs into music and sound(Bilotta et al., 2000a; 2000b; 2000c; Bilotta & Pantano, 2001; Bilotta et al., 2002a; 2002b; Bilotta & Pantano, 2002a; 2002b; Bilotta & Pantano, 2003; Bertacchini et al., 2006; Bilotta et al., 2005; Bilotta et al., 2008). It is as if natural and artificial organisms share nested patterns of growth and synchronization providing a natural basis for the generation of music. Exploiting these patterns we can generate "two or three part inventions" that reflect fundamental harmonies in nature (Hofstadter, 1999). In its simplest form, a canon (term derived from old French, which in turn was taken from the Greek term which means law) is a contrapuntal composition that employs a melody with one or more imitations of the played melody after a given duration. While the initial melody is called the leader, the imitative melody is called the follower, recurring in infinite loops, starting at different times or at different points in the melody, played by different instruments or voices. Canons can be classified by various qualities such as the number of voices, the interval at which each successive voice is transposed in relation to the preceding one, whether voices are inverse, retrograde, or both retrograde and inverse, the temporal distance between each voice, whether the intervals of the second voice are exactly those of the original or if they are varied to fit the musical scale or the tempo of successive voices.. Canons may use more than one of the above mentioned methods.

Applying simple procedures we can generate multi-faceted structures, which we would never be able to recognize if we did not know how they had been created.

We observe that each "voice" (each structure in the CA) has its own structure and organization in time. In this way the music reflects the complex organizational laws governing the behavior of the CA.

CAs can be also transformed into sound. The encoding process we have used makes it possible to visualize the internal structure of 2-D CAs in a two dimensional representation. Similar or repeating elements of fractal nature at different time scales are visually distinct, allowing the identification of structural and rhythmic characteristics of the CAs. By varying the genetic code for these systems, we obtain a huge range of different patterns and consequently of sounds. In these simple artificial systems, we glimpse the harmony of the Universe.

REFERENCES

Aristotle. *Metaphysics,* Translated by Ross, W.D. (2007). ebooks@Adelaide, 2007. http://ebooks.adelaide.edu.au/a/aristotle/metaphysics/.

Bertacchini, P. A., Bilotta, E., & Pantano, P. (2005). Modelli matematici, linguaggi, musica . *Sistemi Intelligenti, 3*, 487–527.

Bilotta, E., Cupellini, E., Pantano, P. (2008). *Sintetizzatore Caotico*. [Patent n° CS2008A00010].

Bilotta, E., Gervasi, S., & Pantano, P. (2005). Reading Complexity in Chua's Circuit by Music. Part I: A new way of understanding chaos. *International Journal of Bifurcation and Chaos in Applied Sciences and Engineering, 15*(2), 253–382. doi:10.1142/S0218127405012156

Bilotta, E., Grob, D., Smith, T., Lenaerts, T., Bullock, S., Lund, H. H., et al. (2002b). *ALMMA II. Artificial Life Models for Musical Applications II: Searching for musical creativity*. Workshop of the 8th International Conference on the Simulation and Synthesis of Living Systems (*ALife VIII*). Sydney NSW, Australia: University of New South Wales.

Bilotta, E., Miranda, E. R., Pantano, P., & Todd, P. M. (2002a). Artificial Life Models for Musical Applications: Workshop Report . *Artificial Life, 8*(1), 83–86. doi:10.1162/106454602753694774

Bilotta, E., & Pantano, P. (2001). Artificial Life Music Tells Complexity. *Proc. of ALMMA I-Artificial Life Models for Musical Applications I, 6th European Conference on Artificial Life*. Prague. pp 17-28.

Bilotta, E., & Pantano, P. (2002a). Self reproducers use contrapuntal means, *Proc. of ALMMA II-Artificial Life Models for Musical Applications II: Searching for musical creativity, Workshop of the 8th International Conference on the Simulation and Synthesis of Living Systems (ALife VIII)*, University of New South Wales, Sydney, Australia, 3-8.

Bilotta, E., & Pantano, P. (2002b). Synthetic Harmonies: Recent results. [MIT Press, Cambridge, MA.]. *Leonardo, 35*(2), 35–42.

Bilotta, E., & Pantano, P. (2003). Matematica, Musica e Tecnologie: un trinomio possibile. *Paper presented at the Conference Matematica senza Frontiere*. Quaderno n. 2. Lecce, IT: Department of Mathematics, University of Lecce.

Bilotta, E., Pantano, P., & Talarico, V. (2000a). Evolutionary Music and Fitness Function, in Progress in Industrial Mathematics at ECMI 2000, A.M.Anile, V.Capasso, A. Greco (Eds), Series: Mathematics in Industry. Vol.1, pp. 118-125, Springer-Verlag, Heidelberg.

Bilotta, E., Pantano, P., & Talarico, V. (2000b). *Music Generation through Cellular Automata: How to Give Life to Strange Creatures. Generative Art GA2000*. Italy: Politecnico di Milano University of Milan.

Bilotta, E., Pantano, P., & Talarico, V. (2000c). Synthetic harmonies: an approach to musical semiosis by means of cellular automata. In M. A. Bedau, J. S. McCaskill, N. H. Packard & S. Rasmussen, (Eds), Artificial Life VII, 537-546. Cambridge, MA: MIT Press.

Bolognesi, T. (1983). Automatic Composition: Experiments with self-similar music. *Computer Music Journal, 7*(1), 25–36. doi:10.2307/3679916

Cope, D. (1991). *Computers and Musical Style*. Madison, WI, USA: A-R Editions.

Cupellini, E., Rizzuti, C., Bilotta, E., Pantano, P., Wozniewski, M., & Cooperstock, J. R. (2008). Exploring Musical Mappings and Generating Accompaniment with Chaotic Systems. *International Computer Music Conference (ICMC)*. Belfast, Ireland.

Dodge, C. (1988). Profile: A musical Fractal. *Computer Music Journal, 12*(3), 10–14. doi:10.2307/3680332

Frova, A. (1999). Fisica nella Musica. Bologna, IT: Zanichelli.

Geymonat, L. (1971). *Storia del pensiero filosofico e scientifico*. Milano: Garzanti.

Hiller, L. (1970). Music Composed with Computers – A Historical Survey. In H. B. Lincoln (Ed.), The Computer and Music (42–96). Ithaca, NY: Cornell University Press.

Hiller, L. (1981). Composing with Computers: A Progress Report. *Computer Music Journal*, *5*(4), 7–21. doi:10.2307/3679501

Hofstadter, D. R. (1999). *Gödel, Escher, Bach: an Eternal Golden Braid*. New York: Basic Books.

Huffman, C. A. (2005). *Archytas of Tarentum, Pythagorean, philosoper and Mathematician King*. Cambridge: Cambridge University Press.

Kramer, G. (Ed.). (1994). Auditory Display: Sonification, Audification and Auditory Interfaces. *Paper presents at the First International Conference on Auditory Display*. Reading, MA: Addison-Wesley.

Leach, J., & Fitch, J. (1995). Nature, Music and Algorithmic Composition . *Computer Music Journal*, *19*(2), 23–33. doi:10.2307/3680598

Manaris, W., Vaughan, D., Wagner, C., Romero, J., & Davis, R. (2003). Evolutionary Music and the Zipf-Mandelbrot Law: progress towards Developing Fitness Functions for Pleasant Music. *Paper presented at EvoWorkshops 2003, Lecture Notes in Computer Science Series*, Berlin: Springer-Verlag.

Mandelbrot, B. (1975). *Les Objects Fractals: Forme, Hasard et Dimension*. Paris: Flammarion.

Miranda, E. R. (2001). *Composing Music with Computers*. Oxford, UK: Elsevier – Focal Press.

Murali, K., Lakshmanan, M., & Chua, L. O. (1995). Controlling and synchronization of chaos in the simplest dissipative non-autonomous circuit. *International Journal of Bifurcation and Chaos in Applied Sciences and Engineering*, *5*(2), 563–571. doi:10.1142/S0218127495000466

Nierhaus, G. (2008). *Algorithmic Composition: Paradigms of Automated Music Generation*. New York, NY: SpringerWien.

NIME. 8th International Conference New interfaces for Musical Expression. (5 - 7 JUNE 2008) Retrieved the conference proceedings http://nime2008.casapaganini.org/.

Pikovsky, A., Rosenblum, M., & Kurths, J. (2001). *Synchronization - A Universal Concept in Nonlinear Science*. Cambridge: Cambridge University Press.

Roads, C. (1996). *The Computer Music Tutorial*. Cambridge, MA: The MIT Press.

Robert, A. Freitas, Jr., & Merkle R. C. (2004). Kinematic Self-Replicating Machines. Georgetown, TX: Landes Bioscience.

Schwanauer, S. M., & Levitt, D. A. (1993). *Machine Models of Music*. Cambridge, MA: The MIT Press.

Stewart, I. (2007). *Why Beauty Is Truth: A History of Symmetry*. New York: Basic Books.

Strogatz, S. H. (2000). *Nonlinear dynamics and chaos: With applications to physics, biology, chemistry, and engineering*. New York: Perseus Books Group.

Strogatz, S. H. (2003). *Sync: The Emerging Science of Spontaneous Order*. New York: Hyperion.

Voss, R. F., & Clarke, J. (1975). 1/F Noise in Music and Speech. *Nature*, *258*(5533), 317–318. doi:10.1038/258317a0

Voss, R. F., & Clarke, J. (1978). 1/f Noise in Music: Music from 1/f Noise. *The Journal of the Acoustical Society of America*, *63*(1), 258–263. doi:10.1121/1.381721

Winfree, A. W. (2001). *The Geometry of Biological Time* (2nd ed.). New York: Springer.

Witten, M. (1996). The sounds of science: II. Listening to dynamical systems - Towards a musical exploration of complexity . *Computers & Mathematics with Applications (Oxford, England)*, *32*(1), 145–173. doi:10.1016/0898-1221(96)00097-1

About the Authors

Eleonora Bilotta is a full professor of General Psychology at the Department of Linguistics, Faculty of Humanities, University of Calabria, Italy, where she is the coordinator of the Ph.D Course in Psychology of Programming and Artificial Intelligence. Her research interests cover a broad range of topics including Cognitive Psychology and Intelligent systems in Education, Animal and Human Behaviour, Human Computer Interaction, Psychology of Music, Continuous and Discrete Dynamical Systems, and Artificial Life and Complexity. Eleonora Bilotta has published more than 250 scientific papers and is co-author of several books, among which, "A Gallery of Chua Attractors", published in 2008 by World Scientific Publishing. Images from Eleonora Bilotta's papers have been used as covers for the journals "Complexity" and the "International Journal of Bifurcation and Chaos. She is currently co- director of the Evolutionary systems Group (ESG) at the University of Calabria. She is co-author of a patent for a sound synthesizer exploiting chaotic dynamics.

Pietro Pantano is a Full Professor of Mathematical Physics at the Faculty of Engineering of the University of Calabria. His research interests are multidisciplinary, covering several different areas of research. These include Theory of Complexity and self-Organizing Systems, Discrete and Continuous Dynamical Systems, Chua's oscillator, Biological Models based on Cellular Automata, Artificial Life, and Generative and Evolutionary Music. Prof. Pantano has published more than 300 scientific papers and has co-authored several books. These include "Ray methods for nonlinear waves in fluids and plasmas" published in 1993 by Longman Scientific and "A Gallery of Chua Attractors" published in 2008 by World Scientific Publishing. Images from papers co-authored by Prof. Pantano have been use as covers for the journals "Complexity" and the "International Journal of Bifurcation and Chaos."

Prof. Pantano is currently co-director of the Evolutionary systems Group (ESG) at the University of Calabria. He is co-author of a patent for a sound synthesizer exploiting chaotic dynamics.

Index

M

N

O

P

Q

R

S

T

U

V

W